T0258764

The Rice-Wheat Cropping System of South Asia: Efficient Production Management

The Rice-Wheat Cropping System of South Asia: Efficient Production Management has been co-published simultaneously as *Journal of Crop Production*, Volume 4, Number 1 (#7) 2001.

The *Journal of Crop Production* Monographic "Separates"

Below is a list of "separates," which in serials librarianship means a special issue simultaneously published as a special journal issue or double-issue *and* as a "separate" hardbound monograph. (This is a format which we also call a "DocuSerial.")

"Separates" are published because specialized libraries or professionals may wish to purchase a specific thematic issue by itself in a format which can be separately cataloged and shelved, as opposed to purchasing the journal on an on-going basis. Faculty members may also more easily consider a "separate" for classroom adoption.

"Separates" are carefully classified separately with the major book jobbers so that the journal tie-in can be noted on new book order slips to avoid duplicate purchasing.

You may wish to visit Haworth's website at . . .

http://www.HaworthPress.com

. . . to search our online catalog for complete tables of contents of these separates and related publications.

You may also call 1-800-HAWORTH (outside US/Canada: 607-722-5857), or Fax 1-800-895-0582 (outside US/Canada: 607-771-0012), or e-mail at:

getinfo@haworthpressinc.com

The Rice-Wheat Cropping System of South Asia: Efficient Production Management, edited by Palit K. Kataki, PhD (Vol. 4, No. 1 #7, 2001). *This book critically analyzes and discusses production issues for the rice-wheat cropping system of South Asia, focusing on the questions of soil depletion, pest control, and irrigation. It compiles information gathered from research institutions, government organizations, and farmer surveys to analyze the condition of this regional system, suggest policy changes, and predict directions for future growth.*

The Rice-Wheat Cropping System of South Asia: Trends, Constraints, Productivity and Policy, edited by Palit K. Kataki, PhD (Vol. 3, No. 2 #6, 2001). *This book critically analyzes and discusses available options for all aspects of the rice-wheat cropping system of South Asia, addressing the question, "Are the sustainability and productivity of this system in a state of decline/stagnation?" This volume compiles information gathered from research institutions, government organizations, and farmer surveys to analyze the impact of this regional system.*

Nature Farming and Microbial Applications, edited by Hui-lian Xu, PhD, James F. Parr, PhD, and Hiroshi Umemura, PhD (Vol. 3, No. 1 #5, 2000). *"Of great interest to agriculture specialists, plant physiologists, microbiologists, and entomologists as well as soil scientists and evnironmentalists. . . . very original and innovative data on organic farming." (Dr. André Gosselin, Professor, Department of Phytology, Center for Research in Horticulture, Université Laval, Quebec, Canada)*

Water Use in Crop Production, edited by M.B. Kirkham, BA, MS, PhD (Vol. 2, No. 2 #4, 1999). *Provides scientists and graduate students with an understanding of the advancements in the understanding of water use in crop production around the world. You will discover that by utilizing good management, such as avoiding excessive deep percolation or reducing runoff by increased infiltration, that even under dryland or irrigated conditions you can achieve improved use of water for greater crop production. Through this informative book, you will discover how to make the most efficient use of water for crops to help feed the earth's expanding population.*

Expanding the Context of Weed Management, edited by Douglas D. Buhler, PhD (Vol. 2, No. 1 #3, 1999). *Presents innovative approaches to weeds and weed management.*

Nutrient Use in Crop Production, edited by Zdenko Rengel, PhD (Vol. 1, No. 2 #2, 1998). *"Raises immensely important issues and makes sensible suggestions about where research and agricultural extension work needs to be focused." (Professor David Clarkson, Department of Agricultural Sciences, AFRC Institute Arable Crops Research, University of Bristol, United Kingdom)*

Crop Sciences: Recent Advances, Amarjit S. Basra, PhD (Vol. 1, No. 1 #1, 1997). *Presents relevant research findings and practical guidance to help improve crop yield and stability, product quality, and environmental sustainability.*

The Rice-Wheat Cropping System of South Asia: Efficient Production Management

Palit K. Kataki, PhD
Editor

The Rice-Wheat Cropping System of South Asia: Efficient Production Management has been co-published simultaneously as *Journal of Crop Production*, Volume 4, Number 1 (#7) 2001.

Food Products Press
An Imprint of
The Haworth Press, Inc.
New York • London • Oxford

Published by

Food Products Press®, 10 Alice Street, Binghamton, NY 13904-1580 USA

Food Products Press® is an imprint of The Haworth Press, Inc., 10 Alice Street, Binghamton, NY 13904-1580 USA.

The Rice-Wheat Cropping System of South Asia: Efficient Production Management has been co-published simultaneously as *Journal of Crop Production*, Volume 4, Number 1 (#7) 2001.

The development, preparation, and publication of this work has been undertaken with great care. However, the publisher, employees, editors, and agents of The Haworth Press and all imprints of The Haworth Press, Inc., including The Haworth Medical Press® and Pharmaceutical Products Press®, are not responsible for any errors contained herein or for consequences that may ensue from use of materials or information contained in this work. Opinions expressed by the author(s) are not necessarily those of The Haworth Press, Inc.

Cover design by Thomas J. Mayshock Jr.

Library of Congress Cataloging-in-Publication Data

The rice-wheat cropping system of South Asia: efficient production management/Palit K. Kataki, editor.
    p. cm.
    Co-published simultaneously as Journal of Crop Production, vol. 4, n. 1(#7) 2001.
    Includes bibliographical references (p.).
    ISBN 1-56022-086-4 (alk. paper)–ISBN 1-56022-087-2 (pbk.:alk. paper)
    1. Rice–South Asia. 2. Wheat–South Asia. 3. Cropping systems–South Asia. I. Kataki, Palit K.
SB191.R5 R5354 2001
631.5'8–dc21
                    2001033161

Indexing, Abstracting & Website/Internet Coverage

This section provides you with a list of major indexing & abstracting services. That is to say, each service began covering this periodical during the year noted in the right column. Most Websites which are listed below have indicated that they will either post, disseminate, compile, archive, cite or alert their own Website users with research-based content from this work. (This list is as current as the copyright date of this publication.)

(continued)

*Special Bibliographic Notes related to special journal issues
(separates) and indexing/abstracting:*

- indexing/abstracting services in this list will also cover material in any "separate" that is co-published simultaneously with Haworth's special thematic journal issue or DocuSerial. Indexing/abstracting usually covers material at the article/chapter level.

- monographic co-editions are intended for either non-subscribers or libraries which intend to purchase a second copy for their circulating collections.

- monographic co-editions are reported to all jobbers/wholesalers/approval plans. The source journal is listed as the "series" to assist the prevention of duplicate purchasing in the same manner utilized for books-in-series.

- to facilitate user/access services all indexing/abstracting services are encouraged to utilize the co-indexing entry note indicated at the bottom of the first page of each article/chapter/contribution.

- this is intended to assist a library user of any reference tool (whether print, electronic, online, or CD-ROM) to locate the monographic version if the library has purchased this version but not a subscription to the source journal.

- individual articles/chapters in any Haworth publication are also available through the Haworth Document Delivery Service (HDDS).

The Rice-Wheat Cropping System of South Asia: Efficient Production Management

CONTENTS

ABOUT THE EDITOR

Palit K. Kataki, PhD, received his doctorate at Cornell University. He has a joint appointment as a Cornell University Coordinator and a CIM-MYT (Centro Internacional de Mejoramiento de Miaz y Trigo) Adjunct Scientist. He has been involved in the planning, implementation, and coordination of research on rice and wheat in South Asia (Bangladesh, India, Nepal, and Pakistan) for several years. Dr. Kataki's research emphasizes tillage and crop establishment, micronutrients, field crop sterility, the establishment of legumes in the rice-wheat cropping system, and seed physiology.

Dr. Kataki has served as Assistant Manager of a tea estate (Tata Tea, Ltd.), Research Associate at ICAR (Indian Council of Agricultural Research), and Assistant Professor at Assam Agricultural University. He has been the recipient of several awards and scholarships, and has authored or co-authored several scientific papers.

Preface

Only a few cropping systems satisfy the calorie needs of over a billion people within a sub-continent, the "Rice-Wheat Cropping System" practiced in the South Asian Indo-Gangetic Plains (IGP) region being one of them. The pros and cons of the past three decades of the largely successful "Green Revolution" era of the IGP are still being debated at various levels of society, from the philosophers to the active field agricultural scientist. Meanwhile, almost as if silently, new sets of problems and concerns have crept into the adoption and spread of this cropping system in the IGP, for both the rice and the wheat crop. These concerns include the need for changes in crop establishment techniques and for crop diversification, declining soil fertility, changes in the pest scenario, a host of issues relating to water management, and a need for policy redirection to sustain productivity growth in the IGP.

At the core, is a search for answers to the questions: "Is the productivity (and sustainability) of the rice-wheat cropping system stagnating OR declining?" AND "Do we have any evidence for it?" for the vast IGP region which encompasses four countries (Bangladesh, India, Nepal and Pakistan) across different agro-ecological zones. Therefore, the theme of this publication on the Rice-Wheat Cropping System of South Asia is centered on these questions. Answering on the affirmative for both these questions would mean understanding the constraints to system productivity and then seeking solutions. The topics in this publication therefore discusses the trends of rice-wheat over time, analyzes rice-wheat productivity changes, followed by specific issues on rice and wheat crop production, which are important in general for the entire IGP region. Topics important in certain areas of the IGP but couldn't be discussed in this special issue are–arsenic contamination of ground water in the eastern IGP, boron toxicity especially to animals in certain districts of Punjab, sulfur deficiency in soils, and weed management. A common denominator for all the issues discussed here is that of varieties, in other words, crop improvement strategies to breed for varieties will be issue and region specific, once the issues (problems) are clearly understood.

[Haworth co-indexing entry note]: "Preface." Kataki, Palit K. Co-published simultaneously in *Journal of Crop Production* (Food Products Press, an imprint of The Haworth Press, Inc.) Vol. 4, No. 1 (#7), 2001, pp. xvii-xx; and: *The Rice-Wheat Cropping System of South Asia: Efficient Production Management* (ed: Palit K. Kataki) Food Products Press, an imprint of The Haworth Press, Inc., 2001, pp. xi-xiv. Single or multiple copies of this article are available for a fee from The Haworth Document Delivery Service [1-800-342-9678, 9:00 a.m. - 5:00 p.m. (EST). E-mail address: getinfo@haworthpressinc.com].

xi

The rice-wheat cropping system is here to stay but there are signs of its productivity growth stagnating or declining. The total production of rice and wheat in the IGP has been increasing annually (chapter by P. K. Kataki, P. Hobbs and B. Adhikary, volume 1 of the series). Under this scenario, it is often difficult to provide hard evidence of stagnating or declining productivity trends given the size (acreage) of the IGP region and the agro-climatic diversity within. The three data sources that provide evidence of productivity changes are from the farmer's field and benchmark surveys, the on-going location specific long-term monitoring (8 to 9 years) surveys, and research station managed long-term soil fertility experiments. The article by P. K. Kataki, P. Hobbs and B. Adhikary (volume 1 of the sereis) also compiles results and discusses the farmer's field and benchmark surveys and Total (TPF) and Partial Factor Productivity (PFP) analysis done for the region. Evidence till date indicate declining productivity. The chapter on the analyses of long-term soil fertility experiment station yields trends by J. M. Duxbury (volume 1 of the sereis) further supports this argument. Analysis of results from the long-term monitoring survey is not complete and is therefore not being discussed here. Focus has shifted from a "Commodity" type research program to diagnosing and understanding the "System" requirements (article by L. Harrington, volume 1 of the sereis). This has encouraged greater inter-disciplinary research in the region and is evident from the topics covered in this special issue.

Agroclimatic characterization of the IGP is important for the diagnosis and analyses of "System" requirement (chapter by J. W. White and A. Rodriguez-Aguilar, volume 1 of the sereis). Options for tillage and crop establishment techniques of rice and wheat have been experimented with and field-tested and are at various stages of adoption and spread in the farming community (chapter by P. Hobbs). These tillage options include region specific solutions based on the land holdings, level of mechanization, soil and climatic conditions, and socio-economic constraints, which varies widely within the IGP. The rice-wheat system has literally reduced the desire to maintain crop diversification within the IGP and has pushed the cultivation of beneficial legumes to more marginal areas in the region (chapter by J. G. Lauren, R. Shrestha, M. A. Sattar and R. L. Yadav, volume 1 of the sereis). Legumes have been an integral part of South Asian diet and culture. This chapter discusses the options available to reintroduce legumes into the IGP to more than its present level of cultivation and to have a greater impact on the soil "health" and on human nutritional benefits.

Soils of the IGP are inherently low in soil fertility, and the further decline in soil fertility in the region is a regional concern. The primary reason for this decline is due to the present soil management practices, or rather its mismanagement in general. Except for the introduction of high yielding, dwarf, fertilizer responsive and disease resistant rice and wheat varieties and increased availability of irrigation water, the "culture" of growing rice and wheat in the region has in general, not adjusted with the demands of the new "commodity" in the last three decades. Higher the crop yield, greater is the nutrient uptake by plants for meeting the "sink" needs, and the requirement for available soil nutrients to sustain this yield jump increases. Higher crop yields (productivity, and therefore the total quantity or production) is absolutely essential to meet the calorie requirement of an ever-increasing population of the region and therefore adjusting crop management practices is important "in making both ends meet." Yadvinder-Singh and Bijay-Singh discusses the management of the primary nutrient requirement for rice and wheat in the IGP. Nitrogen is ubiquitously deficient and deficiencies of P and K are increasing in the region. Fertilizer policy has encouraged imbalance use of N:P:K in the region. Use of organic sources of nutrients has decreased in the IGP. Intensive cropping also increases micronutrient deficiencies. V. K. Nayyar, C. L. Arora and P. K. Kataki discuss the changes in the soil micronutrient status in the IGP, and its management in relation to rice and wheat. In the eastern IGP, soil Boron deficiency is widespread, reducing crop yields (chapter by P. K. Kataki, S. P. Srivastava, M. Saifuzzaman and H. K. Upreti).

The spectrum of pests on rice and wheat has changed since the adoption and spread of the "Green Revolution" in the IGP (article by M. Sehgal, M. D. Jeswani and N. Kalra). Pest attack can be visual (if above ground) or hidden (e.g., damage caused by nematodes, article by S. B. Sharma). The integrated management (and constraints to its adoption and spread) of pests are discussed in these chapters in relation to the IGP.

Inadequate water management practices in the rice-wheat cropping system of the IGP are a widely acknowledged constraint, but the desire to overcome these problems at various levels have fallen short of expectations. Appropriate water management has to be dealt with starting at the macro level through active policy changes, to the micro or field level in terms of crop water needs and its management. Water policy and use is too big a subject to cover all aspects in this publication. Therefore a case study of water management from the Sindh-zone of Pakistan by M. Aslam and S. A. Prathapar, and use and management of poor quality water in the high yielding western IGP region (by P. S. Minhas and M. S. Bajwa) is being discussed here. Finally, policy framing and its adjustment (re-directions) to overcome all the constraints mentioned here, are important for sustainable resource use in the IGP in relation to the rice-wheat cropping system (chapter by P. L. Pingali and M. Shah, volume 1 of the series).

All of the above topics are being published in two volumes. Volume 1 of the publication discusses 6 topics on the status of rice-wheat in the IGP and covers the following: Trends of the Rice-Wheat Cropping System, Long-Term

Soil Fertility Experiments, Agro-Climatic Characterization of the IGP, Legumes and Crop Diversification, Policy Re-Directions for Sustainable Resource Use, and the Synthesis of Systems Diagnosis. Volume 2 of this series discusses the major constraints of the rice-wheat cropping system and includes: Tillage and Crop Establishment, Efficient Management of Primary Nutrients and Micronutrients, Sterility in Wheat and Field Crop Response to Applied Boron, Pest Management, Plant Parasitic Nematodes, Water Management in Sindh of Pakistan, and Use and Management of Poor Quality Waters.

The volume of literature cited in preparation for all the chapters in this publication is a testament to the work done over many years by the agricultural scientists within the network of National Agricultural Institutes, Universities, and International Agricultural Centers in the IGP region. This is specifically true for issues related to rice and wheat production within the last three decades. As a result, these Institutes were able to form the Rice-Wheat Consortium (see Foreword by Professor Timothy Reeves, Director General, CIMMYT) to facilitate greater co-operation, exchange of ideas, research results and research cooperation across the vast IGP region. The volume of work done is immense and this publication on rice-wheat attempts to comprehensively discuss the major problems of this important cropping system and strategies to overcome these problems, the target audience being researchers, students, extension personnel and policy makers.

For providing logistical and financial support for work on rice and wheat in the years of my involvement in the region, I would like to acknowledge the following: National Agricultural Institutes of Bangladesh, India, Nepal and Pakistan, the innumerable agricultural scientists working in these Institutes, International organizations (CIMMYT, ICRISAT, IRRI), Cornell University (especially the Department of Crops and Soils, and CIIFAD), US-AID SM-CRSP program, and the Rice-Wheat Consortium for the Indo-Gangetic Plains. Finally to Kalpita, Gemma, Ronnie and Rubin for their love and patience.

Palit K. Kataki

Foreword

The Rice-Wheat Consortium for the Indo-Gangetic Plains (RWC) was founded in 1993 to foster sustainable productivity in rice-wheat systems. The impetus for creating the RWC was a concern over declining productivity in rice-wheat systems, in which rice and wheat are grown in sequence on the same plot of land. This system meets the food needs of about one billion people: faltering productivity will exact a heavy toll. The rice-wheat sequence occupies some 12 million hectares in Bangladesh, India, Nepal, and Pakistan and accounts for about 90% of food production over this area. Future food security for the region's expanding population is threatened by a range of difficulties, including slower growth in rice and wheat yields, few options for expanding cropped area, and alarming signs of natural resource depletion and degradation.

The RWC has assembled scientific and technical experts from the national research systems of Bangladesh, India, Nepal, and Pakistan, international research centers, and advanced research institutes. The mission they share is to bring a multidisciplinary, holistic perspective to bear on the problems limiting sustainability in rice-wheat cropping systems. The RWC was established in 1994 as an Ecoregional Initiative of the Consultative Group on International Agricultural Research (CGIAR), and in 1998, the International Maize and Wheat Improvement Center (CIMMYT) was assigned convening and leadership responsibilities.

The priorities and research themes of the RWC, which are explored in detail in this volume, reflect the broad systems perspective of the research team:

- long-term experiments to understand why productivity is declining in rice-wheat systems;
- alternative tillage and crop establishment practices (zero tillage, surface seeding of wheat, direct seeding of rice, bed planting, small-scale

[Haworth co-indexing entry note]: "Foreword." Reeves, Timothy G. Co-published simultaneously in *Journal of Crop Production* (Food Products Press, an imprint of The Haworth Press, Inc.) Vol. 4, No. 1 (#7), 2001, pp. xxi-xxiv; and: *The Rice-Wheat Cropping System of South Asia: Efficient Production Management* (ed: Palit K. Kataki) Food Products Press, an imprint of The Haworth Press, Inc., 2001, pp. xv-xviii. Single or multiple copies of this article are available for a fee from The Haworth Document Delivery Service [1-800-342-9678, 9:00 a.m. - 5:00 p.m. (EST). E-mail address: getinfo@haworthpressinc.com].

mechanization) to increase productivity, foster diversification in farming systems, and reduce production costs;

- integrated nutrient management practices (balanced nutrient use, use of chlorophyll meters, studies of micronutrients, new rotations, crop residue management) to obtain higher yields with less chemical fertilizer and rebuild the soil resource base;
- practices to stabilize yields in cropping systems and improve system ecology (alternative crops and rotations, integrated pest management, studies of long-term effects of new tillage practices) to foster long-term productivity, profitability, and diversity in cropping systems;
- integrated water management (practices to reduce salinity and sodicity, groundwater depletion, waterlogging) from the field level to the water basin level, to address projected water shortages and water management problems in coming decades in South Asia;
- the development of rice and wheat varieties to perform well in conservation tillage and direct sowing systems;
- attention to the socioeconomic and policy issues that affect the profitability of new practices, the adoption of new practices, and the rice and wheat economy of the region;
- data and geographic information systems (GIS) to extend the reach of science over years and across sites;
- networking across borders to unite previously isolated researchers and institutions with similar goals; and
- human resource development enabling researchers and farmers to develop, experiment with, and apply new technologies.

Research priorities are constantly evaluated and revised by the members of the RWC, in recognition of the fact that, to be sustainable, farming systems must be economically viable, environmentally sound, socially acceptable, and politically supportable.

Sustainable farming systems must be *economically viable* at the farm and national levels. Poor farmers cannot invest in systems that will not produce reasonable yields and (even better) cash income, now and in the future. At the national level, agriculture must also earn its keep as a significant contributor to GDP and export earnings. The reality in most countries, including those participating in the RWC, is that economic well being and development are almost invariably based on productive and profitable agriculture, the "engine room" of subsequent industrialization.

Sustainable farming systems must be *environmentally sound* as well. Economic success in agriculture cannot come at the expense of our soils, air, water, landscapes, and indigenous flora and fauna.

Sustainable farming systems must be *socially acceptable*. They must be appropriate to the people who, relying on their own meager resources, are responsible for implementing and managing them. The need for socially acceptable systems implies the need for a better understanding of farmer and community needs and values, as well as better targeting of technology to meet local conditions.

Finally, sustainable farming systems must be *politically supportable.* Political support depends largely on successfully meeting the first three requirements of sustainability. If economic growth is catalyzed by agriculture within an environmentally sound, socially acceptable framework, politicians will continue to view agriculture as justifying support.

All four components combine to form the whole: sustainable agriculture. If one is neglected, it can seriously reduce the rate and extent of progress towards sustainability and food security.

Another important contribution of the RWC to research in South Asia is the recognition that to attain sustainable food securities we must change the way we plan, conduct, and communicate about research. We must blend very specialized research disciplines in teams of scientists seeking appropriate outcomes that have an immediate impact in farmers' fields. The RWC, like CIMMYT, has adopted an integrative research paradigm that brings together the best genotypes (G), in the right environments (E), under appropriate crop management (M), generating appropriate outcomes for people (P). All of us who are dedicated to fostering sustainable agriculture in developing countries recognize the interdependence of these factors. Most organizations alone cannot contribute fully to each aspect of $G \times E \times M \times P$. Partnerships and consortia that assemble the best possible teams to execute the $G \times E \times M \times P$ approach, such as the RWC, will underpin the timely and successful achievement of sustainable farming systems and future food security.

The chapters in this volume constitute a valuable overview of our present knowledge of sustainability problems in the rice-wheat systems of South Asia. They present the results of long-term experiments designed to gauge the nature and extent of sustainability problems. They also describe the practices and technologies that have been developed, tested, and in some cases adopted to overcome constraints to sustainability. Readers should note, however, that this volume is not simply an account of the progress that the RWC has achieved to date, as important as that is. This book is also a guide to what may be achieved by using a research approach in which highly focused, multidisciplinary teams of researchers work closely with farmers. Two examples of the impacts of this research approach will suffice to illustrate this point; readers will find many others in the chapters that follow.

FIRST

Two alternative tillage methods promoted by the RWC–direct drilling and surface seeding–allow farmers to prepare soils and sow wheat in a single tractor operation after the rice harvest. These practices enable farmers to reduce production costs by US$ 65/ha. It reduces fuel use by 75% or more, provides better yields, cuts herbicide use in half, and requires 10% less water. In 2000, farmers used direct drilling with locally manufactured drills to plant 8,000 hectares in Haryana, India, and 5,000 hectares in the Punjab of Pakistan. The area of adoption has increased ten-fold each year for several years. The main constraint on more rapid expansion has been a lack of good quality, fairly priced seed drills, but now small private shops are producing more drills in response to the rising demand.

SECOND

The potential environmental impacts of the practices promoted by the RWC are impressive. By changing to a zero-till system on one hectare of land, a farmer can save 98 liters of diesel and approximately 1 million liters of irrigation water. Using a conversion factor of 2.6 kilograms of carbon dioxide per liter of diesel burned, this represents about a reduction in carbon dioxide emission of a quarter ton. Carbon dioxide is a principal contributor to global warming. These benefits increase dramatically if extended across even a portion of the rice-wheat region's 12 million hectares. Adoption of zero tillage on 5 million hectares would represent a saving of *5 billion cubic meters of water* each year. That would fill a lake 10 kilometers long, 5 kilometers wide, and 100 meters deep. In addition, annual diesel fuel savings would come to *0.5 billion liters*–equivalent to a *reduction of nearly 1.3 million tons in CO_2 emissions* each year.

These technologies give an idea of what can be done to achieve more sustainable agriculture in major production systems that are crucial to the welfare of large numbers of people. They also give an idea of the vast challenge taken on by the RWC. The RWC's work is certainly not finished, and readers will appreciate that the chapters in this volume describe a "work in progress" rather than a job completed. The pages that follow reflect the importance of the RWC's work, the dedication of the researchers in the Consortium, and the tremendous human impact of their work. Those of us from CIMMYT who have been privileged to share in this work are proud of the achievements to date. The contents of this volume serve as a testimony to the enduring value of the research and to the great urgency and importance of future support for this group of dedicated researchers.

Professor Timothy G. Reeves
Director General
International Maize and Wheat Improvement Center (CIMMYT)

Tillage and Crop Establishment in South Asian Rice-Wheat Systems: Present Practices and Future Options

P. R. Hobbs

SUMMARY. Tillage is one of the primary activities for crop production. It is a major cost for crop production and has a strong influence on yield through its interaction with timely planting and good plant stand. In the South Asian rice-wheat systems of the Indo-Gangetic Plains the tillage for rice is unique in that wet plowing (locally called puddling) is most commonly used to prepare the field for eventual transplanting of rice seedlings. The field is usually plowed dry, flooded, and then puddled (plowed when flooded) to create a soil with poor soil physical properties and low water infiltration. Most farmers puddle their fields for rice. Wheat and other non-rice crops grown in this system after rice in the same year receive normal dry plowing before the seeds are planted, usually by broadcasting the seed and incorporating by plowing.

Traditional methods of land preparation are either done by animal or by tractor power although there are pockets of land where manual labor provides the power for this work. Animal power is more common in the eastern belt of the Indo-Gangetic Plains in Eastern India, Bangladesh and Nepal. Tractor power has almost replaced animal power in NW India and Pakistan and is gradually replacing animal power in the rest of the region. Land preparation for rice and wheat usually requires multiple passes of the plow and a practice called planking (a wooden plank is drawn over the field) to level the fields and help break the

---

P. R. Hobbs is Regional Agronomist (South Asia), CIMMYT South Asia Regional Office, P. O. Box 5186, Lazimpat, Kathmandu, Nepal (E-mail: p.hobbs@cgiar.org).

[Haworth co-indexing entry note]: "Tillage and Crop Establishment in South Asian Rice-Wheat Systems: Present Practices and Future Options." Hobbs, P. R. Co-published simultaneously in *Journal of Crop Production* (Food Products Press, an imprint of The Haworth Press, Inc.) Vol. 4, No. 1 (#7), 2001, pp. 1-22; and: *The Rice-Wheat Cropping System of South Asia: Efficient Production Management* (ed: Palit K. Kataki) Food Products Press, an imprint of The Haworth Press, Inc., 2001, pp. 1-22. Single or multiple copies of this article are available for a fee from The Haworth Document Delivery Service [1-800-342-9678, 9:00 a.m. - 5:00 p.m. (EST). E-mail address: getinfo@haworthpressinc.com].

clods. When animal power is used this can take many days and lead to delayed planting of the crop. Even with tractor power, land preparation can take several weeks. If the land is too wet or dry this can be even longer. Poor availability of tractor power can also delay planting.

Late planting is a major cause of low wheat yields. If wheat is planted after the end of November, yields can be reduced by one percent or more for every day's delay in planting. In addition, the systems of broadcasting seed to establish wheat can result in poor plant stands and even lower yields than anticipated. In order to overcome the problems of late planting, poor plant stands and to reduce the costs of production research has been evaluating the use of zero- and reduced-tillage systems for wheat in the rice-wheat system. In rice, research has started on systems that do away with puddling the soil since it is hypothesized that by not puddling the soil, the next upland crop will have better soil physical conditions for growth.

This paper first describes the present practices for tillage and crop establishment for rice and wheat and then goes on to describe some of the new tillage and crop establishment options being researched in the region. For wheat, the following systems are described and research results presented:

1. Surface seeding. In this system no land preparation is used and seed are merely broadcast onto saturated soils.
2. Zero-tillage. In this system wheat is planted directly into the rice stubble without land preparation. Different soil coulters are needed depending on the rice residues present at planting
3. Reduced-tillage systems. Usually a rotavator is used to prepare the soil ahead of a seed drill and compaction mechanism. This allows wheat to be planted at the same time as land is tilled.
4. Bed planting systems. These ridge and furrow systems have been introduced to improve water and fertilizer efficiency, allow mechanical weeding and reduced herbicide use and increase yield potential because of less lodging. Beds can be made for every crop or left permanent for succeeding crops.

For rice, the major research presented is in the area of dry seeded rice where puddling is not done. Weeds become a major problem in this system and various options for this problem are discussed. Some time is also devoted to the use of mechanical transplanters and use of direct seeding on puddled soils.

This article discusses several new tillage options for rice and wheat that have great potential in the rice-wheat systems of South Asia. They offer reduced costs of production, the ability to plant on time, give better plant stands and higher crop yields. They also reduce the use of natural resources such as fuel and tractor parts, save water and fertilizer resources and increase the efficiency of agriculture. Also because they can reduce fuel emissions and reduce residue burning they contribute to less global warming. *[Article copies available for a fee from The Haworth Document Delivery Service: 1-800-342-9678. E-mail address: <getinfo@ haworthpressinc.com> Website: <http://www.HaworthPress.com> © 2001 by The Haworth Press, Inc. All rights reserved.]*

KEYWORDS. Bed-planting, cropping system, Indo-Gangetic Plains, reduced-tillage, rice, surface-seeding, tillage, wheat, zero-tillage

INTRODUCTION

Tradition with regards to land preparation says that the more you plough the greater the yield. This age-old concept seems to be no longer valid. In fact some of the results presented in this paper suggest that more yield can be obtained by doing less ploughing and thereby planting on time.

Past experiences with reduced tillage options in the region were hampered by not having suitable equipment for obtaining proper seed placement and getting good germination when tillage was not used. The research also tended to make comparisons of reduced and zero-tillage options with traditional methods using the same planting dates and soil moisture farmers traditionally use. This biased the results in favor of the traditional system since reduced and zero-tillage options require a different soil moisture regime to obtain good germination. In fact, the key to these new reduced tillage options is having higher soil moisture at the time of planting which in turn lowers soil strength to allow plant roots to penetrate the soil. This is essentially what tillage does–it reduces soil strength. It is also the basis for recommending that the first irrigation for wheat be given at the "crown root initiation" stage of growth. It is at this stage that the soil strength needs to be lowered to assist the crown roots to penetrate the soil.

This paper will first briefly describe the present practices for tillage and crop establishment for rice and wheat in the region followed by the research activities being done on new tillage options in the Indo-Gangetic Plains (IGP) of South Asia.

Cropping Systems

The major cropping pattern grown in the IGP is rice-wheat with more than 12 million hectares planted each year to this system. This cropping pattern has become popular among farmers in the last 30 years or since the introduction of the shorter duration, improved "Green Revolution" varieties. These varieties enabled farmers to grow both crops in a single year whereas before, crop duration of rice and wheat varieties didn't match planting needs. Rice is planted from late May to late August and harvested from late September to late November. Later rice planting occurs in the East particularly when spring rice is planted before the main monsoon season crop. Wheat is planted from late October to January with the optimal date for planting in November. Wheat is harvested from early March to May with a shorter season crop found in the warmer eastern areas.

Other crops are also grown in this complex cropping system (Fujisaka, Harrington and Hobbs, 1994). Most of these other crops are temperate oilseed and pulse crops (grain and fodder) that substitute or are used as break crops for wheat. It is harder to find substitute crops for rice because of the flooded and saturated soils in the monsoon season. Sugarcane is grown in a two to three year cycle before planting rice again in parts of western UP in India. Other crops include sunflower, potatoes and various vegetables (peas, tomatoes, cabbage, etc.). Fodder crops like berseem clover are common near villages and cities. Double cropping is the normal practice in the West and triple cropping is possible in the warmer areas of the East.

Description of Farmer Situations

The high population density of the region and the fact that agriculture is the main form of employment and income means that farm size is relatively small. There are some bigger farmers in the area some who use tenants to crop their land. There are many farmers who crop less than one hectare of land and many of these have to rely on off farm employment to feed their families. Mechanization has increased gradually from west to east over the last 30 years. Small 4-wheel tractors in NW India and Pakistan do most land preparation. However, in the east many farmers still rely on animal power for land preparation. This is also changing since it is becoming expensive to keep a pair of bullocks for a year for this purpose. Two-wheel tractors are becoming popular on the small farms and fields in Bangladesh.

Many of the other farm operations like rice transplanting, weeding, fertilizer application, harvesting and threshing are done manually. However, combine harvesters are becoming popular in NW India and Pakistan and wheat threshers have spread from west to east over the past 20 years. Labor constraints, especially at key times like planting and harvesting, are gradually becoming more important for timeliness of operations as industrialization grows and younger people shun the drudgery of farming. Mechanization is therefore bound to increase to fill this power gap.

Farmers Present Tillage Practices

The most common practice for establishing rice in the rice-wheat systems of South Asia is puddling–ploughing soils when they are saturated–before transplanting rice seedlings. Puddling benefits rice by reducing water percolation and controlling weeds. However, puddling also results in degraded soil physical properties, particularly for finer textured soils, and subsequently creates difficulties when it comes to providing a good soil tilth for wheat. This conflicting soil management situation is unique to the rice-wheat cropping system.

Farmers use animals or tractors to till their soil. The use of animals is more prevalent in the eastern rice-wheat areas where yields are lower, although many farmers use a tractor for the first tillage operation and animals for planking and later ploughings. The animal-drawn implement they use usually consists of a wooden stick with a metal point, and with this implement farmers have to make many passes over the land to prepare a reasonably good seedbed for wheat. Six to ten passes of the plough followed by the same number of plankings (a heavy wooden log to compact and break clods) are common on finer textured soils that have poor physical properties after rice has been grown on them (Table 1).

Tractors, mostly four-wheel tractors, are the major power source in the higher potential areas. Even smallholders take advantage of this power source by renting tractors from larger landholders in the village. Tractors will probably eventually replace animal power, as the cost of maintaining a pair of bullocks for a whole year becomes prohibitive. The main implement used for preparing the soil is either a nine-tine cultivator or a disk harrow. Deep ploughing with a mouldboard or disk plough is rare. Although tractors allow land to be prepared more rapidly for wheat after rice, farmers still make many passes of ploughing and planking: six to eight passes of the ploughing implement are common, usually followed by planking. Table 1 illustrates the time and number of operations for ploughing in several areas of the IGP.

TABLE 1. Data on tillage and crop establishment from diagnostic surveys of selected rice-wheat cropping systems, South Asia.

Location	Area planted late (%)	Turnaround time (days)	Average number of passes with plow
Punjab, Pakistan	40	2-10	2-10 (6)
Pantnagar, India	35	15-20	5-12 (8)
Faizabad, India	25	20-45	5-12 (6)
Haryana, India	25	15-35	4-12 (8)
Bhairahawa, Nepal	40	15-35	4-8 (6)

Source: Data for Punjab, Pakistan from Byerlee et al. (1984); for Pantnagar, India, from Hobbs et al. (1991); for Faizabad, India, from Hobbs et al. (1992); for Haryana, India, from Harrington et al. (1993b); and for Bhairahawa, Nepal, from Harrington et al. (1993a).
Note: Late planting is defined as wheat planted after the first week of December.

PROBLEMS SOLVED BY REDUCED
AND ZERO-TILLAGE AND BED PLANTING

Late Planting

Late planting is a major problem in most rice-wheat areas of South Asia, except for the Indian Punjab. To improve the productivity of the rice-wheat system, the wheat crop must be planted at the optimal time. A typical response of wheat to date of planting in South Asia shows an optimum yield towards the end of November followed by a linear decline in yield of 1-1.5% per day after that date (Ortiz-Monasterio, Dhillon and Fischer, 1994; Randhawa, Dhillon and Singh, 1981). Although the slope of the line varies by variety, all show a decline regardless of whether they are short- or medium-duration lines.

Late planting not only reduces yield but also reduces the efficiency of the inputs applied to the wheat crop. Nitrogen response surfaces are much flatter in plots planted late compared to those planted at the optimal time (Saunders, 1990). In other words, applying more nitrogen cannot compensate for the decline in yield from late planting. The reasons for late planting of wheat in the rice-wheat system are many. An obvious cause of late planting is the late harvest of the preceding rice crop or, in some instances, a short-duration third crop planted after rice. Farmers in some parts of the Indo-Gangetic Plains grow long-duration, photosensitive, high-quality *Basmati* rice that matures late. Farmers prefer to grow *Basmati* rice despite its lower yields because of its high market value, good straw quality (livestock feed), and lower fertilizer requirements. *Basmati* varieties cannot be readily replaced by a shorter duration rice variety and late rice harvest results. However, in other areas, high-yielding, modern rice varieties are grown and their planting dates can be manipulated so wheat planting is not delayed. This is the case in the Indian Punjab, where modern rice varieties are planted early, harvested by October, and wheat is planted by November.

The other major cause of late wheat planting is the long turnaround time between rice harvest and wheat planting. Long turnaround time can be caused by many factors, including excessive tillage, soil moisture problems (too wet or too dry), lack of animal or mechanical power for ploughing, and the priority farmer's place on threshing and handling the rice crop before preparing land for wheat.

Plant Stands

Coupled with the problems of late planting of wheat is the problem of poor germination and plant stands. The majority of farmers in IGP plant wheat by broadcasting the seed into ploughed land and incorporating the seed by another ploughing. Part of the reason for this is residue management problems in fields following rice. The loose straw and stubbles are raked and clog the seed drills. Broadcast seed results in seed placement at many different depths and into different soil moistures with resulting variable germination.

Water and Nutrient Efficiency

Water and nutrients are two major factors that can limit yield in the IGP. Water will become a major limiting factor in agriculture in the next decade as competition for water increases between agriculture on one hand and domestic and industrial use on the other. Fertilizers will increase in cost and also is a major factor affecting the cost of production and profit margins. The new tillage options immediately increase the water and nutrient use efficiency by allowing timely planting and higher yields. However, they also increase the efficiency of these inputs in other ways:

1. Zero-tillage can be done in wetter soils, earlier after rice harvest, and so can save the first irrigation.
2. Zero-tillage does not loosen the soil and so the first irrigation flows faster over the land than land prepared by ploughing. Some farmers estimate that it takes 4-5 hours to irrigate an acre when land is prepared and only 1-2 hours when no tillage is done. This saves water and also results in less leaching of nutrients.
3. In bed planting, water is channeled through the furrows and water savings of 30% have been calculated.
4. In bed planting, fertilizer can be managed for higher efficiency by placing the topdress fertilizer in the soil rather than broadcasting on the surface in traditional systems.

Costs of Production

Probably the major benefit of the new reduced tillage options is the reduction in the costs of production. Land preparation is one of the major costs to grow a crop and any reduction in the number of operations obviously results in savings of money. Farmers in the IGP estimate that they pay about US$50-60 per hectare for land preparation. They also use considerable amounts of fuel and there is a lot of wear and tear on tractor engines and accessories. By using reduced tillage systems enormous savings can be obtained by farmers and the country as a whole. In surface seeding which is described below, land preparation costs are entirely eliminated. Since reduced and zero-tillage result in earlier planting and some cases higher yields, the economics of these new tillage options does not require any mathematical calculation.

DESCRIPTION OF THE NEW TILLAGE OPTIONS

Surface Seeding

Surface seeding, the simplest zero-tillage system, is defined here as the placement of seed onto the soil surface without any land preparation. This is a farmer practice for wheat establishment in parts of Eastern India and Bangladesh. Wheat seed is either broadcast before the rice crop is harvested (relay planted) or after harvest. Surface seeding is also commonly used throughout the region for establishing winter pulses and oilseeds after rice.

Farmers have been using surface seeding to plant wheat for several years in areas where the soils are fine textured, drain poorly, and where land preparation is difficult and often results in a cloddy tilth. The key to success with this system is having the correct soil moisture. Too little moisture results in poor germination and too much moisture can cause seed to rot. A saturated soil is best. The roots germinate into the moist soil and follow the saturation fringe as it drains down the soil profile. The high soil moisture reduces soil strength and thus eliminates the need for tillage, but at the same time the moisture level must not be too high, as oxygen is needed for healthy root growth.

As long as soil moisture can be manipulated, surface seeding is also successful on coarser soils. There must be enough surface moisture to germinate the seed (soaking the seed can help), and soil strength at the root penetration stage must be low (moisture dependent) to allow root growth. This may require an early, light irrigation on coarse textured soils. Some farmers who relay wheat into the standing rice crop place the cut rice bundles on the ground after harvest. This allows the rice to dry but also acts as mulch, keeping the soil surface moist and ensuring good wheat rooting. Young seedlings are also protected from birds. However, relay planting can be done only if the soil moisture is correct for planting at this stage.

Zero-Tillage with Inverted-T Openers

For the purposes of this paper, we define zero-tillage as the placement of seed into the soil by a seed drill without prior land preparation. This technology, which has been tested in Pakistan (see Sheikh et al. 1993; Aslam et al. 1993b, 1989), is presently being tested in other areas of the Indo-Gangetic Flood Plains, including India and Nepal. This technology is more relevant in the higher yielding, more mechanised areas of northwestern India and Pakistan, where most land preparation is now done with four-wheel tractors.

The basis for this technology is the inverted-T openers that were developed and imported from New Zealand. This coulter and seeding system places the seed into a narrow slot made by the inverted-T as it is drawn through the soil by the four-wheel tractor. The coulters can be rigid or spring-loaded depending on the design and cost of the machine. This type of seed drills work very well in situations where there is little surface residue after rice harvest. This usually occurs after manual harvesting.

Combine harvesting of wheat is becoming popular among farmers in northwestern India and Pakistan. In this combine system loose straw and residue is commonly left after harvest. The inverted-T opener may not work well where combines are used, since the opener acts as a rake for the loose straw. In this case other zero-tillage coulter designs will be needed.

Leaving the straw as mulch on the soil surface has not been given much thought in Asian agriculture. However, results from rainfed systems and some preliminary results in Asia suggest that this may be very beneficial to establishment and vigour of crops planted this way. Studies are needed to explore the benefits and longer-term consequences of this idea.

Reduced Tillage

The Chinese have developed a seeder for their 12 horsepower, two-wheel diesel tractor that prepares the soil and plants the seed in one operation. This system consists of a shallow rotovator followed by a six-row seeding system and a roller for compaction of the soil. Funding from the Department For International Development (DFID) (UK) and CIMMYT made it possible to import several tractors and implements from Nanjing, China, into Nepal, Pakistan, and India, where they have been tested over the past few years with positive results. Soil moisture was also found to be critical in this reduced tillage system. The rotovator fluffs up the soil, which then dries out faster than with normal land preparation. The seeding coulter does not place the seed very deep, so soil moisture must be high during seeding to ensure germination and root extension before the soil dries appreciably. Modification of the seed coulter to place the seed a little deeper would help correct this problem.

The main drawback of this technology is that the tractor and the various implements are not easily available. It would help if the private or public sector in South Asian countries could import this machinery or develop a local manufacturing capability. As it becomes more costly to keep and feed a pair of bullocks for a year, more farmers in the region are turning to significantly cheaper mechanised options of land preparation. One of the benefits of this tractor is that it comes with many options for other farm operations; it includes a reaper, rotary tiller, and a mouldboard plough, and it can also drive a mechanical thresher, winnowing fan, or pump. However, most farmers are attracted to the tractor because it can be hitched up to a trailer and used for transportation. For smaller-scale farmers who cannot afford their own tractors, custom hiring is a common alternative. In India, a four-wheel tractor version of the above two-wheel technology is available.

Bed Planting Systems

In bed planting systems, wheat or other crops are planted on raised beds. This practice has increased dramatically in the last decade or so in the high-yielding, irrigated wheat growing area of northwestern Mexico (Meisner et al. 1992; K. Sayre, pers. comm.). Bed planting rose from 6% of farmers surveyed in 1981 to 75% in 1994, and farmers have given the following reasons for adopting the new system:

1. Management of irrigation water is improved.
2. Bed planting facilitates irrigation before seeding and thus provides an opportunity for weed control prior to planting.
3. Plant stands are better.
4. Weeds can be controlled mechanically, between the beds, early in the crop cycle.
5. Wheat seed rates are lower.
6. After wheat is harvested and straw is burned, the beds are reshaped for planting the succeeding soybean crop. Burning can also be eliminated.
7. Herbicide dependence is reduced, and hand weeding and rouging is easier.
8. Less lodging occurs.

This system is now being assessed for suitability in the Asian Subcontinent. At Punjab Agricultural University, two bed widths and two or three rows of wheat planted per bed were compared with conventional flat bed planting. Two of the major constraints on higher yields in north-western India and Pakistan are weeds and lodging. Both can be reduced in bed planting systems. The major weed species affecting wheat, *Phalaris minor,* is normally controlled using the herbicide Isoproturon, which is not always effective. Farmers do not always apply Isoproturon well or on time; in addition, recent reports have confirmed that *P. minor* has developed Isoproturon resistance (Malik and Singh 1995; Malik 1996). Alternative integrated weed strategies must be developed to overcome this problem. Preliminary observations indicate that *P. minor* is less prolific on dry tops of raised beds than on the wetter soil found in conventional flat bed planting. Cultivating between the beds can also reduce weeds. Thus bed planting provides farmers with additional options for controlling weeds.

Lodging is also less of a problem on raised beds. Additional light enters the canopy and strengthens the straw, and the soil around the base of the plant is drier. Reduced lodging can have a significant effect on yield, since many farmers in the Punjab do not irrigate after heading precisely because they want to avoid lodging. As a result, water can become limiting during grain filling, resulting in lower yields. On raised beds this irrigation need not be avoided for the reasons stated.

Permanent Bed Systems

An additional advantage of bed planting becomes apparent when beds are "permanent"–that is, when they are maintained over the medium term and not broken down and re-formed for every crop. In this system, wheat is harvested and straw is left or burnt. Passing a shovel down the furrows reshapes the beds. The next crop (soybean, maize, sunflower, cotton, etc.) can then be planted into the stubble in the same bed. Research into permanent bed systems has started at CIMMYT, Mexico, where the results look encouraging. A longer-term trial has also started, for which preliminary results are available. The advantages of this system are reduced costs, erosion control, reduced soil compaction and, it is hoped, better soil physical structure over time. More research is needed to quantify these benefits.

MACHINERY USED AND DEVELOPED

Pantnagar and Other Zero-Till Drills

The potential benefits of zero-till technology may be attractive, but unless farmers can obtain a suitably priced drill, the benefits will remain hypothetical (Aslam et al. 1993b). At Pantnagar University, India, engineers have modified the old *rabi* wheat seed drill by replacing the old seed coulters with the new inverted-T openers that had been tested in Pakistan. This seeder is now being produced locally in India at a fraction of the cost of the similar, imported New Zealand drill, and it is becoming very popular with farmers. The Pantnagar drill uses a rigid fixed arrangement for attaching the coulters to the seed drill frame. In Pakistan, engineers are also experimenting with simple spring arrangements for the coulters. The original New Zealand drill has a spring loaded coulter that is difficult and expensive to duplicate. The drills being produced in India and Pakistan have seed and fertilizer bins that allow fertilizer to be drilled at the time of planting to the side of the wheat seed.

As mentioned above combine harvesting is becoming popular in the high production areas of NW India and Pakistan. In this case various options need to be considered to overcome the residue problem:

1. Stubble can be burnt, as is presently done in most conventional systems. However, this creates environmental problems of air pollution and also results in a loss of organic matter.
2. A suitable trash drill, using some form of disk opener, can be developed. It could either take the form of disk cutters or "rakers" running ahead of the inverted-T openers or a new system of disk planters could be developed and tested. The rakers push the loose straw away allowing the inverted-T planters to plant the seed unobstructed. All these modifications raise the weight of the drill and the cost of the seeder but it might still be within reach of some farmers, particularly those using combines for harvesting.

The combine could be modified to cut the straw into small pieces before it is ejected out of the back of the machine. This small straw would not interfere with the inverted-T openers and would leave stubble mulch on the soil surface.

Strip-Till Drill

Engineers at Punjab Agricultural University, Ludhiana, India, have developed a "strip-till drill," which uses the same rotary land preparation and seeder combination described earlier for the Chinese hand tractor seed drill. It differs by tilling the soil in a strip into which the seed is planted and not the whole area. The results have been encouraging and the drill is likely to become more popular in the future. It also comes with a fertilizer hopper so that this input can be placed in the soil at planting time.

Bed Planters

A local manufacturer in the Indian Punjab has built a prototype low cost planter that enables three or two rows of wheat to be planted on 70-cm beds at the same time as the beds are formed. This should make it more possible for farmers to adopt this technology. It also has a fertilizer hopper. The original bed maker cum seed drill was modified using the inverted-T openers found on the Pantnagar drill. This allows another non-rice crop to be planted into the permanent bed system after wheat harvest. However, more work is needed on developing better drills for planting on permanent beds.

Chinese Hand Tractors

The Chinese hand tractors and an array of accessories are made in China and have been imported into South Asia. However, attempts are now being made to produce the accessories locally. In Bangladesh, a local engineering workshop has made the seed drill and reaper. The Chinese tractor can also be used to power other equipment such as irrigation pumps, threshers, winnowing fans and mills. It is also very popular attached to a trailer and used for transport.

RESEARCH RESULTS INCLUDING ECONOMICS

Surface Seeding

Results from trials on experiment stations and in farmers' fields in Nepal illustrate the benefits of surface seeding. A station experiment in the Tarai area of Nepal examined three wheat establishment methods following three rice cultural practices (Table 2). Among the rice cultural practices, there were no significant differences in yields. Among wheat establishment methods, however, there were large and significant yield differences. Surface seeding gave significantly higher yields than the farmers' practice, and, because the cost of land preparation was zero, surface seeding also generated higher net benefit. The thousand-grain weight of wheat was also significantly better in the surface seeded plots as a result of planting 15 days earlier. Farmers in this area of Nepal receive a premium for bold seeded grain–an added bonus for the farmer.

Surface seeding technology was demonstrated to farmers in the surrounding district in Nepal with success, and data are presently being collected on the extent of adoption. Many farmers experimented with surface seeding by planting wheat before and after rice harvest, with or without soaked seed, at different seed and fertilizer rates, on different seeding dates, and with different soil moisture levels at the time of planting. One farmer even broadcast seed onto puddled soil. Yields exceeding 3 t/ha were obtained by many of the

TABLE 2. Data from a trial on establishment of wheat following rice, Bhairahawa Agricultural Farm, Nepal, 1993-94.

Method	Wheat yield (kg/ha)	1000-grain weight	Cost to plough (Rs/ha)	Net benefit (Rs/ha)	Extra days needed to plant[1]
Surface seeding	2,775a	46.11a	0	11,485a	0
Chinese seed drill	2,831a	45.43b	600	12,090a	8
Farmers' practice	2,314b	40.87c	2,300	8,065b	15

Note: Figures followed by the same letter are not significantly different at 5% probability using DMRT.
[1]: Number of extra days needed for land preparation before seeding compared to the surface seeding.

farmers who planted into the proper moisture, at the proper time. Heavy rains in November 1996 damaged some of the surface seeded wheat, and the yields were lower that year than in 1995, though still higher than yields from ploughed fields.

In another experiment, rice straw was left covering the relay-planted wheat for different lengths of time after harvest. Slightly better yields (but not significantly so) were obtained by keeping the rice straw mulch on the germinating wheat for nine days. However, this occurred in plots without bird problems. The results suggest that where soil moisture is sufficient for surface seeding before harvest, it is better for farmers to relay plant and use the harvested rice straw as mulch. Keeping the rice straw mulch not only protects the seed from birds but also maintains better soil moisture for germination. Many farmers in fact used this method during their experimentation with surface seeding. Fertilizer becomes an issue in surface seeding and other zero-tillage options. When should fertilizer be applied for maximum efficiency? Results in India show that phosphorus and potassium fertilizers can be broadcast onto the soil surface in irrigated wheat systems without a loss in efficiency (Singh, 1980). Nitrogen, on the other hand, is likely to sustain losses if it is applied to the soil surface. Verma, Srivastava, and Verma (1988) in Bihar showed that more nitrogen is required for wheat under zero-tillage. Similarly, a 20-25% loss in nitrogen efficiency was observed in on-farm trials in Pakistan, where nitrogen response trials on zero-tillage plots with surface-applied nitrogen were compared with conventionally tilled plots where the fertilizer was incorporated (Aslam et al. 1993a, 1989). This could be avoided by band placement at planting or top-dressing the fertilizer at a later date in zero-tilled plots. Improved yields were obtained by delaying the application of nitrogen. Results from Mexico also show a benefit to delayed nitrogen application to the first node stage of wheat in conventional tillage (Ortiz-Monasterio et al. 1994a). The benefits included higher efficiency of applied nitrogen, yield and grain protein content. However, in conventional tillage, basal nitrogen can be incorporated to improve efficiency, but in zero-tillage where banding is not possible, this option of incorporation is not possible and surface applied nitrogen losses are high.

Pantnagar and Other Zero-Till Drills

In the late 1980s, 34 zero-tillage trials were conducted on farmers' field over three years in the rice-growing belt of the Pakistan Punjab (Aslam et al. 1993b) (Table 3). The implement used in these trials was a tractor-pulled seed and fertilizer drill with inverted T-openers. With this equipment, farmers could place the seed directly into the standing rice stubble without any land preparation. Any land preparation resulted in loose stubble and a problem of drill clogging. Interestingly, significantly fewer weeds were found under zero-tillage compared to conventional tillage (Verma and Srivastava, 1994; Singh, 1995)), which is the opposite of the experience with zero-tillage systems in developed countries (Kuipers 1991; Christian and Bacon, 1990). This observation has been confirmed in many other locations. Results from 336 on-farm trials in Haryana are shown in Figure 1 where significantly lower weed counts were found in fields with zero-tillage either before or after herbicide applications. This difference can be explained by the nature of the weeds found in the rice-wheat cropping system. Most of the weeds affecting the wheat crop germinate only in the wheat season, and since the soil is disturbed less under zero-tillage, fewer weeds are exposed and fewer germinate. Weed problems typically are more severe under conventional tillage than under zero-tillage, at least in the near term. Longer-term research is needed to anticipate future consequences on tillage changes on weed populations (for example, to quantify likely shifts in weed species and the effects if those shifts on yield stability).

TABLE 3. Effects on grain yield of zero-tillage and farmers' practice for establishment of wheat after rice, Punjab, Pakistan, 1985-88.

Year	Number of locations	Grain yield (kg/ha)	
		Zero-tillage	Farmers' practice
1985-86	15	3,600	3,516
1986-87	13	3,791a	3,509b
1987-88	6	4,279a	3,560b
Pooled data	34	3,890a	3,528b

Adapted from Aslam et al. (1993b).
Note: Means followed by the same letter do not differ significantly at the 5% level using DMRT.

FIGURE 1. *Phalaris minor* populations before herbicide spray and 120 days after seeding in zero-till and normal tilled plots in Haryana, India in 1998-99 (n = 336).

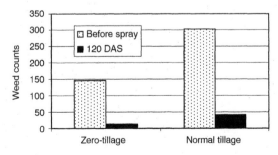

Adapted from Malik, R.K., HAU, Hisar, personal communcation.

Earlier planting is the main reason for the additional yields obtained under zero-tillage (Table 4). Zero-tilled plots were planted as close to as possible 20 November, the optimum date for planting wheat in Pakistan. The results of many trials suggest that the longer the farmer delays planting the lower the yield.

In the 1998-99 wheat season, attempts were made in Haryana and the Punjab of Pakistan to introduce zero-tillage to farmers. A thousand hectares were planted in both places with very good results. Farmers were very impressed with the germination and subsequent growth of the wheat using this technology. Before planting many farmers were sceptical that anything would happen, but afterwards they were pleased they had taken this risk and will be willing to purchase equipment and try again next year.

Chinese Hand Tractor

Table 2 shows that this system generates yields equal to those obtained through surface seeding, and the net returns are superior to farmers' current practice because of the lower land preparation costs. The Chinese tractor can also be used with a rotovator (Bangladesh) to quickly prepare the soil and incorporate the seed after a second pass. This speeds up the planting and results in better stands with less cost than traditional methods. However, the Chinese seeder attachment does a better job because the seed are placed at a uniform depth in the single pass. Engineers are experimenting with removing some of the blades that rototill the soil. In this way, a strip of soil is cultivated rather than the whole area. This reduces the power needs and costs and makes it easier for farmers to manage the tractor.

TABLE 4. Comparison of zero-tillage and farmers' practice for establishment of wheat after rice in Punjab, Pakistan, at locations where planting dates for the two methods differ.

Location	Wheat yield (kg/ha)		Days difference
	Zero-tillage	Farmers' practice	
Daska, site 2	3,143	3,209	10
Daska, site 1	3,842	2,735	13
Ahmed Nagar	4,308	3,526	20
Maujianwala	2,689	2,198	22
Mundir Sharif	4,245	2,660	33
Daska, site 3	3,838	3,420	44
Average	3,677a	2,598b	24

Adapted from Aslam et al. (1993b).
Note: Means followed by the same letter do not differ significantly at the 5% level using DMRT.

Bed Planting

Preliminary results show that there was no significant difference between the methods, which means that yield was not sacrificed by moving to a bed system (Table 5). In fact, with the proper variety (PBW 226), the yield from bed planting can surpass that of planting on the flat. Upright varieties such as HD 2329 perform poorly on beds and cannot compensate for the gap between the beds. Work is just beginning on this new system. Several acres were planted to wheat on beds in Haryana this year with success (Figure 2). This is a rice-wheat area with herbicide resistant *Phalaris*. Bed planting is an option that could help since mechanical weeding of this grass is possible. Bed planting will probably become popular in the next 5 years in rice-wheat and

TABLE 5. Effects of bed size configuration on wheat yield, Punjab Agricultural University, Ludhiana, India, 1994-95.

Variety	Sowing Method					
	On the flat	75 cm beds		90 cm beds		
	25 cm row	2 rows/bed	2 + 1 rows	3 rows/bed	3 +1 rows	Mean
	Wheat yield (kg/ha)					
PBW 226	5,740	6,170	6,390	6,160	6,320	6,160a
WH 542	6,290	5,830	6,360	6,000	6,040	6,110a
CPAN 3004	6,020	5,530	6,140	5,630	5,600	5,780b
PBW 154	5,460	5,110	6,000	5,930	5,880	5,680b
HD 2329	5,770	4,660	6,190	5,580	5,810	5,600b
PBW 34	5,650	5,610	5,800	5,580	5,630	5,650b
Mean	5,820	5,490	6,150	5,810	5,880	

Source: Unpublished data from S.S. Dhillon, wheat agronomist, Punjab Agricultural University.
Note: Means followed by the same letter do not differ significantly at the 5% level using DMRT.

FIGURE 2. Comparison of bed versus flat planting in farmer fields in Pantnagar (n = 3) and Karnal (n = 8), India in 1998-99.

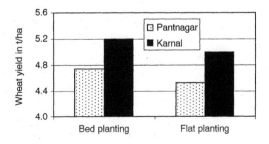

other wheat based systems of Asia. It increases water and fertilizer efficiency (N topdress can be placed), allows mechanical weeding, has less lodging and therefore higher yield potential (more nitrogen and the last irrigation can be given) and by using permanent beds reduces the cost of land preparation.

Rice Establishment Methods

The rice crop in South Asia is mostly transplanted into soil that has been puddled (ploughed when saturated). This is done to reduce water percolation and to control weeds. Research has been done to look at establishment of rice on puddled soils but with direct seeding of sprouted seeds. This has become an alternative for rice establishment in some parts of SE Asia where labor for transplanting is not available. However, in this paper, results will be presented on research where rice is planted dry into soils that were not puddled.

Data in Figure 3 from Nepal shows the yield data for rice established with and without puddling. This is data from a heavy silty clay soil with a high water table. It shows that there is no significant difference between treatments in the rice crop. The figure also shows the yield of wheat following the puddled and un-puddled soils. The wheat yield is significantly better after the dry seeded rice. The total productivity of the non-puddled treatment is also better than where the soil was puddled. Figure 4 shows data from another experiment that was conducted over three years in Pantnagar in India on two different soil types. Yields were only significantly different in the first year in the sandy loam soil and only different in the second year on the finer textured soil. Figure 5 does show that infiltration rates were significantly higher in both soil types when puddling was not used, especially in the coarser textured soil.

Another problem encountered in dry seeding of rice is weeds. Without

FIGURE 3. Rice and wheat yields following with (TPR) and without (DSR) puddling in Nepal in 1998-99.

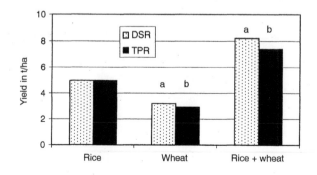

FIGURE 4. Performance of direct seeded and transplanted rice on two soil types in Pantnagar, India.

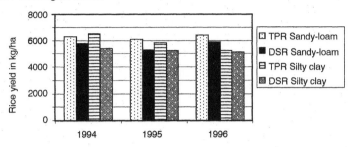

FIGURE 5. Infiltration rate of direct seeded and transplanted rice fields on two soil types at Pantnagar, India.

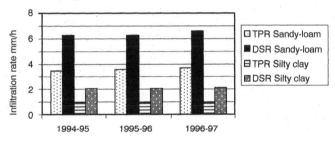

controlling weeds in non-puddled soils this method of rice establishment will not show the expected benefits to the total system productivity. Weeds were important for these different soils since infiltration rates were high and control of weeds was more difficult in the sandy soils. The data from the two sites, however, do indicate that dry seeding of rice is possible, especially where high water tables occur and water is not a problem. Dry seeding of rice deserves more attention in future research agendas.

Farmer Participatory Methods Used

The participation of farmers and local manufacturers in the experimentation and development of these new tillage options was and will continue to be critical for its success. The farmers have to be convinced that the technology will provide substantial benefits to him. Once he gets to this situation he can put pressure on local manufacturers to provide him with the equipment he needs. The local manufacturers have to be involved from the start so that prototypes can be made and tested in the field and modifications made to the design.

In Asia, a farmer participatory approach has figured strongly in the implementation of research. Different methods have been used. At the start, the traditional method of demonstration is used. In this case farmers are asked to cooperate and the fields he selects are planted with the help of engineers and researchers. Later, as the farmer becomes more convinced of the benefits, he is loaned the equipment and plants himself. This provides very valuable feedback. It provides information on how the farmer utilizes the equipment but also feedback on what modifications are needed. Work has also involved organizing farmer groups. This has been successful in Nepal and Bangladesh. In this case, farmers in a village are approached and organized into a group for experimenting with the equipment. Operators are trained and farmers organize themselves for timing of planting, collection of costs and maintaining the equipment. Help is also given to the farmers for obtaining spare parts, repairing the equipment and helping them get loans from local banks for procuring their own machinery.

CONCLUSIONS

South Asia has a booming human population that will not significantly slow down for several decades. In order for the region to be self sufficient in food, farmers will have to increase production by 2.5% per year over the next 20 years (Hobbs and Morris, 1996). This growth in production will have to come from yield growth. It will also have to come by increasing efficiency and reducing the costs of production because it would be politically volatile to increase food production by raising urban food costs.

The new tillage options described in this paper are one way to achieve higher yields and at the same time increase efficiency and reduce costs of production. They can also enlarge the windows between succeeding crops by timeliness that allows other crops to be grown and the system diversified. There are few traditional mechanisms for yield growth that have not already been tapped to their potential. It is therefore imperative that researchers are encouraged to pursue these new tillage options to their fullest in the next decade. The tillage systems need to be adapted to local conditions. Equipment has to be developed, fine tuned and manufactured to meet farmer demands. By adopting this strategy it is possible to meet the food needs of this continent in the years to come.

REFERENCES

Aslam, M., N.I. Hashmi, A. Majid, and P.R. Hobbs. (1993a). Improving wheat yield in the rice-wheat cropping system of the Punjab through fertiliser management. *Pakistan Journal of Agricultural Research* 14: 1-7.

Aslam, M., A. Majid, N.I. Hashmi, and P.R. Hobbs. (1993b). Improving wheat yield in the rice-wheat cropping system of the Punjab through zero tillage. *Pakistan Journal of Agricultural Research* 14: 8-11.

Aslan, M., A. Majid, P.R. Hobbs, N.I. Hashmi, D. Byerlee. (1989). Wheat in the rice-wheat cropping system of the Punjab: A synthesis of on-farm research results 1984-88. PARC/CIMMYT Paper No. 89-3. pp 1-58.

Byerlee, D., A. Sheikh, M. Aslam, and P.R. Hobbs. (1984). *Wheat in the Rice-Based Farming System of the Punjab: Implications for Research and Extension.* PARC/ CIMMYT Wheat Paper No. 84. Islamabad, Pakistan: Pakistan Agricultural Research Council (PARC) and CIMMYT.

Christian, D.G., and E.T.G. Bacon. (1990). A long-term comparison of ploughing, tyne cultivation and direct drilling on the growth and yield of winter cereals and oilseed rape on clayey and silty soils. *Soil and Tillage Research* 18: 311-331.

Fujisaka, S., L.W. Harrington, and P.R. Hobbs. (1994). Rice-wheat in South Asia: System and long-term priorities established through diagnostic research. *Agricultural Systems* 46: 169-187.

Harrington, L.W., S. Fujisaka, P.R. Hobbs, C. Adhikary, G.S. Giri, and K. Cassaday. (1993a). *Rice-Wheat Cropping Systems in Rupandehi District of the Nepal Terai: Diagnostic Surveys of Farmers' Practices and Problems, and Needs for Further Research.* Mexico, DF: CIMMYT.

Harrington, L.W., S. Fujisaka, M.L. Morris, P.R. Hobbs, H.C. Sharma, R.P. Singh, M.K. Chaudhary, and S.D. Dhiman. (1993b). *Wheat and Rice in Karnal and Kurukshetra Districts, Haryana, India: Farmers' Practices, Problems and an Agenda for Action.* Mexico, DF: Haryana Agricultural University (HAU), Indian Council for Agricultural Research (ICAR), CIMMYT, and the International Rice Research Institute (IRRI).

Hobbs, P.R., G.P. Hettel, R.P. Singh, Y. Singh, L.W. Harrington, and S. Fujisaka (eds.). (1991). *Rice-Wheat Cropping Systems in the Terai Areas of Nainital, Rampur, and Pilibhit Districts in Uttar Pradesh, India: Sustainability of the Rice-Wheat System in South Asia. Diagnostic Surveys of Farmers' Practices and Problems, and Needs for Further Research.* Mexico, DF: CIMMYT, ICAR, and IRRI.

Hobbs, P.R., G.P. Hettel, R.K. Singh, R.P. Singh, L.W. Harrington, V.P. Singh, and K.G. Pillai. (1992). *Rice-Wheat Cropping Systems in Faizabad District of Uttar Pradesh, India: Exploratory Surveys of Farmers' Practices and Problems, and Needs for Further Research.* Mexico, DF: ICAR, NDUAT, CIMMYT, and IRRI.

Hobbs, P.R., and M.L. Morris. (1996). *Meeting South Asia's Future Food Requirements from Rice-Wheat Cropping Systems: Priority Issues Facing Researchers in the Post Green Revolution Era.* NRG Paper 96-01. Mexico, DF: CIMMYT.

Kuipers, H. (1991). Agronomic aspects of ploughing and non-ploughing. *Soil and Tillage Research* 21: 167-176.

Malik, R.K., and S. Singh. (1995). Littleseed canarygrass resistance to isoproturon in India. *Weed Technology* 9: 419-425.

Malik, R.K. (1996). *Herbicide Resistant Weed Problems in Developing World and Methods to Overcome Them.* Proceedings of the 2nd International Weed Science Congress. Copenhagen, Denmark. Pp. 665-673. FAO, Rome.

Meisner, C., Acevedo, E., Flores, D., Sayre, K, Ortiz-Monasterio, I., Byerlee, D.

(1992). *Wheat Production and Grower Practices in the Yaqui Valley, Sonora, Mexico*. Wheat Special Report #6. CIMMYT. Mexico, DF.

Ortiz-Monasterio, J.I., S.S. Dhillon, and R.A. Fischer. (1994). Date of sowing effects on grain yield and yield components of irrigated spring wheat cultivars and relationships with radiation and temperature in Ludhiana, India. *Field Crops Research* 37: 169-184.

Ortiz-Monasterio, J.I., K.D. Sayre, J. Peña, and R.A. Fischer. (1994a). *Improving the Nitrogen Use Efficiency of Irrigated Spring Wheat in the Yaqui Valley of Mexico.* Transactions 15th World Congress of Soil Science 5b: 348-349.

Randhawa, A.S., S.S. Dhillon, and D. Singh. (1981). Productivity of wheat varieties, as influenced by the time of sowing. *Journal of Research, Punjab Agricultural University* 18: 227-233.

Saunders, D.A. (1990). Crop management research summary of results. *Wheat Research Centre Monograph 5*. Nashipur, Bangladesh: WRC.

Sheikh, A.D., M.A. Khan, M. Aslam, Z. Ahmad, and M. Ahmad. (1993). *Zero Tillage Wheat Cultivation Technology in the Rice-Wheat Areas of the Punjab.* Pakistan, Islamabad: National Agricultural Research Council (NARC).

Singh, R.P. (1980). Nitrogen and phosphorus management of wheat under the condition of delayed availability of fertilizers. *Indian Journal of Agronomy* 25(3): 433-440.

Singh, Y. (1995). *Rice-Wheat Cropping System NARP Basic Research Sub-Project. Progress Report 1994-95*. G. B. Pant University of Agriculture and Technology, Pantnagar, U.P., India.

Verma, U.N., and V.C. Srivastava. (1989). Weed management in wheat under zero and optimum tillage conditions. *Indian Journal of Agronomy* 34: 176-179.

Verma, U.N., and V.C. Srivastava. (1994). Production technology of non-tilled wheat after puddled transplanted rice. *Indian Journal of Agricultural Science* 64: 277-284.

Verma, U.N., V.C. Srivastava, and U.K. Verma. (1988). Nitrogen management in wheat under conventional and no-tillage conditions in rice-wheat sequence. *Indian Journal of Agronomy* 33: 37-43.

Efficient Management of Primary Nutrients in the Rice-Wheat System

Yadvinder-Singh
Bijay-Singh

SUMMARY. A better understanding of the nutrient economy in relation to productivity is immensely important to develop a sustainable rice-wheat cropping system in the Indo-Gangetic Plain (IGP). Fertilizer (N, P and K) use pattern for the rice-wheat system in the IGP varies greatly at the country, state, village, and farmer field level. In general, fertilizer use is conspicuously more in the western compared to the eastern IGP. Continuous government subsidies on nitrogenous fertilizers have helped create imbalanced N, P, and K fertilizer use by farmers. Removal of P and K by the rice-wheat system far exceeds its additions through fertilizers and recycling.

Nitrogen from rice-wheat systems can often be lost via ammonia volatilization, denitrification and leaching. Nitrification acts as a key process in determining fertilizer-use-efficiency by crops as well as N losses from soils. Loss of N via ammonia volatilization can be substantial when urea is top-dressed. Placement of fertilizer N beneath the soil surface or transportation of N to subsoil layers along with irrigation water can be a useful management option to reduce NH_3 volatilization losses. Water management in rice fields influences the extent of N losses due to nitrification-denitrification. Up to 50% of the applied N can be lost through denitrification when alternating aerobic-anaerobic conditions prevail in rice fields. Leaching losses of N as NO_3 are minimal under wheat but can be substantial under rice grown in readily percolating coarse textured soils found in northwestern India. Irrespec-

Yadvinder-Singh and Bijay-Singh are Senior Soil Chemists, Department of Soil Science, Punjab Agricultural University, Ludhiana, Punjab–141004, India.

[Haworth co-indexing entry note]: "Efficient Management of Primary Nutrients in the Rice-Wheat System." Singh-Yadvinder, and Bijay-Singh. Co-published simultaneously in *Journal of Crop Production* (Food Products Press, an imprint of The Haworth Press, Inc.) Vol. 4, No. 1 (#7), 2001, pp. 23-85; and: *The Rice-Wheat Cropping System of South Asia: Efficient Production Management* (ed: Palit K. Kataki) Food Products Press, an imprint of The Haworth Press, Inc., 2001, pp. 23-85. Single or multiple copies of this article are available for a fee from The Haworth Document Delivery Service [1-800-342-9678, 9:00 a.m. - 5:00 p.m. (EST). E-mail address: getinfo@haworthpressinc.com].

tive of its source, application of fertilizer N at 120 kg N ha^{-1} has been the recommended level for rice and wheat in most of the Indo-Gangetic Plain. However, recent trend indicates the use of more than 150 kg N ha^{-1}, particularly to rice grown in northwestern Indian states, where more than 10 t ha^{-1} grain yield per annum is being harvested from the rice-wheat system. Efforts to increase the notoriously low fertilizer N use efficiency in rice through modifications in N sources have been promising for slow release N fertilizers. In spite of the distinct advantage of using supergranules of urea in fine textured soils, its use is not popular amongst farmers due to the lack of a suitable mechanical applicator. Application of fertilizer N in three and two equal split doses to rice and wheat, respectively, has proved to be an efficient management option. Need based fertilizer N applications to rice using chlorophyll meter is promising for increasing the fertilizer use efficiency. In northwestern India, nitrate content in ground water bodies is continuously increasing and it has been linked to the inefficient fertilizer N use in the rice-wheat system.

In the rice-wheat system, response of wheat to fertilizer P application is more common than its application to rice. Moreover, the residual response of P applied to wheat on rice is greater than that applied to rice on wheat. The availability of soil and fertilizer P increases under submergence and at high temperatures prevailing during the rice season. Thus rice could meet its P requirement from the soil and the residual fertilizer P, when the recommended P fertilizer has been applied to the preceding wheat crop. Improved chemical tests for predicting indigenous soil P supply need to be developed to predict response of rice and wheat to fertilizer P in the cropping system. In neutral and alkaline soils, fertilizers containing P in water-soluble forms are more effective than those containing P soluble in citric acid. Nitrophosphates containing 60% or more of total P in water-soluble forms have been found to be as efficient as diammonium phosphate and superphophates in wheat. In some neutral and acidic soils, a part of the total fertilizer P supplied through diammonium phosphate and superphosphates can be replaced by rock phosphate.

Most of the soils in the Indo-Gangetic Plain contain illite as dominant clay mineral and are medium to high in ammonium acetate (1 M, pH 7.0) extractable K. Therefore, response of rice and wheat to applied K are generally small. Farmers apply very small quantities of K fertilizers to rice and wheat. Total annual K removal by the rice-wheat system is quite large causing depletion of soil K supply. The suitability of ammonium acetate extractable K as an index of plants available K for different soils varying in texture and clay mineralogy remains controversial. Dynamic soil test using resin capsule, which integrates intensity, quantity and delivery rate measures of P and K to rice and wheat can

overcome many of the theoretical limitations associated with rapid chemical extraction. *[Article copies available for a fee from The Haworth Document
Delivery Service: 1-800-342-9678. E-mail address: <getinfo@haworthpressinc.
com> Website: <http://www.HaworthPress.com>* © 2001 by The Haworth Press,
Inc. All rights reserved.]*

KEYWORDS. Fertilizers, nitrogen, phosphorous, potassium, rice, wheat

INTRODUCTION

Rice followed by wheat is a dominant cropping sequence under a range of
management regimes in the Indo-Gangetic Plains of South Asia. It covers
more area in semi-arid ecoregions than in humid or subhumid regions and is
practiced essentially under irrigated conditions. Productivity of this intensive
agricultural production system spread over 12 million ha in India, Pakistan,
Nepal and Bangladesh (Huke and Huke, 1992) varies greatly due to variations in soil, climate, and nutrient and water management. Since, more than 1
billion people rely on this cropping system for a large share of their daily
calorie intake, income, and employment, the importance of studying sustainability of rice-wheat systems in terms of nutrient balance and productivity is
immense. The present population growth rate of 1.8% is estimated to create a
shortfall of 20 million tons of cereal grains by the year 2020. The case of
efficient nutrient management in this context is very well illustrated from
observations on rice. To obtain the projected grain yield of 8.0 t ha^{-1} in
irrigated rice by the year 2025, it is necessary to apply 280 kg N ha^{-1} at 33%
fertilizer N recovery efficiency (Cassman and Pingali, 1995). This means that
urea fertilizer applied to irrigated rice in Asia would increase from 15.5 to
43.6 million tons–nearly a 300% increase in N use for a 63% increase in
yield. By increasing fertilizer N recovery efficiency by 50%, it is possible to
reduce N application rate from 280 to 187 kg ha^{-1} and urea fertilizer need
from 43.6 to 29.1 million–still a 200% increase in fertilizer N application for
a 63% increase in yield (Cassman and Pingali, 1995).

During 1960 to 1990, genetic improvements leading to development of
fertilizer responsive rice and wheat varieties and improved management
strategies resulted in a dramatic rise in productivity and production of the
rice-wheat system. Both rice and wheat are exhaustive feeders, and the
double cropping system is heavily depleting the soil of its nutrient content. A
rice-wheat sequence that yields 7 t ha^{-1} of rice and 4 t ha^{-1} of wheat
removes more than 300 kg nitrogen, 30 kg phosphorus, and 300 kg ha^{-1} of
potassium from the soil. Even with the recommended rate of fertilization of
this system, a negative balance of the primary nutrients still exists, particular-

ly for nitrogen and potassium. The system in fact, is now showing signs of fatigue and is no longer exhibiting increased production with increases in input use.

Soil degradation is caused by nutrient imbalance (deficiency or toxicity), salinity/alkalinity, water logging, subsoil compaction, and declining organic matter quality and soil N supply (Kundu and Ladha, 1995). Nutrient management needs are predominantly farmand field-specific; knowledgeable decisions on the quantity and time of nutrient application are a necessity. The success of these decision-making processes is expressed as high crop yields. For both the low- and high-resource input conditions, both the crops (rice and wheat) must use the nutrients contained in the soil and that applied through fertilizers and manures efficiently, to reach their yield potential under this system. The requirements of nitrogen, phosphorus, and potassium for high rice and wheat yields vary, depending upon the soil and climatic conditions.

Governments in Asia are moving away from fertilizer subsidies–a practice initiated in the 1970s. While the subsidy structure may have accounted for excessive growth of fertilizer inputs to intensive rice systems, the farmers' current decisions on what can profitably be used in crop production may be guided more by fertilizer market prices. Efficient fertilizer use will produce healthy plants that are less vulnerable to pests, diseases, and to lodging. Optimal crop management requires farmer knowledge of matching inputs to crop production needs (Pingali et al., 1995). The overuse or improper user of nutrient inputs is highly damaging to crops and the environment. Excess nitrates pollute not only the soil and groundwater but also the produce itself. Nitrous oxide released from denitrification of nitrates pollutes the air. Therefore, to minimize health risks, new methods and/or products must be developed to increase nutrient use efficiency. Efficient fertilizer use will minimize water pollution by nitrates, phosphates, and will reduce accumulation of free nitrates in food.

SOIL FERTILITY AND FERTILIZER USE
IN SOUTH ASIAN RICE-WHEAT SYSTEM

Diagnostic surveys (Yadav et al., 2000) have indicated that farmers apply high doses of N (130 to 195 kg N ha^{-1}) to non-basmati rice in Punjab, Haryana and western Uttar Pradesh states of India (Trans-Gangetic plain and parts of Upper Gangetic plain). In most cases, diammonium phosphate (DAP) is applied as a basal dressing and urea, is top dressed in 2 to 3 equal split doses. Application of 11 to 14 kg P ha^{-1} is common in these areas, but K fertilization is limited. In Middle Gangetic plain (eastern Uttar Pradesh and Bihar), 50 to 100 kg urea ha^{-1} 50 to 120 kg DAP ha^{-1} are applied as basal dressings, and 50 to 100 kg urea ha^{-1} is top dressed in 2-3 splits. In Lower

Gangetic plain (West Bengal) fertilizer use is very low. Rice and wheat receive a mixture of 75-100 kg ha^{-1} urea and 60-80 kg ha^{-1} DAP as basal dressing and only 50 kg ha^{-1} urea is applied as top dressing. In wheat, use of 95 to 200 (average 153) kg N ha^{-1} and 13 to 24 (average 17) kg P ha^{-1} is common throughout the rice-wheat system in South Asia. In western and central parts of Uttar Pradesh, 100 to 125 kg DAP ha^{-1} is applied as a basal dose and urea N is top dressed. In Bihar surveys of farmer's fields revealed that 50-100 kg urea ha^{-1} and 75-125 kg DAP ha^{-1} is applied to wheat grown after rice. Potassic fertilizers are rarely used in these areas. Rice-wheat farmers in the IGP seldom adopt recommended fertilizers.

Fertilizer use pattern for the rice-wheat system varies greatly within the Indo-Gangetic Plains. For example, of the 30 districts of Punjab and Haryana states of northwestern India, more than 150 kg (N + P_2O_5 + K_2O) ha^{-1} was applied in 18 districts (Table 1). But 82 of 120 districts of the eastern Indian states of Uttar Pradesh, Bihar and West Bengal applied less than 100 kg (N + P_2O_5 + K_2O) ha^{-1}. Fertilizer use, N/P_2O_5 and N/K_2O ratios have simultaneously widened in most of the Indo-Gangetic Plains (Table 2). Reduction in subsidies of phosphate and potash in India adversely affected their use, while nitrogen remained heavily subsidized. This resulted in continued imbalance fertilizer use, the N:P_2O_5:K_2O ratio in 1996/97 was 9.9:2.9:1 compared to 6.3:2.3:1 in 1970/71 (FAO, 1998).

The productivity of both rice and wheat in the Indian states where these crops are grown sequentially is determined by the total consumption of N + P_2O_5 + K_2O (Table 3). Fertilizer N-use constitutes the major component of the total fertilizer use pattern, in the IGP. Interestingly N-use data for the state of Bihar was higher than that of W. Bengal, but the rice and wheat productivity trends were opposite (Table 3). It seems that an N/P_2O_5 ratio of around 2 in West Bengal compared to around 6 in Bihar proved favorable for better

TABLE 1. Classification of districts according to the range of annual nutrient consumption (N + P_2O_5 + K_2O) in different states of India in the Indo-Gangetic Plains in 1996-97.

State	Number of districts	Range of nutrient consumption (kg ha^{-1})				
		> 200	200-150	150-100	100-50	< 50
Punjab	14	1	10	5	1	–
Haryana	16	3	4	5	3	1
Uttar Pradesh	64	2	7	25	14	15
Bihar	39	3	1	9	18	24
West Bengal	17	1	1	4	11	–

Source: Fertiliser Association of India (1999)

TABLE 2. Ratios of $N:P_2O_5:K_2O$ and nutrient consumption per hectare of cropped area in India.

States	1984-85		1991-92		1997-98	
	$N:P_2O_5:K_2O$	$N + P_2O_5 + K_2O$ (kg ha^{-1})	$N:P_2O_5:K_2O$	$N + P_2O_5 + K_2O$ (kg ha^{-1})	$N:P_2O_5:K_2O$	$N + P_2O_5 + K_2O$ (kg ha^{-1})
Punjab	33.9:11.9:1	151.2	51.5:17.7:1	168.4	45.2:12.9:1	170.9
Haryana	35.8:7.4:1	57.7	62.6:31.8:1	112.8	171.0:47.8:1	140.1
Uttar Pradesh	14.7:3.4:1	65.1	16.8:4.4:1	88.7	26.0:6.3:1	117.7
Bihar	7.8:1.8:1	35.9	9.1:2.8:1	57.9	11.5:2.9:1	87.2
West Bengal	3.9:1.4:1	54.8	2.5:1.3:1	90.5	3.2:1.5:1	112

Source: Fertiliser Association of India (1999).

TABLE 3. Productivity of rice and wheat in relation to fertilizer use, irrigation and N:P ratio in the Indo-Gangetic Plains of India in 1996-97.

States	Nutrient use, kg ha^{-1}		% of gross irrigated area of gross cropped area	N/P_2O_5		Productivity (t ha^{-1})	
	$N + P_2O_5 + K_2O$	N		Rabi	Kharif	Wheat	Rice
Punjab	157.0	125.0	93.7	3.5	5.4	4.24	3.40
Haryana	127.7	103.8	76.4	3.0	12.5	3.88	2.97
Uttar Pradesh	107.6	86.3	67.4	4.0	6.1	2.66	2.12
Bihar	79.6	64.8	47.9	6.1	6.6	2.17	1.43
West Bengal	102.8	60.6	35.0	2.2	2.6	2.39	2.18

Source: Fertiliser Association of India (1999).

productivity in the former. The data shown in Table 3 indicate that percentage of irrigated area in different states substantially influenced the productivity of rice and wheat. A positive interaction between water and nutrient use in the rice-wheat system is well documented.

Soils in the Indo-Gangetic Plain generally contain sufficient exchangeable K and K bearing minerals (illite) that release exchangeable K to meet crop requirements. Total K in alluvial soils of IGP ranges from 1.28 to 2.77% and exchangeable K contents of 78-273 mg kg^{-1} soil (Tandon and Sekhon, 1988). Most of the soils in Bangladesh are low in exchangeable K. Of the 73 samples analyzed, 52% contained 0.15 cmol of K kg^{-1} (Kawaguchi and Kyuma, 1977). Potassium is now becoming a limiting factor in crop production in intensively cropped areas of rice and wheat in many south Asian countries. Deficiencies of N, P and K are also common in Bangladesh (Islam, 1995). All the soils under rice-wheat system are deficient in N. Phosphorus is deficient in 57%, and K in 85% of the sampled soils (Islam and Hossain, 1986). Most of the soils in Nepal are low in N and P status. The most extensive nutrient deficiencies are those of N, P and K because these are the nutrients which rice and wheat absorb in large amounts. However, long-term fertilizer experiments conducted on the rice-wheat system in the Indo-Gangetic Plain indicates that the availability of nutrients can be maintained with continuous use of fertilizers (Table 4) (Yadav et al., 2000).

NUTRIENT REMOVAL BY THE RICE-WHEAT SYSTEM

Rice-wheat cropping system annually removes 270-680 kg ha^{-1} of N, P and K (Table 5). The nutrient removal depends on the production level, soil type and whether crop residues are removed or recycled into the soil. When crop residues are retained in the field, large amounts of nutrients are recycled. Average N, P and K uptake per ton of grain is about 24.5, 3.8 and 27.3 kg for wheat and 20.1, 4.9 and 25.0 kg for rice, respectively (Tandon and Sekhon, 1988).

TABLE 4. Soil fertility changes in long-term rice-wheat rotation experiments under continuous application of recommended NPK doses (120 kg N + 26.2 kg P + 33 kg K ha^{-1}) in the Indo-Gangetic Plains.

Fertility Parameter	Initial	After 11 years	Change (%)
Trans-Gangetic Plain (Ludhiana)			
Organic C (%)	0.31	0.43	36.7
Available P (mg kg^{-1})	5.09	10.09	98.2
Available K (mg kg^{-1})	46.00	42.00	−8.7
Upper Gangetic Plain (Pantnagar)			
Organic C (%)	1.43	0.95	−33.6
Available P (mg kg^{-1})	9.09	9.68	6.5
Available K (mg kg^{-1})	65.00	66.00	1.5
Middle Gangetic Plain (Faizabad)			
Organic C (%)	0.37	0.40	8.1
Available P (mg kg^{-1})	6.27	9.27	47.8
Available K (mg kg^{-1})	161.00	133.00	−17.4
Lower Gangetic Plain (Kalyani)			
Organic C (%)	0.92	0.94	2.2
Available P (mg kg^{-1})	7.27	6.77	−6.9
Available K (mg kg^{-1})	36.00	33.00	−8.3

Adapted from Yadav et al., 2000.

TABLE 5. Nutrient (N, P and K) removal by rice-wheat cropping system.

Cropping system	Total productivity (t ha^{-1})	Nutrient uptake (kg ha^{-1})			References
		N	P	K	
Rice-wheat	13.2	278	53	287	1
Rice-wheat-cowpea	9.6 + 3.9 (fodder)	272	67	324	2
Rice-wheat-jute	6.9 + 2.3 (fibre)	170	33	212	2
Rice-wheat	10.7	185	38	271	3
Rice-wheat	8.8	235	40	280	4
Rice-wheat	5.7	124	17	130	5
Rice-wheat-rice	8.1	164	34	185	5

1. Kanwar and Mudahar (1986); 2. Nambiar and Ghosh (1984); 3. Saggar et al. (1985); 4. Sharma and Prasad (1980); 5. Saunders (1990).

Removal of P and K by the rice-wheat system far exceeds its additions through fertilizers and recycling. Total P uptake by rice-wheat system ranges from 17 to 67 kg ha^{-1} (Table 5). The amount of K removed by rice-wheat cropping system is quite high and ranges from 212 to 363 kg ha^{-1}. Dobermann, Cruz and Cassman (1995) reported estimates of K removal by modern rice varieties ranging from 20-50 kg K t^{-1} of rough rice in India. Optimum application of N increased K uptake by 57% over control plots and N and P application increased K uptake by 145% (Tandon and Sekhon, 1988). Long-term studies have indicated that continuous rice-wheat cropping will lead to depletion of K in soil even when optimum levels of fertilizer K have been applied (Tandon and Sekhon, 1988; Meelu et al., 1995). The quantities of nutrient removed by rice and wheat crops are important parameters for evaluating the fertilizer requirement.

NITROGEN

Transformations and Losses of Nitrogen from the Soil-Plant System

In any given ecosystem, different amounts of various forms of N form products that become substrates for other N transformations. Because N supply, transformations and transfers are linked, understanding the N economy of a rice-wheat system and its surroundings requires accurate understanding of the rate limiting steps of different N turnover processes. Even under the best circumstances, no more than two third of the N added as a fertilizer or as biological N_2 fixation, can be accounted for as being utilized by crop plants or to be present in the soil. Losses of one-half of the amount of N applied are commonly observed. Nitrogen from rice-wheat systems can more often be lost via ammonia volatilization, denitrification and leaching. Since the major portion of total fertilizer N used in the rice-wheat cropping system in the Indo-Gangetic Plains is urea, the rate of nitrification is a primary determinant of N losses. Nitrate ions are a substrate for denitrification. Nitrification rates in excess of NO_3 utilization by plants can result in increased levels of NO_3-N in run-off and groundwater rendering it unsafe for human consumption. On the other hand, slow rates of nitrification would result in accumulation of NH_4-N, which may enhance fertilizer-use efficiency by reducing denitrification and leaching losses. However, reduced nitrification could also increase N losses via NH_3 volatilization. Thus nitrification acts as a key process in determining fertilizer-use-efficiency by crops as well as N losses from soils.

Nitrification

The major factors governing nitrification include soil aeration status, concentration of NH_4-N, temperature, soil pH and soil texture. Bhupinderpal-Singh, Bijay-Singh and Yadvinder-Singh (1993) observed that the rate of nitrification in several semiarid subtropical soils was optimum at $35°C$ compared to the optimum temperature of 20 to $25°C$ in temperate soils, thereby indicating adaptation of soil nitrifiers to a high temperature optimum and tolerance. Kuldip-Singh et al. (1996) observed that under upland conditions, rapid nitrification in neutral soils (7 mg N kg^{-1} d^{-1}) resulted in complete oxidation of applied 100 mg NH_4-N kg^{-1} within 10 days. However, nitrification potential as affected by soil pH, was modest in alkaline soil with pH 9.8 (3 mg N kg^{-1} d^{-1}) and lowest in the acidic soil with pH 4.8 (1 mg N kg^{-1} d^{-1}). Nevertheless, substantial nitrification observed in acidic and alkaline soils suggested adaptation of nitrifiers to the prevailing soil pH. Soil texture has no effect on nitrification in upland conditions provided aeration status is comparable (Kuldip-Singh, 1993).

In several regions of the Indo-Gangetic Plains, continuous flooding for rice usually cannot be maintained either due to the shortage of water or due to high water percolation rates of the soil, leading to alternating flooded (reduced or anaerobic) and drained (oxidized or aerobic) conditions. Therefore, nitrification occurs during "dry spells" and nitrates thus produced are subsequently reduced to N_2 and N_2O through denitrification when soils are re-flooded. Although there is much variation in the data on nitrification rates estimated at different locations in the Indo-Gangetic Plain, a trend with respect to soil temperature and pH is easily discernible (Aulakh and Bijay-Singh, 1997). Incorporation of crop residues can also influence nitrification in flooded soils. While incorporation of wheat residues did not influence the rate of nitrification in the upland soil, its successive levels in flooded soil showed profound effect on the rate of nitrification (Aulakh, 1989).

Ammonia Volatilization

To wheat, urea is generally applied after the irrigation event to minimize leaching and denitrification losses. However, placement of urea on the wet surface of alkaline porous soils promotes ammonia volatilization. For example, in a field study with wheat (percolation rate 78 mm d^{-1}) comparing ^{15}N labeled (i) urea, (ii) urea plus dicyandiamide (DCD), a nitrification inhibitor, (iii) urea plus phenylphosphorodiamidate (PPD), a urease inhibitor, and (iv) KNO_3 applied before or 20 h after the irrigation event, Katyal et al. (1987) completely recovered top dressed KNO_3 in the soil/plant system after the harvest of the crop (Table 6). Thus, leaching and denitrification were not involved as significant mechanisms under soil and water management systems for crops like wheat. On the other hand, losses from urea-based fertilizers were substantial with 30-35% loss from urea-PPD and 21% from urea-DCD. With negligible leaching and denitrification, volatilization of NH_3 seems to be the most likely mechanism of N loss. The timing of fertilization and irrigation could further influence the losses of urea applied to porous soils. If applied on the wet soil surface following irrigation, as much as 42% of the applied ^{15}N was lost (Table 6), most likely due to volatilization. Shankaracharya and Mehta (1971) earlier espoused the notion that irrigation following fertilizer application reduces volatilization of NH_3 based on research conducted on sandy soils. Deep placement of urea due to its application before irrigation and resultant reduction in losses of applied ^{15}N from 42 to 15% (Table 6) is very well demonstrated from the work of Katyal et al. (1987). This finding has an important bearing on the losses of applied N via this mechanism in soils under rice-wheat system. Hence, where NH_3 volatil-

TABLE 6. ^{15}N balance in a field study for basal and top dressed applications of N to wheat grown on a porous soil.

N application	N source	^{15}N recovered (% of applied N)			Losses
		Straw	Grain	Soil	
Basal	Urea	7.2de*	28.0c	34.7bc	30.1cd
	Urea + PPD	8.4cde	28.5c	28.5cd	34.6de
	Urea + DCD	6.8de	30.9c	41.6a	20.8bc
	KNO$_3$	14.4ab	50.3a	30.2b	5.1a
Top dressed	Urea	6.7de	23.8c	27.4cd	42.2ef
after	Urea + PPD	8.1de	31.7bc	27.0d	33.2de
irrigation	Urea + DCD	6.2e	23.1c	24.5d	46.2f
Top dressed	Urea	11.5bc	40.3ab	32.4bcd	15.8b
before	Urea + PPD	9.2cd	30.5c	27.4cd	32.9de
irrigation	Urea + DCD	11.3bc	43.0a	32.0b	13.7b
	KNO$_3$	15.0a	49.6a	37.1ab	−1.7a

Figures in the same column with a common letter are significantly different (p = 0.05).
Reproduced with permission from Katyal et al., 1987.

ization is likely to be a problem, placement of fertilizer N beneath the soil surface could be a useful management technique to reduce NH_3 volatilization losses and improve utilization of N by crops.

When factors such as pH, temperature, wind velocity and CEC of the soil are favorable for ammonia volatilization, loss of N will be determined by the concentration of NH_4 in the floodwater of flooded rice soils. However, in coarse textured soils, NH_4-N in floodwater is readily transported to subsurface soil layers along with the percolating water. Ammonium-N placed at depth cannot move back up to the soil surface because of regular downward flux of percolating water under repeatedly irrigated rice culture. Thus, in contrast to ideal rice soils, highly permeable soils under wetland rice do not favor substantial losses of N via ammonia volatilization. For instance, in a field experiment with rice conducted on highly percolating porous soils, Katyal et al. (1985b) found that at mid day when temperature of the flood water was in the range of 28 to 39°C, and pH and alkalinity were very conducive for loss of N via ammonia volatilization, NH_4-N concentration in the floodwater even one day after application of urea was negligible (< 5 mg N 1^{-1}). But when leaching in porous soils was checked in microplots, NH_4-N concentrations in the floodwater increased rapidly from 2 to as high as 40 mg N 1^{-1} (Figure 1) thereby creating conditions for substantial losses through NH_3 volatilization.

Denitrification Under Wetland Rice

In submerged soils under rice, nitrification in oxidized soil zones and floodwater converts the ammonical N formed by ammonification and hydrolysis of urea into nitrate N. The nitrate N can thereafter move into reduced soil zones where it is readily denitrified. Thus, water management in lowland

FIGURE 1. NH_4-N concentration in the floodwater of leached (L) and unleached (NL) microplots after application of prilled urea and urea supergranules (1 g size) (adapted from Katyal et al., 1985b).

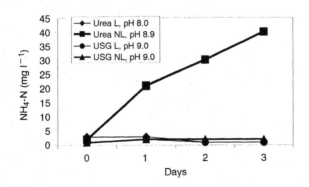

rice fields can influence the extent of N losses due to nitrification-denitrification. Alternating aerobic-anaerobic conditions in soils under rice could result in greater total N loss from the soil than would be found under continuous anaerobic condition (Aulakh, Khera and Doran, 1998). Organic N, including urea N is converted to NH_4 under both flooded and drained conditions, though at different rates. Under aerobic condition, NH_4-N thus formed is oxidized to NH_3 and NO_3-N. When the soil is reflooded NO_3-N is denitrified under anaerobic conditions.

Depending upon the soil texture, water management and profile characteristics of a porous soil, length of each aerobic and anaerobic period in the soil may vary. Reddy and Patrick (1975, 1976) observed a 24.3% total N loss in a soil system that underwent maximum numbers of alternate aerobic and anaerobic periods of 2 days each during a 4-month incubation period. Increasing the duration of aerobic and anaerobic periods reduced total N loss. In lysimeters packed with 30 cm deep sandy loam soil with a bulk density of 1.35 g cm^{-3}, rice plants were grown for 60 days and 30 mm depth of irrigation water was applied every 24 h (Bijay-Singh et al., 1991b). Using NO_3 as an N source and estimating N-uptake by plants, leachate analysis revealed 40% N-loss via denitrification. Under field conditions too, by applying ^{15}N-labeled urea to rice grown in a porous soil Katyal et al. (1985a) could show, though indirectly, that around 50% of the applied N was lost via nitrification-denitrification. In this experiment, irrigation to a depth of 5 cm was applied daily for the first month after transplanting of rice and every alternate day thereafter. During the first 30 days of crop growth, the water disappeared within 8-10 hours after irrigation.

Very limited information is available on the direct measurement of gaseous N losses via denitrification in porous soils and in semiarid subtropical regions. Aulakh et al. (1998) has conclusively shown that in partially anaerobic soil system simulating highly percolating soil under irrigated rice cultivation, occurrence of concurrent nitrification-denitrification resulted in 50% loss of applied ammoniacal fertilizer N.

Leaching

Leaching losses of N mainly occur as NO_3. The leaching of NH_4 in soils may not be a problem except when applied in very large quantities on coarse textured soils having low cation exchange capacity. Leaching of unhydrolyzed urea, particularly when applied as large granules can also occur in coarse textured soils under wetland rice. Under ideal lowland rice conditions generally achievable on nonporous fine textured soils, nitrification proceeds very slowly leading to production of a small amount of nitrates. Due to restricted downward percolation, even the possibility of nitrates to leach beyond the rice root zone is very small.

In a highly percolating soil under rice-wheat system, Katyal et al. (1987) studied distribution of ^{15}N in soil profiles under wheat to which urea, urea-PPD, urea-DCD and KNO_3 were either basally applied or top-dressed before or after irrigation. Despite a high percolation rate for the soil (78 mm d^{-1}), leaching of applied N was not found to be significant. In fact, the applied nitrate was fully recovered between the plant and soil. In case of urea-based sources, about 75% of the soil ^{15}N was located in the top 15 cm, reflecting little or no leaching of N. In contrast when applied before irrigation, ^{15}N derived from urea or urea-DCD moved down to 15- to 30-cm layer. The most pronounced movement was observed for KNO_3 applied before irrigation with 47% of the soil ^{15}N found between 50 and 90 cm depth.

Application of 100 kg urea-N ha^{-1} to wetland rice grown in highly percolating soil in 10 equal split doses at weekly intervals, rather than 3 equal split doses at 7, 21 and 42 days after transplanting did not increase rice grain yield and N uptake significantly (Bijay-Singh et al., 1991b). Results suggested that leaching losses of N were not substantial. In lysimeters planted with rice, leaching losses of N as urea, NH_4, and NO_3 beyond 30 cm depth of a sandy loam soil for 60 days were about 6% of the total urea-N and 3% of the total ammonia sulphate-N applied in three equal split doses. Application of urea even as a single dose at transplanting resulted in substantially lower N leaching losses (13%) compared to those observed from KNO_3 (38%) applied in three split doses (Figure 2). In lysimeters, unlike in field situations, soil solution percolates down under saturated conditions, which favor denitrification. Recovery of around 60% KNO_3-N in leachate and in the plants suggested that the remaining 40% was apparently lost via denitrification. Supply of

FIGURE 2. Cumulative leaching losses of N as urea, NH_4, and NO_3 and apparent fertilizer N recovery by rice grown in lysimeters. In each lysimeter a total of 848-mg N equivalent to 120 kg N ha^{-1} was applied (adapted from Bijay-Singh et al., 1991).

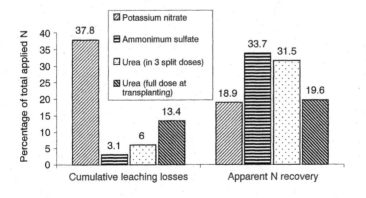

organic carbon to the denitrifiers in the coarse textured soil seemed to determine the amount of N lost through denitrification. Prevalence of frequently alternating aerobic-anaerobic soil conditions should have favored NO_3 formation under the experimental conditions.

In a field study using labeled urea, Katyal et al. (1985a) observed that less than 4% of the applied ^{15}N was recovered from the soil below 15 cm at maturity of rice in a permeable (percolation rate 4.5 mm h^{-1}) loamy sand soil. The leaching losses of applied N below 30 cm were found to be negligible. These observations along with data pertaining to ^{15}N balance in rice with different N sources applied at 116 kg N ha^{-1} and grain yield of rice (Figure 3) throw more light on N loss mechanisms operating in highly percolating porous soils. The fate of urea applied as supergranules weighing 1 g each (USG) merits particular attention in this regard.

Urea super granules were developed for easy application of urea in the reduced zone of puddled rice soils so as to minimize N loss through NH_3 volatilization (Savant, DeDatta and Craswell, 1982). On the basis of data on floodwater parameters (pH, alkalinity and NH_4-N concentration) in the plots to which USG was applied, Katyal et al. (1985a) reported that NH_4 volatilization was negligible even in highly percolating soils. Katyal, Bijay-Singh and Craswell (1984) have observed that irrespective of leaching restritons in field microplots, deep placement of USG did not allow a significant build up of NH_4-N in the flood water and thus could effectively reduce ammonia volatilization losses. Field experiments conducted by Bijay-Singh and Katyal (1987) and Rana et al. (1984), however, revealed that USG applied in the reduced zone of coarse textured soils proved to be least efficient source of N for increasing rice yield. Katyal et al. (1985a) found that percent recovery of ^{15}N at maturity of rice was only 9% from USG as compared to 27% from

FIGURE 3. [15]N balance in a field study at maturity of rice for different N sources applied at 116 kg N ha^{-1} (adapted from Katyal et al., 1985a).

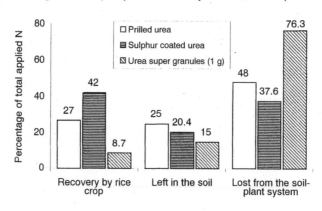

prilled urea (Figure 3). Total ^{15}N recovery from soil and plant was 24% from USG as compared to 52% under prilled urea treatment. Leaching of N from USG is a major cause of its poor efficiency in wetland rice grown in highly percolating porous soils. Distribution pattern of ^{15}N in the profile showed that substantial losses of N from USG should have occurred via leaching below 70 cm. In a lysimeter study Katyal et al. (1985a) found that a substantial portion of USG-N both as urea and NH_4 was recovered in the leachate within 2 days of its application. Vlek, Byrnes and Craswell (1980) showed that at a percolation rate of 8.1 mm d^{-1} an average of 5% of USG-N was leached as urea during the first 7 days. But at a percolation rate of 18.3 mm d^{-1}, an average of 72% of the USG-N was leached as urea by fourth day. Apparently, at high percolation rates, leaching of fertilizer urea is rapid enough to escape hydrolysis by soil urease. The unchanged nature of urea thus renders the soil retention capacity virtually ineffective.

NITROGEN FERTILIZER MANAGEMENT

Amount of N

Irrespective of its source, application of fertilizer N at 120 kg N ha^{-1} has been the recommended level in most of the rice growing regions. In eastern parts of the Indo-Gangetic Plains, lower N rates are being practiced (Table 7). Response of rice to application of N at different locations is shown in Table 8. In general, responses to application of N up to 120 kg N ha^{-1} are significant (Meelu et al., 1987) and economical keeping in view the cost of the fertilizer and value of the produce. A large number of experiments conducted on

TABLE 7. Recommended dose of N for the rice-wheat crop rotation in different Indian states and in Bangladesh.

States/Country	Rice	Wheat
Punjab	120	120
Haryana	150	150
Uttar Pradesh	100 (120)*	120
Bihar	75-90	100 (80)**
West Bengal	50-60	100
Bangladesh	T. Aus 60, T. Aman 80	80[§]

*Medium late maturing varieties, **Late sown wheat
Source: Package of practices for different crops/Handbooks of Agriculture issued by different State Agricultural Universities
[§]Bhuiyan et al. (1993).

farmer's fields under the All India Coordinated Agronomic Research Project confirm these observations (Narang and Bhandari, 1992). In soils testing low in organic carbon, significant and economical responses of rice have been recorded up to 150 kg N ha^{-1} (Bijay-Singh, Yadvinder-Singh and Maskina, 1987) and many farmers follow this recommendation.

Experiments conducted on research stations and farmers' fields reveal that wheat responded significantly to N application up to 120 kg N ha^{-1} with occasional responses observed up to 180 kg N ha^{-1} under intensive cropping (Table 9). Tandon (1980) reviewed work on fertilizer use in India and concluded that the optimum dose of N for wheat was in the range of 140-150 kg N ha^{-1}. The response to N seems to depend on soil characteristics such as organic carbon content, cropping intensity, length of the growing season, irrigation and weed management practices. Application of 25% higher dose

TABLE 8. Grain yield of rice (t ha^{-1}) as a function of N application levels at different locations in India.

Location	N levels (kg N ha^{-1})						References
	0	40	80	120	150	180	
Pura farm (U.P.)	3.22	3.74	4.19	4.47	–	–	Anonymous (1979)
Pantnagar (U.P.)	4.42	5.64[a]	6.26	–	–	5.85	Singh and Sharma (1979)
Ropar (Punjab)	3.41	6.01[a]	7.59	–	–	8.59	Thind et al. (1984)
Ludhiana (Punjab)	3.79	5.56[a]	–	6.69	–	6.64	Beri and Meelu (1983)
Ludhiana (Punjab)	3.50	5.10	–	6.30	6.70	6.90	Bijay-Singh et al. (1987)
Hissar (Haryana)	3.57	4.49	5.79	6.38	–	–	Singh et al. (1985)
Karnal (Haryana)	3.60	4.29	5.39	–	5.96	–	Singandhupe and Rajput (1990)
Ranchi (Bihar)*	1.38	2.16	2.58	–	–	–	Mahapatra and Srivastava (1984)

* Upland rice; [a] At 60 kg N ha^{-1}.

TABLE 9. Grain yield of wheat (t ha^{-1}) as a function of N application levels at different locations in the Indo-Gangetic Plains.

Location	N levels (kg N ha^{-1})					References
	0	40	80	120	160	
Pantnagar (U.P.)	1.26	2.15	3.01	3.93	4.71	Rajput et al. (1984)
Ropar (Punjab)	2.47	3.36	4.05	4.55	3.99	Thind et al. (1984)
Ludhiana (Punjab)	2.57	3.61	4.82	5.65	5.36	Kapur et al. (1982)
Haryana	2.15	3.25*	–	4.03	–	Singh and Singh (1992)

* At 60 kg N ha^{-1}.

of N over rates applied to normal soils is a common recommendation for wheat grown in alkali soils.

Due to the adoption and spread of this intensive rice-wheat cropping system, the need for fertilizer use has increased. Recent data indicates that the application of more than 150 kg N ha^{-1} to rice following wheat in the north-western Indian states of Punjab, Haryana and western Uttar Pradesh. In this region, grain yields of 10 to 12 t ha^{-1} per annum is being harvested from the rice-wheat system. Soil N supply also influences responses of rice and wheat to applied N. In an investigation carried out by Adhikari et al. (1999), soil N-supplying capacities (SNSC), measured as total N accumulation from the zero-N plots, and grain yields without N additions were greater for rice than for wheat in both Nepal and Bangladesh. Higher SNSC in rice was probably due to greater mineralization of soil organic N in the warm, moist conditions of the monsoon season than in the cooler, drier wheat season. However, SNSC was not correlated with total soil N, two soil N availability tests (hot KCl-extractable NH_4 or 7-day anaerobic incubation), exchangeable NH_4/ or NO_3. Wheat in Nepal had greater N-recovery efficiency, agronomic efficiency of N and physiological efficiency of N than rice. Nitrogen internal-use efficiency of rice for all treatments in both countries was within published ranges of maximum sufficiency and maximum dilution. In wheat, the relationship between grain yield and N accumulation was linear, indicating that the mobilization of plant N to the grain was less affected by biotic and abiotic stresses than in rice.

N Sources

Urea is the main source of fertilizer N for the rice-wheat system in the Indo-Gangetic Plains of South Asia. Small quantities of calcium ammonium nitrate (CAN), ammonium chloride (AC), and ammonium sulphate (AS) are also available to farmers growing rice and wheat. Review of work done on sources of N in India suggests that urea, CAN, AC and AS are generally equally effective in wheat (Yadvinder-Singh, Meelu and Bijay-Singh, 1990; Meelu, Yadvinder-Singh and Bijay-Singh, 1990). CAN performed better than other sources on calcareous soils of Bihar (Gupta and Narula, 1973) and alkaline soils of Punjab (Thind, Singh and Gill, 1984) and Delhi (Prasad and Prasad, 1992). The fact that CAN contains both NH_4 and NO_3 should be a reason for its favorable effect on wheat.

The relative performance of N sources in rice have been found to be affected by soil conditions and fertilizer management practices, particularly the time and method of its application. In most instances, ammonium-forming fertilizers exhibited similar efficiency in increasing grain yield of rice. Nitrate containing fertilizers such as CAN when applied to rice proved less efficient (Table 10) because nitrate is prone to be lost via denitrification and

TABLE 10. Comparative efficiency of different sources of N in rice grown in the Indo-Gangetic Plains of India.

Location	Grain yield of rice (t ha^{-1})				References
	Calcium ammonium nitrate	Urea	Ammonium chloride	Ammonium sulphate	
Himachal Pradesh	3.72	3.54	3.50	3.63	Kanwar and Joshi (1964)
Punjab	6.42	7.48	7.16	–	Thind et al. (1984)
Punjab	5.10	5.31	–	–	Meelu et al. (1987)
Delhi	4.51	4.60	–	4.83	Prakasa Rao and Prasad (1980)
Haryana	4.80	5.36	–	5.51	Singh et al. (1985)

leaching under submerged soil conditions (Meelu et al., 1987). Ammonium sulphate, due to its strong acidic property, has been more efficient as a N source than urea on salt-affected soils (Yadvinder-Singh, Meelu and Bijay-Singh, 1990). Nitant and Dargan (1974) have shown that as compared to CAN and urea, ammonium sulphate was better source of N for wheat grown in alkali soils.

During the last two decades, substantial efforts have been made to increase the notoriously low fertilizer N use efficiency in rice and wheat through modifications in N sources. The two most studied categories of modified N sources are urea supergranules (USG) and coated slow release urea. Urea supergranules are granules each weighing 1 g, and are placed in the soil at 5-7 cm depth. This modified source of N has been extensively tested for rice in South Asia. It has been observed that depending upon the soil and agroclimatic conditions, one time application of deep placed USG produced on an average, yield benefit of 15-20% over that obtained by the same amount of N applied in split doses through prilled urea (Table 11). Deep placed USG, however, did not perform better than prilled urea in coarse textures soils with high percolation rates as are commonly found in the Indian Punjab (Table 11). In spite of distinct advantage of USG in fine textured soils, it is not popular amongst farmers due to the lack of a suitable mechanical applicator.

Slow release N fertilizer developed by coating urea granules with sulfur has been tested vis-à-vis ordinary urea in a large number of field trials on rice in the Indo-Gangetic Plains (Table 11). This material out performed prilled urea in almost all soil types. However, due to the exorbitant price of sulfur in South Asia (there are no S reserves in this region), yield benefits obtained by using sulfur coated urea is not economical. Recently, new slow release materials with coatings of different types of biodegradable polymers have been introduced. These modified N sources are yet being tested at different locations.

TABLE 11. Relative performance of modified urea sources in rice grown in the Indo-Gangetic Plains.

Location	Grain yield of rice (t ha^{-1})				References
	Urea	Urea super granules	Sulphur coated urea	Neem cake coated urea	
Ludhiana (Punjab)	6.27	4.97	7.80	–	Bijay-Singh and Katyal (1987)
Ludhiana (Punjab)	5.65	4.06	6.36	4.96	Meelu et al. (1983)
Gurdaspur (Punjab)	6.38	5.97	6.31	5.94	Meelu et al. (1983)
Varanasi (U.P.)	3.02	3.44	3.39	3.23	Singh et al. (1990)
Kanpur (U.P.)	2.90	3.20	–	2.94	Agrawal et al. (1980)
Pantnagar (U.P.)	5.80	6.40	6.40	–	Pandey et al. (1989)
Delhi	4.17	4.60	–	–	Prasad et al. (1989)
Karnal (Haryana)	5.32	5.10	–	–	Singandhupe and Rajput (1990)

Time and Method of N Application

Fertilizer N, applied at a time when crop needs are high, reduces N losses from the soil-plant system, resulting in higher N-use efficiency. Application of fertilizer N in three equal split doses at transplanting, tillering (21 days after transplanting [DAT]) and panicle initiation (42 DAT) is more efficient in terms of grain yield compared to its application in one or two doses irrespective of the N-source and the soil type (Meelu et al., 1987). Application of one-third to one-half dose of N at transplanting was found to enhance overall fertilizer N use (Meelu and Bhandari, 1978). Application of N at 7 DAT proved more beneficial than application of the first dose at transplanting, as crop needs of N during the first seven days are likely to be very small.

Improved fertilizer N management should aim at maximum N absorption during those crop growth stages when N is most efficiently translated into grain yield. In a field study conducted on non-traditional rice soil (Bijay-Singh et al., 1986), path coefficient showed that the rice grain yield response to applied N was due to higher panicle density and spikelet number. The 1000-grain weight was only slightly influenced by N fertilization. Since panicle density and spikelet number are known to be determined within 70 days after transplanting of rice, N uptake during this period was found to be critical in achieving a maximum grain yield response to applied N.

Fertilizer N use efficiency in wheat can be increased by N-application at appropriate crop growth stages, and in a manner that applied N is not prone to losses from the soil-plant system. Application of fertilizer N in two equal split doses–half at sowing and half at crown root initiation stage (along with first irrigation) is beneficial in increasing grain yield and N uptake of wheat and therefore is a general recommendation over a vast area under wheat. Wheat yields differences were not signficant for N-application in two or three equal split doses (Meelu et al., 1987). Application of the first half N dose with pre-sowing irrigation resulted in significantly higher wheat yield than its application at sowing (Sidhu, Sur and Aggrawal, 1994). Possibly, N applied along with pre-sowing irrigation was transported to lower soil depths and therefore was not prone to losses via ammonia volatilization. On coarse-textured soils, application of N in three equal split doses at sowing, along with first and second irrigation has been recommended. Chaudhary and Katoch (1981) observed 14% higher grain yield of wheat when N was applied in three equal split doses rather than two.

Drilling of half N at sowing of wheat has been shown to perform better than broadcasting (Khanna and Chaudhary, 1979). Top dressing of second dose of N after first irrigation increased N use efficiency compared to its application before the irrigation (Verma and Srivastva, 1995).

However, in highly percolating soils, application of N before irrigation resulted in the transportation of N to a depth, hence losses of N via ammonia volatilization were curbed. Therefore, better fertilizer N use efficiency and higher grain yields of wheat were observed (Katyal et al., 1987). Results of a number of field experiments showed that 33% of total N applied as basal and rest in two equal split doses at 21 and 42 days after sowing of wheat gave the best results in alkali soils.

Nitrification Inhibitors

A large number of chemicals have been proposed and tested for controlling nitrification of NH_4 in the soil. Nitrification inhibitors are particularly effective in reducing losses of N via leaching and denitrification in soils under rice and wheat where aerobic and anaerobic regimes frequently alternate. Dicyandiamide (DCD) and calcium carbide coatings of urea have significantly reduced losses of applied N via leaching and denitrification and thus increased N recovery and use efficiency in rice-wheat cropping system (Banerjee et al., 1990; Singh and Prasad, 1992). In a field experiment on a sandy loam soil in New Delhi, wheat yields with 80 kg urea-N ha^{-1} + DCD (10%) were higher than the application of 120 kg N ha^{-1} without DCD (Singh and Prasad, 1992). Neem (*Azadirachta indica*) seed extract contains compounds, which have nitrification inhibition properties. Grain yield increases of 4-12% for wheat and 10-20% for rice using neem coated urea have been demonstrated in several places in India (Agrawal, Shanker and Agrawal, 1980).

Need Based Fertilizer N Management in Rice Using Chlorophyll Meter

Nitrogen applied to soils when crop demand is low is subject to losses from the soil-plant system in large quantities leading to low fertilizer N use efficiencies. Recently, Peng and Cassman (1998) demonstrated that recovery efficiency (by difference method) of top dressed urea during peak demand periods of rice, such as panicle initiation, could be as high as 78%. Need based fertilizer N applications to rice therefore seem promising, provided the critical time of need for top dressing can be determined rapidly and with a low cost. Visual assessments of green leaf color by farmers for determining the needs for N fertilization of rice plants is subjective, and non-quantitative, and leaf color is influenced by sunlight variability. Soil tests for N fertilizer recommendations in flooded rice soils have proved elusive (Stalin et al., 1996; Adhikari et al., 1999). Recent research shows a close link between leaf chlorophyll content and leaf N content. Thus, chlorophyll meter, for example, SPAD 502 model developed by Minolta Camera Company, can quickly and reliably assess the leaf N status during different stages of crop growth. It has been used successfully by soil scientists in Nebraska, USA, for corn, by applying in-season N fertilizer doses. This has been done using sufficiency indices of chlorophyll meter readings of the plot in question divided by a well-fertilized reference plot or strips falls below 0.95 (Peterson et al., 1993; Varvel, Schepers and Francis, 1997). This method has the advantage of being self-calibrating for different soils, seasons and varieties.

For rice in Asia, the approach of using the chlorophyll meter has focused on determining a fixed critical reading that farmers can easily refer to in the field. Peng et al. (1996) suggested the use of 35 as a critical chlorophyll meter reading for a recent IRRI variety, IR72 in the dry season, when 30-kg N ha^{-1} top-dressings could be applied as SPAD readings fell below this number. The 35 reading was based on a correspondence to 1.4-g N m^{-2} leaf area, a number that was found to be fairly stable for a high yielding IR72 crop during the growing season. During field trials in the Philippines, the use of the 35 critical reading was found to result in higher AE (i.e., similar yields with less N fertilizer applied) compared to fixed split-timing schemes (Peng et al., 1996).

In a recent study by Bijay-Singh et al. (1999), use of critical chlorophyll meter reading (SPAD) of 35 to match N top-dressing with plant demand was not satisfactory for the two rice cultivars–PR 106 and PR 111. This method resulted in the application of only 60 kg N ha^{-1} resulting in low rice yields. On the other hand, applying 30 kg N ha^{-1} when SPAD reading fell below the critical value of 37.5 resulted in net saving of 30 kg N ha^{-1} to produce grain yields of rice equivalent to those obtained by applying 120 kg N ha^{-1} in fixed-time splits. This saving was observed only in 1997. Therefore use of critical SPAD values to manage N in rice may not be very reliable, as it seems to be influenced by the rice variety grown and by the climatic conditions that prevail. However, basal application of 30 kg N ha^{-1} has a distinct advantage over its non-application. Based on sufficiency index criteria, a total of 90 kg

N ha^{-1} was applied to both the rice cultivars in 1997 and 1998 and grain yields were statistical similar to those recorded with 120 kg N ha^{-1} applied in three fixed-time splits. Two applications of 30 kg N ha^{-1} were applied to correct for slight deficiencies below the 90% sufficiency index. These studies indicated that the use of a well-fertilized reference plot and the chlorophyll meter have potential to be used as in-season N status indicator for irrigated rice in northwest India. Yields similar to the "best management practice," usually 120 or more kg N ha^{-1} split three or four times at fixed growth stages were achieved with the chlorophyll meter using less N fertilizer. Therefore, the sufficiency index approach seem to adapt to different seasons and varieties.

ENVIRONMENTAL ASPECTS OF FERTILIZER NITROGEN USE

A number of environmental issues linked to N use in rice-wheat system can be identified. These issues include: nitrates in drinking water, nutrient enrichment, eutrophication and deterioration of surface water quality due to nutrient leaching and/or run off. The significance of N use in agriculture in polluting the air is being increasingly appreciated. Nitrous oxide originating from soils under rice can damage the ozone layer in the stratosphere. Except for nitrate pollution of ground water, investigations on other N related environmental issues are not available from the rice-wheat regions of South Asia.

In extensively irrigated regions of northwestern India, particularly in the state of Punjab, fertilizer application rates to rice-wheat system exceed 200 kg N ha^{-1} year^{-1}. In some central districts, fertilizer levels exceed 300 kg N ha^{-1} year^{-1} and possibly at several farms fertilizers are poorly managed. The soils in this region are predominantly coarse textured (sandy loam and loamy sand as the dominant textural classes) and about 75% of the total annual rainfall of more than 600 mm falls between July to September. NO_3 concentration of water from shallow (4 to 10 m deep) open wells located in cultivated areas decreased significantly with depth from the water table, and correlated positively with amount of fertilizer N applied ha^{-1} year^{-1} on the farms located in the vicinity of wells (Bijay-Singh and Sekhon 1976c). The amount of NO_3-N contained in the soil profile down to 210-cm depth in June correlated significantly with the NO_3 concentration of well water in September. Nitrates therefore reach the water table during the rainy season (July-September). Between 1975 to 1988, average fertilizer N consumption in the Indian Punjab increased from 56 to 188 kg N ha^{-1} year^{-1} (on net area sown basis). Monitoring of the NO_3-N concentrations in the shallow well waters in 1982 and 1988 (Bijay-Singh, Sadana and Arora, 1991a) revealed that the increase in fertilizer consumption correlated with an increase of NO_3-N by almost 2 mg l^{-1}. Bajwa, Bijay-Singh and Parminder-Singh (1993) investigated 236 samples from 21 to 38 m deep tube wells in different blocks of Punjab where fertilizer N consumption ranged from 151 to 249 kg N ha^{-1} year^{-1}. No significant correlation was observed between fertilizer consumption and NO_3 content of well waters, possibly because samples were taken from deep wells. Nevertheless, percentage of groundwater samples containing more than 5 mg NO_3-N l^{-1} from tube wells located in vegetable growing areas was 17 as compared to 3 and 6% from wells located in regions of rice-wheat and potato-wheat growing areas, respectively.

Vegetation retards NO_3 leaching from root zone by absorbing nitrates and water. Rooting habits of plants exert a profound influence on NO_3 mobility in the root zone. Bijay-Singh and Sekhon (1977) showed that maximum leaching of NO_3-N below the root zone occurred from crop rotations with heavily fertilized shallow rooted crops. Wheat, when grown in a rotation absorbed a large fraction of the applied N due to its deep rooting system. Bijay-Singh and Sekhon (1976a) were able to show that balanced application of N, P and K can significantly reduce the amount of unutilized nitrates in the root zone. When 120 kg N ha^{-1} was applied to wheat, a substantial portion was recoverable as NO_3 from the soil profile up to 200-cm depth. However, when 120 kg N, 26 kg P and 25 kg K ha^{-1} were applied, there was little NO_3-N in the soil profile, which could potentially leach below the root zone. Increasing the rate/amount of irrigation water and decreasing its frequency increased the amount of NO_3-N leaching to deeper soil layers (Bijay-Singh and Sekhon, 1976b). With lighter and more frequent irrigation schedules, NO_3-N from only the third or fourth split dose of fertilizer remained unutilized in the root zone. Ensuring a deep and extensive root system in the early stages of crop growth by adopting proper irrigation and fertilizer application schedule can considerably reduce the leaching of NO_3-N beyond the root zone. These results indicated the need to delay the application of large amounts of nitrogenous fertilizers until the crop can utilize it and also to avoid irrigation when a large amount of NO_3-N is present in the root zone.

INTEGRATED NUTRIENT MANAGEMENT USING ORGANIC MANURES AND FERTILIZERS

In the coming decades, a major issue in designing sustainable agricultural system will be the management of soil organic matter and the rational use of organic inputs such as animal manures, crop residues, green manures, sewage sludge and food industry wastes. Among organic manures, farmyard manure (FYM) is the most commonly used manure by the rice-wheat farmers in the Indo-Gangetic Plain. The rate and periodicity of application of FYM, however, depends on its availability with the farmers. Diagnostic surveys (Yadav et al., 2000) reveal that 15-20 t FYM ha^{-1} is applied at an interval of 3-5 years in Trans Gangetic Plain and Upper Gangetic Plain, while the average rates of FYM application are low in the Middle Gangetic Plain (5-6 t ha^{-1} at an interval of 2-3 years) and Lower Gangetic Plain (2-4 t ha^{-1} at 2-3 years interval). Since organic manures cannot meet the total nutrient needs of modern agriculture, integrated use of nutrients from fertilizers and organic resources is essential. The basic concept underlying the integrated nutrient management system (INM), nevertheless, is the maintenance and possible improvement of soil fertility status for sustained crop productivity on a long-term basis and also to reduce fertilizer input cost. The different components of integrated nutrient management is diverse in terms of chemical and physical properties and nutrient release patterns.

Animal Manures

Studies conducted on farmers fields under All-India Coordinated Agronomic Research Project have shown that combined use of 12 t FYM ha^{-1} and 60 kg N ha^{-1} produced rice grain yield equivalent to that obtained by the application of 120 kg N ha^{-1} (Table 12) (Kulkarni, Hukeri and Sharma 1978). Experiments conducted in Ludhiana showed that the application of 12 t FYM ha^{-1} in combination with 80 kg N ha^{-1} produced the same rice yield as that obtained from 120 kg N ha^{-1} (Maskina et al., 1988). In the succeding wheat crop, significant residual effect equivalent to 30 kg N and 30 kg P$_2$O$_5$ ha^{-1} was obtained due to FYM application to rice. At Kaul (Haryana), rice responded significantly to 120 kg N ha^{-1} on FYM amended plots compared to 180 kg N ha^{-1} on unamended plots (Agrawal et al., 1995). Long-term studies on integrated N management of the rice-wheat system conducted at several locations in the Indo-Gangetic Plains show that 25 to 50% of total N to rice can be supplied through FYM (Table 13).

The N mineralization rate in soil from poultry manure containing narrower C/N ratio is substantially higher than that from FYM (Yadvinder-Singh et al.,

TABLE 12. Effect of combined use or organic and inorganic fertilizers on crop yields (t ha^{-1}) in a rice-wheat rotation.

Fertilizer treatment		Bhagalpur alluvial (means of 87 trials)		Manipur red and yellow (means of 96 trials)		Ludhiana loamy sand alluvial soil (means of 5 years)	
Rice	Wheat	Rice	Wheat	Rice	Wheat	Rice	Wheat
F$_0$N$_0$	F$_0$N$_0$	2.18	1.57	3.33	0.57	2.8	1.3
F$_0$N$_{120}$	F$_0$N$_{60}$	4.21	2.75	4.41	1.04	5.6	2.8
[a]F$_{12}$N$_{60}$	F$_0$N$_{60}$	4.14	2.95	5.44	1.33	5.7	3.9

N = Nitrogen. [a] N$_{80}$ to rice and N$_{90}$ to wheat at Ludhiana. F$_0$ and F$_{12}$ refers to Farm Yard Manure at 0 and 12 t ha^{-1}.

TABLE 13. Integrated use of farmyard manure (FYM) and fertilizer N in rice-wheat cropping system.

Treatment		Grain yield (kg ha^{-1})	
Rice (% NPK* + % FYM)	Wheat (% NPK)	Rice	Wheat
Varanasi (Uttar Pradesh) 1985-86 to 1990-91			
100 + 0	100	3924	3968
50 + 50	100	3558	4226
Raipur (Madhya Pradesh) 1988-89 to 1990-91			
100 + 0	100	4528	1678
50 + 50	100	4654	1839
75 + 25	75	4546	1764
R.S. Pura (Jammu and Kashmir) 1985-86 to 1990-91			
100 + 0	100	4616	3445
50 + 50	100	4401	3549
Palampur (Himachal Pradesh) 1985-86 to 1990-91			
100 + 0	100	3136	2076
50 + 50	100	3075	2278
Kalyani (West Bengal) 1986-87 to 1990-91			
100 + 0	100	3573	2638
50 + 50	100	3390	2926
75 + 25	75	3787	2666

* Substitution through FYM based on N requirement.
Adapted from Hegde, 1992.

1988). In Ludhiana, application of 4 t poultry manure ha^{-1} along with 60 kg N ha^{-1} produced rice yield similar to that obtained from 120 kg N ha^{-1} urea (Bijay-Singh et al., 1997). Nitrogen from poultry manure and urea gave similar N use efficiency in rice.

The fertilizer equivalence approach can be effectively used to determine the value of organic manures when used in combination with inorganic fertilizers. Field studies have compared the rates of organic manures needed to obtain a target yield or N uptake by a crop, equivalent to that obtained with fertilizer N application. Yadvinder-Singh et al. (1995) reported that fertilizer N equivalents of FYM in rice ranged from 42 to 52% of the total N applied. Apparent N recovery was 20% from FYM as compared with 35 to 46% from urea. Similar fertilizer N equivalence for poultry manure worked out to be 80 to 127% for rice (Bijay-Singh et al., 1997).

In several studies, combined application of organic manures and chemical fertilizers generally produced higher crop yields compared to their single application. This increase in crop productivity may be due to many components present in the organic manures, their effects on improving soil physical and biological properties and partly due to synergism. The slow release of nutrients from manures and composts provides stable supplies of NH_4^+ thereby increasing crop yields. The effect of integrated use of organic and inorganic sources of nutrients is more spectacular in acidic and sodic soils than in normal soils.

Green Manures

Green manures (GM) constitute a valuable potential source of N and organic matter. *Sesbania aculeata* is a versatile flood tolerant green manure, highly adapted to varying soil and climatic conditions. *Crotalaria juncea* and *Tephrosia purpurea* are important drought tolerant legumes. A 45-60 days old GM crop can accumulate more than 200 kg N ha^{-1}, but 100 kg N ha^{-1} accumulation is more common (Bijay-Singh and Yadvinder-Singh, 1997). The 100-kg N ha^{-1} accumulation correspondence to the amount of mineral fertilizer N generally applied to crops in Asia. Green manuring is better suited for rice-based cropping systems. Integrated use of green manure and chemical fertilizers can save 50 to 75% of N fertilizers in rice (Meelu, Yadvinder-Singh and Bijay-Singh, 1994; Yadvinder-Singh, Meelu and Bijay-Singh, 1991). Green manuring also increases the availability of several other plant nutrients through its favorable effect on the chemical, physical and biological properties of soil (Yadvinder-Singh, Bijay-Singh and Khind, 1992; Meelu, Yadvinder-Singh and Bijay-Singh, 1994). In Bangladesh, N supplied by *Sesbania* green manure was effective for rice grown in coarse textured soils but its residual effect on the following crop of wheat was negligible (Bhuiyan et al., 1993).

The fertilizer equivalence of GM-N, the amount of split-applied urea N required to obtain equivalent yield, is generally higher in coarse textured than in fine textured soils (Becker et al., 1995). Green manure N is as efficient as fertilizer N and with its application at more than N-equivalence of 100 kg N ha^{-1}, no fertilizer N application will be required to raise a crop of rice (Bijay-Singh and Yadvinder-Singh, 1998). At application rates of less than 80 kg N ha^{-1} lowland rice uses GM-N more efficiently than urea-N. However, when GM-N is applied in excess of 100 kg N ha^{-1}, GM-N use efficiency declined more rapidly than that of urea. Integrated use of GM and inorganic fertilizers increased rice grain yield to a level not achieved by fertilizer application alone. This integrated nutrient use allows omission of N-fertilizer application during rice transplaning (Bijay-Singh et al., 1992).

Farmers have opted for fertilizer N-use because technical, socio-economic, and human resource constraints have limited the adoption of green manuring and BNF in cropping systems. Technical problems can be solved by basic and applied research, while socio-economic and human resuource constraints by education and training. Fitting legumes into cropping systems is a task of challenging complexity. Within the major cropping systems, more niches for legumes must be found and associated agronomic limitations must be overcome to exploit the direct and indirect contributions of BNF through green manuring. Sustainable production and availability of seeds for planting GM crops is the most important agronomic constraint. Other agronomic constraints include: tillage and seeding costs, successful establishment of GM crops, insect and pest attack, and the cost and ease of GM crop incorporation into the soil. Environmental constraints include factors that limit legume/rhizobia symbiosis, limited rainfall/irrigation oppurtunities during the dry and hot pre-monsoon North Indian summers, and soil phosphorous deficiency.

In northwestern India, short-duration legume pulses, e.g., mungbean, cowpea, etc., can be grown in the fallow period after the harvest of wheat. In the rice-wheat cropping system, incorporation of mungbean residue after the picking of pods, significantly increased rice yield and economized 60 kg N ha^{-1} (Rekhi and Meelu, 1983). In many experiments, the advantages of legume residues and green manuring to rice, were similar (Singh and Dwivedi, 1996).

Crop Residue Management in Rice-Wheat System

About 400 million tons of crop residues are produced in India alone. Crop residues are good sources of plant nutrients and are important components for the stability of agricultural ecosystems. In areas where mechanical harvesting is practised, a sizable quantity of crop residues are left in the field, which can be recycled for nutrient supply. On an average, 25% of N and P, 50% of S, and 75% of K uptake by cereal crops are retained in crop residues, making them valuable nutrient sources.

Traditionally, wheat and rice straw were removed from the fields for use as cattle feed and for several other purposes in South Asia. Recently, with the advent of mechanized harvesting, farmers have been burning *in-situ* large quantities of crop residues left in the field. This practice has important implications on the organic matter content of the soil. As crop residues interfere with tillage and seeding operations for the next crop, farmers often prefer to burn the residue *in-situ*, causing loss of nutrients and organic matter. Compared to its removal or burning, incorporation of straw builds up soil organic matter, soil N, and increases the total and available P and K contents (Tale 14) of the soil (Beri et al., 1995; Misra, Yadvinder-Singh and Bijay-Singh, 1996).

The major disadvantage of incorporation of cereal straws (wide C:N ratios) is the immobilization of inorganic N and crop growth may be adversely

TABLE 14. Effect of crop residue management on organic carbon (O.C.) and total nitrogen (N) content of soil under rice-wheat cropping system in India and Pakistan.

References	Type of crop residue	Duration of study (years)	Residue management	O.C. (%)	Total N (%)
Beri et al. (1995)	Rice straw in wheat and	10	Removed	0.38	0.051
	wheat straw in rice		Burnt	0.43	0.055
			Incorporated	0.47	0.056
Sharma et al. (1987)	Rice straw in wheat and	6	Removed	1.15	0.144
	wheat straw in rice		Incorporated	1.31	0.159
Zia et al. (1992)	Rice straw in rice	3	Removed	0.53	–
			Incorporated	0.63	–
Yadvinder-Singh et al.	Wheat straw, green manure	6	Removed	0.38	–
(2000a)	and wheat straw + green		Incorporated	0.49	–
	manure (GM) in rice		Green manure	0.41	–
			Straw + GM	0.47	–

affected due to deficiency of N. Incorporation of crop residues into rice or a wheat field significantly lowered crop grain yields (Beri et al., 1995). Wheat yield depression due to straw incorporation decreased from 0.54 t ha^{-1} to 0.08 t ha^{-1} with the application of 60 kg N ha^{-1} and 180 kg N ha^{-1}, respectively. Residue characteristics and, soil and management factors affect residue decomposition in soil. Under optimum temperature and moisture conditions, the period of immobilization can last as long as 4 to 6 weeks (Toor and Beri, 1991, Yadvinder-Singh et al., 1988). The long-term adverse effects of wheat straw incorporation can be averted by combined incorporation of green manure (narrow C:N ratio) and cereal straw (wide C:N ratio) into the soil before rice transplanting (Yadvinder-Singh et al., 2000a).

Incorporation of rice straw before wheat planting compared to wheat straw before rice planting is difficult due to low temperatures and short time interval between rice harvest and wheat planting. However, similar rice and wheat yields under different straw management practices (burning, removal or incorporation) have been reported (Walia et al., 1995; Yadvinder-Singh and associates, personel communitcation). Verma and Bhagat (1992) observed lower wheat yields during the first three years of rice straw incorporation 30 days prior to wheat planting, but in later years, straw incorporation did not effect wheat yields adversely. In contrast, rice straw incorporation gave significantly higher wheat yields of 3.51 t ha^{-1} compared to straw removal (wheat yield of 2.91 t ha^{-1}) in Pakistan (Salim, 1995). Singh et al. (1996)

reported that the incorporation of rice straw three weeks before wheat sowing significantly increased wheat yields on clay loam but not on sandy loam soil. The work of Koyama (1981) suggests that 10-20% of the N supplied through organic materials with a high carbon to nitrogen ratio like rice straw and stubble, is assimilated by the rice crop, 10-20% is lost through various pathways, and 60-80% is immobilized in the soil. In some experiments, addition of 10 t ha^{-1} rice straw at 4-5 weeks before transplanting of rice was equivalent to the basal application of 40 kg N ha^{-1} through urea (Chatterjee et al., 1979). Verma and Mathur (1990) found that the incorporation of rice straw (11.2 Mg ha^{-1}) along with cellulytic microorganisms and rock phosphate 15 days before sowing of wheat significantly increased wheat grain yield compared to the recommended fetilizer practice.

Results from the All India Coordinated Agronomic Research Project showed the beneficial effects of wheat residues when applied as a substitute for chemical fertilizer needs of rice in the rice-wheat cropping system (Singh and Dwivedi, 1996). During a 10-year (1984-1994) long-term field experiment (Rattan et al., 1996), comparisons were made between the application crop residues versus inorganic fertilizers on rice and wheat. In the first year of this study, inorganic fertilizer treated plots of rice and wheat yielded the highest. However, in the 2nd and 3rd year of this study, the combined application of wheat straw and inorganic fertilizer gave results similar to that of the inorganic fertilizer applied plots. Beyond the 4th year of this experiment, combination of wheat straw and inorganic fertilizer applied plots out yielded all the other treatments. Incorporation of wheat straw (10 Mg ha^{-1}) saved 50% of the recommended fertilizer dose (60 kg N + 13.1 kg P + 25 kg K ha^{-1}) and helped achieve higher yield of rice (Rajput, 1995).

The extent of N-immobilization due to soil incorporation of crop residues can be reduced by proper fertilizer management practices. These practices include optimum method, time and rate of fertilizer N application. Therefore, placement of fertilizer below the residue incorporated C-enriched surface soil layer, application of higher than recommended N-fertilizer dose, and delayed crop planting, are strategies to reduce the adverse effects of N-immobilization. The total grain yield of rice and wheat was higher for the combined treatment of 20 kg N ha^{-1} starter dose with straw incorporation compared to either burning of straw or straw incorporation without a starter dose (Misra et al., 1996). Similarly Bali et al. (1986) found that the application of 15 kg N ha^{-1} as starter dose with straw incorporation increased the yields of both the rice and wheat crops. However, using ^{15}N-labeled urea in a field experiment Bijay-Singh et al. (2000) found that the application of a part of the recommended N (25%) at the time of straw incorporation (to hasten decomposition of straw) led to large N losses and low wheat yields. Recovery of ^{15}N by wheat was maximum (41.8%) when rice straw was removed or burnt and minimum (30.4%) when 30 kg of 120 kg N ha^{-1} fertilizer was applied along with straw incorporation at 20 days before sowing of wheat. Yadav (1997) recorded similar observations from long-term field experiments (1990-96) at two locations in the Indo-Gangetic Plain. When an extra 20-kg N ha^{-1} was applied, higher crop yields were recorded. Pathak and Sarkar (1997) reported that at recommended fertilizer N level, rice straw incorporation produced lower rice yields than urea alone. A higher dose of urea-N application with rice straw incorporation was necessary to get good rice yields. The beneficial effect of straw incorporation before rice planting did not carry over to the succeeding wheat crop. In another study, application of 30-kg extra N ha^{-1} than the recommended fertilizer dose increased rice yields only slightly (Agrawal et al., 1995). Thakur and Singh (1987) also observed the benefits by applying higher fertilizer N doses to rice-wheat cropping system in soils amended with crop residues.

PHOSPHORUS

Phosphorus Fertilizer Management

Crop Responses to P Application

Based on data from single crop experiments at research stations and farmers' fields, recommendations were developed for P application to rice and wheat. Rice generally responded up to 13 kg P ha^{-1} while wheat responded to 26 kg P ha^{-1} (Abrol and Meelu, 1998). Rice and wheat crops respond to higher rates of applied P in high P fixing soils (Tandon, 1987). Response of wheat to 26 kg P ha^{-1} varied from 620 kg to 1,500 kg grain yield ha^{-1}.

It is estimated that cereal crops use 20-25% of applied P, the rest remain in the soil as residual P (Jain and Sarkar, 1979). Large build up of available P in soils has been observed with the long-term use of P fertilizers in the rice-wheat system (Prem Narain, Soni and Pandey, 1990; Nambiar and Ghosh, 1984). Therefore, fertilizer P applied to a crop must be considered as a component of the cropping system. It has been advocated that P should be applied to the wheat crop and the rice crop will then benefit from the residual P left in the soil (Meelu, Kulkarni and Bhandari, 1982). As P availability changes with alternate submergence and drying, the P applied to wheat shows greater residual effect on the succeeding rice, while the P applied to rice has less residual effect on the succeeding wheat (Palmer et al., 1990). Application of the total amount of required P to wheat doubled the P uptake and grain yield of wheat and of the succeeding rice crop compared to the total P application to the rice crop in China (Run-Kun, Baifan and Cheng Kwei,

1982). The availability of soil and residual fertilizer P increases under submergence and at high temperatures prevailing during the rice season, and rice has greater ability to utilize the residual P from Fe-P and Al-P fractions than wheat (Goswami and Singh, 1976; Gill and Meelu, 1983). Response of wheat to applied P are, therefore, more common than rice. Better utilization of applied P by the rice-wheat system has been obtained by P application to wheat than to rice (Tiwari et al., 1980). On soils highly deficient in available P, it should be applied to both the crops in the rice-wheat system, but the dose of P can be reduced for rice (Tiwari et al., 1980). In a long-term experiment on Mollisols (Pantnagar, India), Modgal et al. (1995) observed that the response of rice to applied P appeared after 8 years of continuous cropping, whereas in wheat, response to applied P was obtained just 2 years after initiation of the experiment.

Response to fertilizer P application at a given soil test value tends to be greater on sandy soils compared to silty or clayey soils. More fertilizer P is needed on soil with high clay content for obtaining the same yield response due to greater P fixation (Vig et al., 1986). Response to added P varies between crops and varieties of the same crop. This is because of differences in their P requirement and uptake efficiency for a given level of production.

In the sates of Indian Punjab, the rice crop did not respond to P application when 26 kg P ha^{-1} was applied to the preceeding wheat crop (Table 15). However, P balance remained negative with the application of only 26 kg P ha^{-1} to either the rice or the wheat crop. A total application of 39 kg P ha^{-1} to both crops maintained a positive P balance (Table 16). Continuation of these studies over 7 years revealed that current recommendation of applying 26 kg P ha^{-1} only to wheat is insufficient to sustain a neutral or positive P balance, high indigenous P supply, and high yields of wheat and rice over the long term (Yadvinder-Singh et al., 2000b). Furthermore, this study suggests that P should also be applied to rice at rates of >15 kg P ha^{-1} if rice yields of greater than 6 t ha^{-1} are being targeted. Similar conclusions could be drawn from a 4-year study (Faroda, 1992) on a clay loam soil of the Haryana state in India. This study showed that the optimum dose and time of P application to a rice-wheat system at that site was 26 kg P ha^{-1} for wheat and 13 kg P ha^{-1} for rice (Table 15).

Soil Tests for P and Crop Response to P Application

There is a great deal of literature on soil tests for available P. Irrespective of the wide variations in the P status of soils, Olsen-P for alkaline soils and Bray-1 for neutral and acidic soils are the most widely used indices of available soil P for rice and wheat in the Indo-Gangetic Plains of South Asia (Singh and Sharma, 1994). However, interpretation of soil tests for P for both the rice and wheat crops remains controversial. Critical levels of Olsen-P for

TABLE 15. Effect of P application on crop yields (t ha^{-1}) in a rice-wheat system of northwestern India.

Rate of P (kg ha^{-1})		Punjab (7 years average)[1]		Haryana (4 years average)[2]	
Rice	Wheat	Rice yield	Wheat yield	Rice yield	Wheat yield
0	0	4.0	2.3	5.0	3.0
0	26	6.6	4.1	5.9	4.3
26	0	6.5	2.4	5.8	3.8
13	26	6.6	4.2	6.2	4.4
26	26	6.6	4.2	6.3	4.6

[1]Yadvinder-Singh et al. (2000b); [2]Faroda (1992).

TABLE 16. Phosphorus balance over 7 years in a rice-wheat system on a loamy sand soil of Punjab, India.

Fertilizer P applied (kg ha^{-1})		Total P added in 7 years (kg ha^{-1})	Total P recovery by rice and wheat (kg ha^{-1})	P balance after 7 years (kg ha^{-1})
Rice	Wheat			
13	26	273	237	+36a
0	26	182	222	−40b
26	0	182	215	−33b
13	13	182	218	−36b
0	0	0	166	−166c
0	39	273	231	+42a

Values in the column followed by same letter do not differ significantly (p = 0.05).
Adapted from Yadvinder-Singh et al., 2000b.

rice ranges from 4 to 29 mg P kg^{-1} soils (Goswami, 1986; Panda, Panda and Jena, 1981). The behavior of P in flooded soils under rice markedly differs from that of upland soils under wheat. Flooding a dry soil causes an initial increase in the P concentration in soil solution because of reduction of Fe (III) compounds that liberate sorbed and co-precipitated P. The initial P flush, however, is followed subsequently by a decrease due to the resorption or precipitation of Fe (II)-P compounds (Ponnamperuma, 1972; Willet, 1986). Much of the reimobilized P becomes acid-soluble (Kirk and Saleque, 1995). A large proportion of P uptake by rice seems to be drawn from P pools that are soluble under alkaline aerobic conditions [Fe (III)-P and Al-P], but these are transformed into acid-soluble P forms due to submergence and reduction (Jianguo and Shuman, 1991). Thus, a suitable soil test for rice should provide an index of this pool.

Critical levels of Olsen-P for wheat are not unique for different soils. The critical level of Olsen-P for wheat grown in different soils is considered to be

9 mg P kg^{-1} soil, but significant responses to applied P have been obtained in soils with Olsen-P of up to 17 mg kg^{-1} soil (Goswami, Bapat and Pathak, 1971; Gill, 1992). Obviously, the soil tests for P are not providing a correct picture of the P that can be made available to wheat. Soils differ in their chemical and physical properties thereby influencing the availability of native and applied P. Alternatively, there is a need for revising critical thresholds for Olsen-P in wheat for different soil groups varying in clay mineralogy, physical and chemical properties. Therefore, P fertilizer requirement for targeting optimum wheat yields will depend on varietal yield potential, fertilizer effectiveness, and soil characteristics (Choudhary, Arora and Hundal, 1997). P sorption maxima and P buffering parameters may be included in fertilizer recommendation practices for wheat to improve reliability.

Ion-exchange resins can be used to estimate bioavailable nutrients dynamically, because they maintain low ion concentration in solution, thereby stimulating further release from soil solids (Skogley and Dobermann, 1996). Capsules containing a mix of cation and anion resin (H$^+$/OH$^-$ form) are placed in soil paste made from freshly sampled paddy fields or placed in-situ in-between rice and wheat rows (Yadvinder-Singh et al., 2000b; Yang et al., 1991). The adsorption kinetics of anions and cations are then measured in a subsequent 7-day and 14-day anaerobic incubation at 30°C. The capsule provides a strong and finite nutrient sink over long incubation periods and mimics some of the key processes occurring in the root zone of a rice and wheat field, including root nutrient uptake and rhizosphere acidification. Resin capsule sensitively measured the build up or depletion of soil P and K and provides a "soil fertility history" across a number of sites (Dobermann et al., 1996). In the rice-wheat long-term fertility experiment at Quingpa (China), treatments that had received P for a long time had a much greater P release than treatments to which P was not applied. At this site, the treatment differences were mostly reflected by differences in the 'b' coefficient describing slow P release over time. Similarly, in the rice-wheat system at Ludhiana, Yadvinder-Singh et al. (2000b) observed that 'b' coefficients measured 14 days after incubation correlated significantly with grain yield, P uptake by rice and wheat and P-balance in soils. Predicting soil-P status at harvest of a crop based on its initial soil-P value without additional soil chemical analysis, will be a useful tool for making recommendation for the succeeding crop. This approach will enable the soil testing laboratories to recommend fertilizer doses for the whole cropping system based on the initial soil test value.

Fertilizer recommendations based on soil tests have been an important approach for efficient nutrient management in the rice-wheat system. Soil test values for available P and other nutrients are generally classified into low, medium or high category (Table 17). Some workers have classified these ratings into 5 or 6 categories. The response to P application decreases as available P status of soil increases. For soils testing either high or low in its available P status, the fertilizer dose is decreased or increased respectively by 25% of the recommended fertilizer dose. Tandon and Sekhon (1988), however, found that yield response to 26 kg P ha^{-1} on low and medium P testing soils were very similar, suggesting a need for revising the soil fertility categories. Wheat grown in soils testing low or medium in its P status responded to 39 and 26 kg P ha^{-1}, respectively (Table 18). Crop responses to P on high P testing soils are generally considered to be small or non-significant. However, reports of crop response to P application has been observed for soils testing high in available P. There are also cases of lack of response to P application in P deficient soils, particularly in rice (Table 18). Thus, there is a need for revising the soil fertility categories for rice and wheat.

Sources of P

Diammonium phosphate (DAP) is the major source of P used in the rice-wheat system and accounts for nearly 65% of the P used in India. The other sources of P used commonly are single superphosphate (SSP), ammonium nitrophosphates (ANP) and compound fertilizers (12-32-16 grade). The efficiency of a P source varies depending upon the proportion of water-soluble P and soil properties (soil pH, P-fixing capacity and organic matter content). In neutral to alkaline soils, materials containing water soluble P are

TABLE 17. N, P, and K soil test value range limits used in India for rating soils.

Nutrient	Soil test method	Low	Medium	High
Organic C (%)	Walkey and Black	Below 0.5	0.5-0.75	Above 0.75
Available N (kg ha^{-1})	Alkaline KMnO$_4$	Below 280	280-560	Above 560
Available P (kg ha^{-1})	Olsen	Below 10	10-24.6	Above 24.6
Available K (kg ha^{-1})	Ammonium acetate extractable	Below 108	108-280	Above 280

TABLE 18. Response (kg grain kg^{-1} P) of wheat to applied P (26 kg P ha^{-1}) on soils varying in available P status in India.

State	Low P	Medium P	High P
Uttar Pradesh	55	41	25
Haryana	41	32	23
Punjab	30	23	5
Himachal Pradesh	64	50	5

Adapted from Singh and Singh, 1992.

generally more efficient than materials containing citric acid soluble or citric acid insoluble P (Bijay-Singh and Sekhon, 1973a,b; Bijay-Singh et al., 1976; Meelu, Kulkarni and Bhandari, 1982; Saggar, Meelu and Dev, 1985). High P utilization by rice and wheat from DAP, compared to SSP, was reported by Goswami and Kamath (1984). Similarly, several experiments conducted under the All India Coordinated Agronomic Research Project at different locations in India showed that DAP was a better source of P than SSP in almost all the soils, both for its direct effect on rice and its cumulative effect on wheat (Table 19). The efficiency of P sources decreased with decreasing water soluble P content in the fertilizer on highly calcareous soils of north India (Mishra and Mishra, 1992). But in very acidic soils, rock phosphate (RP) was as effective as water soluble P sources for rice and wheat (Mathur and Lal, 1989).

For the rice-wheat system practiced on a clay loam soil of pH 7.1, the relative agronomic efficiency of different P fertilizers were in the decreasing order of diammonium phosphate (100%), SSP (98%), ANP with 70 (90%), 50 (80%) and 30 (65%) percent water solubility, and Jordan RP (50%) as reported by Hundal, Deol and Sekhon (1977). Bijay-Singh et al. (1976) recorded similar observations for rice and wheat not grown in a sequence. In

TABLE 19. Direct and cumulative effect of different P sources on crop grain yield (t ha^{-1}) in the rice-wheat system in India.

Location	Grain yield of	P sources*				
		SSP	DAP	RP	RP + SSP (2:1)	RP + PSBNP
RS Pura	Rice	5.10	5.52	4.90	5.02	5.01
	Wheat	3.52	3.62	3.14	3.30	3.17
Kharagpur	Rice	3.71	4.28	3.99	4.18	–
	Wheat	2.27	2.72	2.64	2.52	–
Kanpur	Rice	4.27	4.44	4.14	4.31	4.34
	Wheat	3.27	3.45	3.12	3.32	3.27
Varanasi	Rice	3.99	4.10	3.68	3.78	3.82
	Wheat	3.62	3.75	3.34	3.42	3.40
Raipur	Rice	4.17	4.36	3.95	4.32	3.93
	Wheat	1.47	1.73	1.35	1.50	1.58
Jabalpur	Rice	2.36	2.44	1.73	2.12	1.64
	Wheat	2.47	3.18	2.09	2.31	2.05

*SSP: Single super phosphate; DAP: Di-ammonium phosphate; RP = Rock phosphate, PSBNP = Phosphorus solubilizing microorganisms.
Adapted from PDCSR, 1998.

view of the high cost of DAP fertilizer, there is an increasing interest to use a relatively cheaper source of P such as ANP, which can have variable proportion of water soluble total P. A recent study (1995-98) on a loamy sand soil (pH 7.6) showed that ANP containing 60% or more of water soluble total P was as efficient as DAP or SSP for increasing crop yields of the rice-wheat system (Bijay-Singh and Yadvinder-Singh, 1999). The type of crop grown, the rate and method of P application, and the soil properties, determine the level of water solubility of P in fertilizers. In general, high levels of water soluble P is required for alkaline and calcareous soils than for acid soils, and for wheat than for rice.

Time and Method of P Application

Due to the rapid fixation of soluble P fractions in the soil, it is important to apply P fertilizers at the right time and to place it in the rightsoil zone. The demand and rate of P uptake is generally high during the vegetative growth of most cereal crops. Available P in the soil generally cannot meet P requirement of crops at this stage, particularly in the case of wheat. As discussed earlier, rice can generally meet its P requirements utilizing the residual P from adequately fertilized wheat. However, there are situations when rice, grown after wheat, may need an additional dose of P fertilizer.

Most of the studies from the IGP suggest the application of the entire P dose at wheat sowing time, rather than its split application during later crop growth stages. Due to low temperatures encountered during the wheat season, more P is required in the early crop growth stages due to low rate of P diffusion in soils. The widely followed recommended practice for rice is to apply P at puddling. However, application of each half dose of P-fertilizer at sowing/harvesting and 21 days later, compared to its full dose application during planting, gave higher rice and wheat yields (Singh, Singh and Seth, 1980). Nevertheless, application of P at sowing of wheat or puddling of rice is more convenient than its top dressing during the later crop growth stages.

The mode of P application will depend on the P fixation, P source, soil P fertility level, and tillage pracices. Broadcasting and incorporation, rather than the banding of water-soluble P fertilizers determine the extent of P fixation. Uptake by crops can be enhanced by banding P-fertilizers below or near the seed, thus making a P-source readily available to the roots. As the soil P fertility level increases, the advantage of banding decreases. Banding is efficient compared to the broadcast method. Closer spaced crops, e.g., wheat and rice, benefits from banding, compared to wider spaced crop, e.g., corn. However, broadcasting and mixing P fertilizers to soil during rice transplanting was also reported to be effective compared to its placement (Abrol and Meelu, 1998). For wheat, results are overwhelmingly in favor of drilling and placing P fertilizers below the soil surface and into the root zone. Under irrigated conditions, yield increases of 400-700 kg ha^{-1} have been recorded when P was placed or drilled compared to its broadcasting (Tandon, 1987). Phosphorus use efficiency due to placement of P in wheat was reported to be 1.5 times greater than its broadcasting (Vig and Singh, 1983). However, in commercial farming this practice is extremely limited because few seed cum fertilizer drills are available. On acid soils, broadcasting finely ground RP or partially acidulated RP followed by its incorporation is recommended.

Development of Low Cost P Fertilizer Use Practices

Limited production of P fertilizers, low P use efficiency, and escalating fertilizer prices, are major constraints to agricultural productivity in South Asia. In this context, several studies on the use of indigenous RP with appropriate management practices such as partial acidulation of RP, reducing particle size, use of a mixture of RP and SSP and pre-incubation of RP with organic manures or pyrite, have been taken up. Research has also been conducted to increase the availability of soil P and to solubilize insoluble inorganic P sources like RP.

The use of a mixture of RP+SSP has been based on the principle that plants absorb P from SSP to meet their initial P requirements and subsequently from RP during later crop growth stages. Jaggi and Luthra (1983) suggested that 25-50% of the P requirement could be supplied through RP. Field studies conducted under All India Agronomic Research Project in India showed that a mixture of SSP+RP (2:1) was better than SSP in slightly alkaline to acid soils of Kharagpur (West Bengal), Raipur (Madhya Pradesh) and Kanpur (Uttar Pradesh) (Table 19). Mishra et al. (1980) reported that in neutral to alkaline soils, a mixture of RP and SSP (1:1) was nearly as effective as SSP. Relative agronomic efficiency of partially acidulated RP is greater in acid soils with high P fixing capacity. However, high rates of P as RP will be needed on acid soils.

A group of heterotrophic microorganisms have the ability to solubilize inorganic P from insoluble P sources (Chhonkar, 1994). Some of these P solubilizing microorganisms (PSM), particularly *Bacillus polymyxa, Pseudomonas striata* and *Bacillus frmus*, appear to be very promising for use as biofertilizers. Adding organic manures to the alluvial soils containing very little organic matter stimulates the ability of P solubilizing microorganisms. An increase in the efficiency of RP applied to rice and wheat has been observed with the inoculation of seed/seedling with PSM along with organic amendments (Gaur, 1990). The Indian Agricultural Research Institute, New Delhi, has developed a P-solubilizing biofertilizer known as 'Microphos' containing *P. striata* and *B. polymyxa*. Application of RP after its inoculation with PSM was found to be as good as SSP at some locations (Table 19). Although, it has been well established that PSM increased the availability of

P to plants, available information on their practical utility is inadequate. Efficient organisms adapted to different soil and climatic conditions should be isolated and conditions for their multiplication in soils should be well defined. Soils where such artificial inoculations are likely to be more successful need to be identified.

Vesicular-arbuscular mycorrhizae (VAM) plays an important role in the absorption and uptake of P besides many other nutrients by plants. Inoculation of soil and roots with VAM enhances rice P use efficiency on high P fixing soils (Chhonkar and Tilak, 1996). High P levels in soils can reduce the colonization, infection or spore formation of VAM. Together with other microorganisms, VAM dissolves the otherwise unavailable P and through their extensive mycelial network, transports P to the host plant. Rice and wheat has responded significantly to VAM. However, technical difficulties for the mass scale VAM production limit its use as a biofertilizer.

Fertilizer P Management Based on Target Yields

N, P and K recommendations for rice and wheat have been made for different yield targets. In this approach, a linear relationship between grain yield and nutrient uptake by the crop is assumed. This strategy takes into account the nutrient removal, soil nutrient supply, and the fertilizer efficiency for a given soil-climatic environmental condition. The fertilizer recommendations based on this concept are more quantitative, precise, and meaningful due to the combination of soil and plant analysis. The essential basic data set required are: (1) nutrient requirement in $kg\ t^{-1}$ of grain production (NR), (2) percent contribution from soil (CS), and (3) percent contribution from fertilizer nutrient (CF). The fertilizer dose (FD) can be calculated as:

$$FD\ (kg\ ha^{-1}) = NR/CF * 100T - CS/CF * STV,$$

where T is yield target $(t\ ha^{-1})$ and STV is soil test value $(kg\ ha^{-1})$.

For making N, P and K fertilizer recommendations based on soil tests, many researchers have developed targeted yield equations for rice and wheat. Dhillon et al. (1987) predicted the application of 22 and 0 kg P ha^{-1} for soils testing 12 and > 20 kg Olsen P ha^{-1} to obtain 7 t ha^{-1} of rice yield. Similarly, when soil test values for P were 10 and 20 kg ha^{-1}, the required fertilizer rates for wheat (target yield of 5 t ha^{-1}) in the Indian Punjab were calculated to be 64 and 44 kg ha^{-1}, respectively. Fertilizer recommendations based on target yields gave higher yield response and required less fertilizer use, compared to general fertilizer recommendations (Table 20). The target yield equations are applicable to similar or related soils in the same agroclimatic region. The targeted yields should be within the range of experimental yields and the equations must be used within the experimental range of soil

TABLE 20. Comparison of fertilizer recommendations based on target yield approach and general fertilizer recommendations (GFD) for rice and water.

Crop	No. of experiments	Fertilizer recommendation approach	Average fertilizer dose (kg ha[*1])			Average yield (t ha[-1])	Response to fertilizer (kg grain kg[-1] nutrient)
			N	P	K		
Rice	10	Target yield	79	14	4	5.10	52.4
	10	GFD	118	20	40	5.27	29.6
Wheat	21	Target yield	97	15	20	3.58	27.0
	21	GFD	110	24	38	3.36	19.6

Adapted from Velayutham et al., 1985.

test values and should not be extrapolated. Moreover, other nutrients should not be limiting crop yields.

Fertilizer P Management Models

General models describing the long-term crop response to fertilizer P have been developed (Wolf et al., 1987; Janssen and Wolf, 1988). These appear to be useful for specifying fertilizer recommendation as the data needed for model calibration can be obtained from a single field experiments. The Wolf-model uses the first season recovery as an input variable, which includes dynamic changes during that period. Therefore, it may also account to some extent for the dynamic changes in the P chemistry of flooded soils (Willet, 1986; Kirk, Yu and Choudhury, 1990) and for possible rhizosphere effects on the solubilization of fertilizer P.

Another basic model for calculating fertilizer P requirement for crops has been proposed by Driessen (1986). The P requirement is the difference in P uptake by unfertilized and fertilized crop divided by an index of fertilizer recovery. The fertilizer requirement (Fp) is calculated as:

$$F_P = (Um - Ua)/R$$

where, Um is the P uptake by the P-fertilized crop at the desired yield level, Ua is the P uptake by the unfertilized crop, and R is the recovery of applied P. Both Um and Ua are determined from field experiments. There is, however, little information on its applicability to commercial production.

The amount of P required to raise the P level of soil to its critical threshold value is calculated from the equation:

$$P \text{ requirement} = \{\text{Critical threshold value} - P \text{ (48 h-resin)}\}/a$$

where, 'a' is constant for each group of soils. The values of 'a' were determined from the slope of desorption isotherm of the soils. Although this system has been tested and found suitable for a wide range of soil types, it relies on a number of generalized assumptions that may reduce the accuracy with which fertilizer requirements are assessed for individual soil types. Incorporating the varying external requirements of different crop species can make improvements in this approach. Fertilizer P requirement for a unit rise in soil test value is a useful parameter for estimating P additions required for improving fertility. In many alluvial soils, it usually takes 20-40 units of fertilizer P to raise Olsen P by one unit under field conditions (Tandon, 1987).

POTASSIUM

Potassium Transformations in Soils

Soil potassium exists in exchangeable and non-exchangeable (fixed and structural K) forms, in solution. The amount of solution and exchange K is usually a small fraction of total K (1-2% and 1-10%); the bulk of soil K exists in K-bearing micas and feldspars (Sekhon, 1995). The amount of K present in the soil solution is often smaller than the crop requirement for K. Thus, continuous renewal of K in the soil solution for adequate nutrition of high yielding varieties of rice and wheat is obvious. Similarly, the exchangeable K component has to be continuously replenished by the release of fixed K and weathering of K minerals. Hence, K availability to crops is a function of the amounts of different forms of K in the soil, their rates of replenishment, and the degree of leaching. The release of K from illitic materials through weathering may account for the apparent lack of crop response to K application in alluvial soils of the Indo-Gangetic Plain. Dynamic equilibrium reactions occurring between different forms of K have a profound effect on the chemistry of soil K. The direction and rate of these reactions determine the fate of the applied K and the release of non-exchangeable K. Under certain conditions, added K is fixed by the soil colloids and is not readily available to plants. Clays of 2:1 type can readily fix K and in large amounts. The mechanisms governing K fixation especially in flooded soils are not clearly understood.

Available K is relatively high in flooded soils because large amount of soluble Fe^{2+} and Mn^{2+} ions brought into solution displaces cations from the clay complex, and exchangeable K is then released into the soil solution. In fields with adequate drainage, K and other basic cations are lost via leaching. Leaching losses of K are a major concern on highly permeable wetland rice soils with low cation exchange capacity. Abedin Mian et al. (1991) reported from Bangladesh that leaching losses could be as high as 0.1-0.2 kg K ha^{-1}

day^{-1}. Leaching losses of K will depend on soil solution concentration and percolation rates. The increase in K availability in flooded soils also enhances K uptake by rice. Water regime is highly dynamic in the rice-wheat system and may influence the availability and fixation of K in soils. In a long-term rice-wheat crop rotation experiment in the Terai plains of Southern Nepal, 46% to 56% of added K was fixed in the soil in a wet/dry equilibration, and fixation was linear with addition rates of up to 25 mM K kg^{-1} soil (Regmi, 1994).

Assessment of Soil K-Supplying Capacity

Soil testing is widely used in the Indo-Gangetic Plain to estimate the amount of K that becomes available to plants during the rice and the wheat cropping seasons. Use of 1 M ammonium acetate at pH 7.0 to extract plant available K (exchangeable + water soluble K) is still widely used as a soil K availability index for rice and wheat. But its suitability as a measure of plant available K remains controversial, especially when soils with different textures and clay mineralogy are considered together (Kemmler, 1980; Dobermann et al., 1996). For example, in Gurdaspur (India) 40% of the soil samples and 7% of the plant samples were deficient in K (Tandon and Sekhon, 1988). Critical levels for 1 M ammonium acetate-extractable K for rice soils varies from 0.17-0.21 cmol K kg^{-1} (Prasad and Prasad, 1992). Soils have been grouped into 3 categories of low, medium and high on the basis of soil test values (Table 17). Usually diverse soils analyzing < 55 mg K and >110 mg K kg^{-1} soils in 1 M ammonium acetate solution are rated as low and high in its available K content, respectively.

Most of the soils in the Indo-Gangetic Plain contain illite as dominant clay mineral. The high root density, relatively high maximum influx and low minimum solution concentrations for K uptake indicate that rice and wheat depends on the non-exchangeable fraction for much of their K supply in such soils (Meelu et al., 1995). Hence, it appears desirable to include a measure of non-exchangeable K in our estimate of plant available K. A measure of non-exchangeable K in soil is determined by boiling 1 M HNO$_3$, but results do not always correlate with grain yield and total K uptake.

Sub-soil K fertility makes a significant contribution to plant nutrient, and differences in the mineralogy and reserve K and relationship between exchangeable K and water soluble K among soil series and soil types suggest the need for different rates of critical limits for different soils (Sekhon, 1995). In addition, ammonium acetate-extractable K for soil testing should include soil properties such as clay content, cation exchange capacity and organic matter content. On alkaline soils, reduced K activity in soil solution due to preferential K adsorption may contribute to low K uptake by rice, when ample K is available (Dobermann et al., 1996). Other methods such as Q/I relationship and electro-ultra filtration are laborious and/or expensive and not used in analysis of soils for the rice-wheat cropping system.

Tandon and Sekhon (1988) suggested that crops grown in soils with low available K (< 100 mg kg^{-1} soil) be expected to readily respond to K application. Soils with low available K but with a high reserve K (> 1000 mg K kg^{-1} soil) status will need lower rates of K application. However, soils with high available K (> 100 mg K kg^{-1} soil) but with a low reserve K status can support crops for some years without K fertilizer application.

All chemical soil tests used for K analysis for rice and wheat production have the following theoretical limitations: (1) nutrient availability in irrigated rice-wheat ecosystem is extremely dynamic and tests on air-dried soil may not fully reflect nutrient status of submerged soils, (2) differences in clay mineralogy and physical properties have a strong impact on desorption characteristics and K availability to plants, (3) unextracted nutrient pools may also contribute to plant uptake, (4) diffusion is a key process for K (and P) transport to the root surface, (5) external mechanisms such as root-induced solubilization of P by acidification contribute significantly to K uptake by rice and wheat roots, and (6) kinetics of nutrient release are not measured. Dynamic soil tests overcome many of the theoretical limitations associated with rapid chemical extractions (Dobermann et al., 1998). The resin capsule, for example, integrates intensity, quantity and delivery rate measures of P and K supply to the rice and wheat roots and provides parameters to assess short- and long-term nutrient supplying power in a dynamic manner. The simultaneous extraction of N, P, K, S, Ca, Mg, Fe, Mn, Zn, Cu and other nutrients across different soil types in situ, without the need for soil sample collection and processing, is a major advantage of this method as a universal soil test procedure.

Though the role of non-exchangeable K in plant nutrition and the effect of soil texture on K release is known, these factors are not considered when K recommendations are made by the soil testing laboratories. The question of critical limits of K in soil continues to be the first problem of interpretation. Further research, to develop field applicable critical limits for diagnosing soil and crop K deficiency is needed.

Fertilizer K Management for Rice-Wheat System

Crop Response to Applied K

Most of the alluvial soils in the Indo-Gangetic Plain are generally medium to high in extractable K. It was thought that additional K supply with irrigation water would make K a rate limiting factor in irrigated rice-wheat system (Kawaguchi and Kyuma, 1977; Bajwa, 1994). But K deficiency is now increasing even in alluvial illitic soils of India (Tiwari, 1985). Yield response to applied K depends upon crop, variety, soil fertility status, soil texture, cation exchange capacity, soil depth, soil moisture and application of other nutrients. Earlier studies conducted on large number of farmers fields in India showed that application of 50 kg K ha^{-1} gave response of 290 and 240 kg grain ha^{-1} for wheat and rice, respectively (Randhawa and Tandon, 1982). Average agronomic response of 6-kg grain kg^{-1} K to the application of 37.5-kg K ha^{-1} was observed in rice and wheat. In later studies, Meelu et al. (1992) reported that the response of rice to 25-50 kg K ha^{-1} ranged from 210-370 kg grains ha^{-1} in the northern Indian states of Punjab, Haryana and Uttar Pradesh. Dobermann et al. (1995) reported significant yield increase of 12% to K application in rice at Pantnagar. From a 5-year field study on a sandy loam soil testing 123 kg available K ha^{-1}, mean increase in the yield of rice and wheat was 280 and 160 kg grain ha^{-1}, respectively, due to application of 25 kg ha^{-1} (Meelu et al., 1995). The low response to fertilizer K application observed in rice and wheat on alluvial soils of the Indo-Gangetic Plain suggests that the release of native K from illitic minerals in these soils could be met by the K needs of these crops.

Using time series analyses, Bhargava et al. (1985) showed that the response to K application has been increasing over time. The response to K in wheat was in the range of 6.7-12.7 kg grain kg^{-1} K during 1977-1992 as against 2.0-5.0 kg grain kg^{-1} K during 1969-1971. The corresponding values for rice were 5.3-10.7 kg and 1.8-8.0 kg grain kg^{-1} K. The increasing trend in response to K over the years suggests the need for its application in intensive rice-wheat cropping system. In another long-term expriment in India, response to K application increased with time, while that of N decreased (Table 21). During 1975-1979, Kemmler (1980) reported that the average increase in rice yield > 0.4 t ha^{-1} was due to K application in 400 out of 500 trials in Bangladesh.

Time and Method of K Application

It is recommendation that the full dose of basal-K be applied at puddling for rice and at sowing of wheat. When cation exchange capacity of soil is low and drainage in soil is excessive, basal application of K to rice should be avoided. Because rice and wheat require large quantities of K, its supply is necessary up to the reproduction stage. On coarse textured soils, split application of fertilizer K to both rice and wheat may give higher nutrient use efficiency than its single application, due to reduction in leaching losses and luxury consumption of K (Tandon and Sekhon, 1988). In Punjab, Kolar and Grewal (1989) reported a yield advantage of 250 kg grains ha^{-1} by split application of K (half at transplanting + half at active tillering stage) compared to its single application at transplanting. Other studies also showed the distinct benefit of K application in split doses (Tandon and Sekhon,

TABLE 21. Response of crops to N, P, and K application over time in long-term experiments in India.

Crop	Year	Control yield (kg ha^{-1})	Response to applied nutrient (kg ha^{-1})		
			N120	P23	K33
Location: Faizabad (PDCSR, 1990)					
Rice	1977-78	1008	2905	500	50
	1989-90	820	2642	925	231
Wheat	1977-78	833	2625	617	35
	1989-90	602	2141	1169	398
Location: RS Pura (Hegde and Sarkar, 1992)					
Rice	1981-82	1490	2499	654	-250
	1989-90	1550	880	989	666
Wheat	1981-82	980	873	458	66
	1989-90	730	423	1191	2250

1988). Similarly, in sandy loam soils of Uttar Pradesh, Singh and Singh (1987) reported a wheat yield increase of 440-490 kg grain ha^{-1} by split application of K compared to its single application. Fertilizer K at sowing of wheat and transplanting of rice is normally applied by drilling, placement or broadcast followed by incorporation.

Muriate of potash (KCl) is a principal fertilizer source of K for rice and wheat because of its low cost and high K analysis. However, its use in salinity affected areas is discouraged. Potassium sulphate may be used in areas with S deficiency.

Interactions of Potassium with Other Nutrients

The interaction among plant nutrients is a common feature of crop production. Potassium plays an important role in ensuring efficient utilization of N. Large quantities of N applied to the intensive rice-wheat cropping system encourage crop uptake of N and K and in turn depletes soil K. Application of N and P resulted in 145% increase in K uptake compared to the control treatment (Tandon and Sekhon, 1988). If insufficient N and P or other essential plant nutrients restrict the crop development, the amount of K present even at low soil test values may be sufficient to meet crop needs. Tiwari et al. (1992) reported that the response of rice to K application increased with increasing rates of N application. To obtain high yields, application of higher levels of K along with higher N rates, was suggested.

Effect of K Fertiliry Status of Soils and Response to Potassium

It is a common finding that responses of rice and wheat to K application are higher on soils testing low in 1 M ammonium acetate-extractable K than on high K soils (Tandon and Sekhon, 1988; Yadvinder-Singh and Khera, 1998). Significant responses of wheat to applied K were observed up to 25 kg K ha^{-1} on soils testing low in available K in Punjab, but no significant increase in wheat yield was observed on soils testing medium and high in available K (Sharma et al., 1978). From another study at Gurdaspur in the Indian Punjab, Azad, Bijay-Singh and Yadvinder-Singh (1993), reported that wheat yield increased significantly up to 75 kg K ha^{-1} on soils testing low in available K. However, significant wheat yield increase was observed only at 25 kg K ha^{-1} on soils testing medium as well as high in available K. Rana et al. (1985) observed that rice responded to 50 kg K ha^{-1} on soils testing low and medium in available K, but no significant response to applied K was observed on soils testing high in available K. Tandon and Sekhon (1988) concluded that response of high yielding varieties of rice and wheat to K application in soils rated medium in available K were only marginally lower than responses in low K soils. Such results emphasize the need for a fresh look at soil fertility limits used for categorizing soils into low, medium and high with respect to available K, particularly for highly productive rice-wheat cropping system.

Field experiments conducted at different locations in the Indian Punjab showed that rice responded more to applied K in the north-eastern districts (Gurdaspur, Amritsar, Kapurthala and Hoshiarpur) than in the central and south-western districts (Ludhiana, Bathinda, Sangrur, Ferozepur) (Singh and Bhandari, 1995). The values of available K in soil ranged from 150-180 kg K ha^{-1} in northwestern districts and 112-165 kg K ha^{-1} in central and south-western districts. The lower rates of K release from clay minerals could be a possible reason for the greater responses to applied K in northeastern districts as compared to the central and southwestern districts. A recent study (1997-2000) conducted at two locations in Punjab showed that the rice and the wheat crops responded significantly to the application of 50 kg K ha^{-1} on loamy soils but not on sandy soils (Yadvinder-Singh and associates). Soils at both these locations tested low in available K. These studies suggest that the same test values represent varying K supplying capacity of different soils.

Potassium Balance Under Rice-Wheat Cropping System

Introduction of modern production technologies for rice and wheat with N responsive, high yielding varieties, has resulted in increased annual crop demand for K by the crops. From the nutrient removal data (Table 5) it is evident that in the rice-wheat system, annual removal of K equals or exceeds that of N, while the replacement of K by fertilizer represents only a fraction of N. Furthermore, most of the K uptake in rice and wheat crops is stored in the straw, which is mostly removed from the field as animal feed and are not recycled into the soil. Long-term studies have shown that K balance in the rice-wheat system is highly negative and application of recommended doses of K has only slightly improved the K balance (Nambiar and Ghosh, 1984). In a long-term experiment at Ludhiana, net negative K balance of about 250 kg K ha^{-1} year^{-1} was observed and it decreased available K level of the soil (Table 22). There was considerable K removal from the non-exchangeable K reserve of the soil, which contributed about 95% towards the K nutrition of rice and wheat crops. Tiwari (1985) observed a decline in available and non-exchangeable K by 17% and 2.8% after two cropping cycles measured on 14 fields at Kanpur (Uttar Pradesh). Complete straw removal from the rice and wheat fields at harvest, leads to K mining at an alarming rate since it contains 80-85% of the K uptake by crops. The K rates applied by most farmers are lower than that used in the long-term experiments. The K balance in the farmer's fields in Haryana showed a negative K balance of 170 kg K ha^{-1} after an annual cropping cycle of rice and wheat.

The negative K balances mean that it will be impossible to maintain the present production levels. Results from long-term fertility experiments in India show that crop response to K application start appearing over a period of time in soils which were initially well supplied with K (Nambiar and Ghosh, 1984). Such responses to K started appearing after 3 years in rice and 11 years in wheat at Pantnagar (Uttar Pradesh) and after 3 and 7 years, respectively, at Barrackpore (West Bengal). Long-term studies suggest that application of FYM and recycling of crop residues can help improve the K balance in the rice-wheat cropping system (Yadvinder-Singh and Khera, 1998). Studies on K balance help to understand the K dynamics with special reference to K build up or depletion under different levels of K management.

TABLE 22. Potassium balance under different rates of K application after 4 cycles of rice-wheat crop rotation at Ludhiana, India.

K rate (kg ha^{-1})	Total K applied (kg ha^{-1})	Total K uptake (kg ha^{-1})			Net K balance (kg ha^{-1})	Depletion of availabe K (kg ha^{-1})	Contribution from non-exchangeable K (%)
		Rice	Wheat	Total			
0	0	535	348	883	−883	34	96
25	200	563	360	931	−731	28	96
50	400	548	384	932	−532	26	95
75	600	551	404	955	−355	23	94

Depletion of available K was calculated from 0-15 cm soil layer.
Adapted from Meelu et al., 1995.

Presently little data is available on the recovery of fertilizer K applied to the rice-wheat cropping system. The long-term fate of added K will depend on the soil texture (leaching), clay mineralogy (fixation), and crop removal.

Potassium Uptake and K Requirements in the Future

Dobermann et al. (1995) showed that the relationship between rice grain yield and K uptake is scattered and nonlinear. There is a wide range of K uptake (16-50 kg K) for one ton of rough rice, which makes it difficult to estimate the K requirement for a given yield target over a large area. The total K content in the grain of modern rice and wheat varieties is fairly constant between 0.25-0.33% for rice and between 0.30-0.40% for wheat. The relationship between yield and K uptake showed that the K requirement of rice in the yield range of 4-8 t ha^{-1} varied from 17 to 30 kg K ha^{-1} of grain. Using this relationship, rice yield of 5-8 t ha^{-1} would require a total K uptake of 105-237 kg ha^{-1}, because K is absorbed at a rate that parallels dry matter accumulation between tillering and early grain filling stage, a period as short as 60 days. Thus large amounts of K must be easily available in order to meet daily K uptake rates of 1.4-3.6 kg K ha^{-1} day^{-1} over that period. Similarly, peak K uptake rate of wheat is also quite high.

Straw management will have a large impact on the overall K balance. Removal of crop residues for use as forage or fuel will increase depletion of K from croplands. On the other hand, returning all or most of the crop residues will reduce K removal and K fertilization. Dobermann et al. (1995) suggested that yields of irrigated rice in Asia must be raised to 8.0 t ha^{-1} from the present level of 4.9 t ha^{-1}. Hence large quantities of fertilizer K will be needed in the future to achieve such high yields of rice and wheat.

NUTRIENT MANAGEMENT FOR RICE-WHEAT SYSTEM IN SALT AFFECTED SOILS

A large area (about 2.8 M ha) in the Indo-Gangetic Plains is highly alkaline (pH > 8.5) and contains excessive concentration of soluble salts, high exchangeable sodium percentage (> 15%), and $CaCO_3$. Salt-affected soils are low in organic matter and have poor air-water relations. The above characteristics of these soils influence the transformation and availability of native and applied nutrients. The recovery efficiency of fertilizer N by both rice and wheat crops grown on salt-affected soils is even lower than on the normal soils. Excessive losses of fertilizer N from soil-plant system via NH_3 volatilization occur on sodic soils (Bajwa and Singh, 1992). Therefore, general fertilizer N recommendations for rice and wheat are normally 25% higher than for normal soils. Leaching loss of N is not important during initial years on sodic soils. Nitrogen remains within the root zone. Therefore, N fertilizer (urea) applied in a single dose can be as good as that applied in split doses (Tiwari et al., 1998). However, when permeability of the soil improves after a few years of reclamation, split application of N is useful.

Both rice and wheat grown on freshly reclaimed sodic soils (i.e., during initial 3 years) do not respond to applied P due to the relatively high content of available P in the soil (Swamp and Singh, 1989). Long-term studies have shown significant response of rice and wheat to applied P when Olsen-P was less than 12.7 and 8.7 kg ha^{-1} in 0-15 cm soil depth, respectively. Application of 26 kg P to wheat and 13 kg P ha^{-1} to rice has produced maximum yields on these soils. On sodic soils, rice responds more to applied P than wheat. Swamp and Singh (1989) found that the application of fertilizer K did not significantly increase crop yields in the rice-wheat rotation on reclaimed sodic soils in Haryana, even after 12 years of continuous cropping. However, in salt affected soils of Kanpur, application of 25 kg K ha^{-1} to both crops produced additional grain yield of 0.50 and 0.61 t ha^{-1} of rice and wheat, respectively (Tiwari et al., 1998).

FUTURE RESEARCH NEEDS

Nitrogen

Precise and accurate N budgets need to be prepared for the rice-wheat system practiced in different agroclimatic zones in the Indo-Gangetic Plains of South Asia. These should take into account various N inputs (atmospheric depositions, N fixation, and N fertilization) and outputs via crop removal and losses through mechanisms like ammonia volatilization, nitrification-denitrification and leaching. Better understanding of N transformation processes controlling losses of N via mechanisms such as nitrification-denitrification and leaching should be obtained in soils under rice-wheat cropping system. Along with rate limiting steps for different N transformation processes, this information will be utilized for working out better N management options for rice-wheat system in South Asia. Where wetland rice is grown in porous soils subjected to alternate flooded and drained conditions, there is an urgent need to estimate total N_2O emissions and of the relative predominance of N_2O in N gas emissions. It will help in assessing the contribution of the rice-wheat system in the destruction of stratospheric ozone. The effect of changing land use patterns (from wheat to wetland rice) on soils need to be studied in terms of sustainable crop production and pollution of the environment. Long-term investigations need to be initiated to monitor leaching of nitrate N beyond the crop root zone and its movement in the unsaturated zone before it reaches groundwater. These studies become all the more important because porous soils are increasingly being brought under irrigation and are producing high yields due to application of heavy doses of fertilizers. Development of quantitative predictive models for different N loss mechanisms and N uptake by the rice-wheat system should be attempted. Running these models with the available data would help identify the information that needs to be obtained for a more reliable prediction of losses of N and fertilizer N use efficiency in soils under rice-wheat cropping system.

To better adjust fertilizer rates, simple models describing recycling of N, P and K from crop residues and organic manures are needed. This information particularly relates to the question of how different straw management practices affect the amount of nutrient recycled and also the dynamics of nutrient release. The integration of N_2-fixing species in the rice wheat system enables a reduction in N fertilizer use. There has been limited research comparing the efficiency of fixed N and fertilizer N at similar inputs. A few studies have indicated that N losses from the fertilizer N were greater than or similar to those from fixed N. There is a need to develop simple models relating nutrient release from organic inputs to management practices. The existing knowledge based on optimum nutrient dynamics for high yielding rice (exceeding 8 t ha^{-1}) and wheat (exceeding 5.5 t ha^{-1}) is inadequate. There is a need to determine fertilizer application rates to balance nutrient removal and losses from the system for different yield targets in different regions.

Phosphorus and Potassium

The soil moisture regimes in the rice-wheat system varies from a flooded-wet to a dry state, but the effects of such moisture changes on P and K availability and its associated crop response, has not received much attention. Research is needed to clarify processes of fixation and release of fixed P and K during drying and wetting cycles, and the extent of impeded K diffusion in the rhizosphere of rice and its effect on K uptake at very high yield levels. A better understanding of the processes affecting the long-term fate of fertilizer P and K in irrigated rice-wheat system, including the influence of wetting and drying cycles on the recovery of P and K is needed (Dobermann et al., 1998). Research may be initiated to predict the time taken by soils well supplied with K to become deficient in K. Information on rate of P adsorption and sorption isotherms for different soil is required. Greater predictive understanding of the P sorption and release characteristics of soils under rice-wheat cropping system would help in developing improved nutrient management practices.

Applied research must provide the tools necessary for practical use of long-term strategies for N, P and K management based on nutrient balance concept. Tools such as the resin capsule, models for estimating crop nutrient requirements based on the interactions of N, P and K, and models for predicting the long-term fate of added P and K fertilizers should be fully developed for the rice-wheat system. These models may provide a basis for introducing farm or field-specific nutrient management practices. A systems approach in research on nutrient management of the rice-wheat system must be encouraged. There is a need to develop improved nutrient management practices that will enhance fertilizer use efficiency, reduce nutrient losses, and increase benefits to the farmers; i.e., time and method of fertilizer application, exploitation of indigenous and alternative P and K sources, and use of biofertilizers.

Dynamic soil test methods assaying the N, P and K·supplying power of soils under rice-wheat system should be developed. An innovation in agronomic soil testing for P would be to use indices or models that combine soil test P values with other soil properties related to P intensity and buffering capacity (e.g., soil pH, organic matter content, clay content and cation exchange capacity), thereby improving the prediction and accuracy of P requirement. A simple modeling approach to study the P requirement of the rice-wheat system including trends in extractable soil P values over time will be of considerable value. Soil testing protocols should be developed for an agro-climatic region rather than for different states. An integrated approach towards K fertilizer recommendation may be developed based on site specificity, laboratory analysis, weather and soil survey records (e.g., clay mineralogy, soil depth), target yields, and crop response information. Computer models can then be used to process the database for making fertilizer recommendations.

Recycling of crop residues, organic inputs, time of land preparation, and fertilizer management, influence nutrient supplying capacity of the soil. Quantification of the effects of crop management variables and climatic factors on nutrient supplying capacity of the soil, will lead to better models of soil-nutrient cycling and nutrient losses as well as to improved practical soil testing methods. A better understanding of the effects of crop residue management on nutrient cycling, soil nutrient pools, and on crop response is essential. This will facilitate the development of budgets to balance nutrient removal with nutrient application for different yield targets which seems necessary for a sustainable, high yielding, rice-wheat production system. Medium and long-term field experiments on well-characterized sites should be initiated and should be carefully monitored for changes in soil fertility and crop productivity.

REFERENCES

Abedin Mian, M.J., H.P. Blume, Z.H. Bhuiyan, and M. Eaqub. (1991). Water and nutrient dynamics of a paddy soil of Bangladesh. *Zeitschrift für Pflanzenernaehr und Bodenkunde* 154: 93-99.

Abrol, I.P. and O.P. Meelu. (1998). Phosphorus in rice-wheat cropping systems in

South Asia. In *Nutrient Management for Sustainable Crop Production in Asia*, eds. A.E. Johnston, and J.K. Syers, London: CAB International, pp. 211-218.

Adhikari, C., K.F. Bronson, G.M. Panuallah, A.P. Regmi, P.K. Saha, A. Dobermann, D.C. Olk, P.R. Hobbs, and E. Pasuquin. (1999). On-farm soil N supply and N nutrition in the rice-wheat system of Nepal and Bangladesh. *Field Crops Research* 64: 273-286.

Agrawal, R.P., V.K. Phogat, T. Chand, and M.S. Grewal. (1995). Improvement of soil physical conditions in Haryana. In *Research Highlights*, Hisar, India: Department of Soils, C.C.S. Haryana Agricultural University, pp. 69-75.

Agrawal, S.R., H. Shankar, M.M. Agrawal. (1980). Effect of slow-release nitrogen and nitrification inhibitors on rice-wheat sequence. *Indian Journal of Agronomy* 35: 337-340.

Anonymous. (1979). *Annual Report of the Coordinated Agronomic Research Project*, New Delhi, India: Indian Council of Agricultural Research. 282 pp.

Aulakh, M.S. (1989). Transformations of ammonium nitrogen in upland and flooded soils amended with crop residues. *Journal of the Indian Society of Soil Science* 37: 248-255.

Aulakh M.S. and Bijay-Singh. (1997). Nitrogen losses and N-use efficiency in porous soils. *Nutrient Cycling in Agroecosystems* 47: 197-212.

Aulakh, M.S., T. Khera, and J.W. Doran. (1998). Effect of nitrate and ammoniacal N on denitrification in upland, nearly-saturated and flooded subtropical soils. In *Proceedings of the 16th World Congress of Soil Science*, Symposium 26, Montpellier, France: International Society of Soil Science.

Azad, A.S., Bijay-Singh, and Yadvinder-Singh. (1993). Response of wheat to graded doses of N, P and K in soils testing low, medium and high with respect to P and K in Gurdaspur district of Punjab. *Journal of Potassium Research* 9: 266-270.

Bajwa, M.I. (1994). Soil potassium status, potash fertilizer usage and recommendations in Pakistan. *Potash Review No. 3.*, Basel, Switzerland: International Potash Institute. 4 pp.

Bajwa, M.S., and Balwinder Singh. (1992). Fertiliser nitrogen management for rice and wheat crop grown in alkali soils. *Fertiliser News* 37(8): 47-59.

Bajwa M.S., Bijay-Singh and Parminder-Singh. (1993). Nitrate pollution of groundwater under different systems of land management in the Punjab. In *First Agricultural Science Congress-1992 Proceedings*, New Delhi, India: National Academy of Agricultural Sciences, pp. 223-230.

Bali, S.V., S.C. Modgal, and R.D. Gupta. (1986). Effect of recycling of organic wastes on rice-wheat rotation under alfisol soil conditions of North-Western Himalayas. *Himachal Journal of Agricultural Research* 12: 98-107.

Banerjee, N.K., A.R. Mosier, K.S. Uppal, and N.N. Goswami. (1990). Use of encapsulated calcium carbide to reduce denitrification losses from urea fertilized flooded rice. Proceedings International Denitrification Workshop, Giessen, FRG, March 1989. Mitt. Deutsch Boden Des. 60: 245-248.

Becker, M., J.K. Ladha, and M. Ali. (1995). Green manure technology: potential, usage, and limitations. A case study for lowland rice. *Plant and Soil* 174: 181-194.

Beri, V., B.S. Sidhu, G.S. Bahl, and A.K. Bhat. (1995). Nitrogen and phosphorus transformations as affected by crop residue management practices and their influence on crop yield. *Soil Use and Management* 11: 51-54.

Beri, V. and O.P. Meelu. (1983). Integrated use of inorganic and biofertilizers in rice. *Fertiliser News* 28(11): 33-38.

Bhargava, P.N., H.C. Jain, and A.K. Bhatia. (1985). Response of rice and wheat to potassium. *Journal of Potassium Research* 1: 45-61.

Bhuiyan, A.M., M. Badaruddin, N.U. Ahmed, and M.A. Razzaque. (1993). Rice-wheat system research in Bangladesh: A review. Wheat Research Center, Bangladesh Agricultural Research Institute, Dinajpur, Bangladesh. 96 pp.

Bhupinderpal-Singh, Bijay-Singh and Yadvinder-Singh. (1993). Potential and kinetics of nitrification in soils from semiarid regions of North-Western India. *Arid Soil Research and Rehabilitation* 7: 39-50.

Bijay-Singh, and G.S. Sekhon. (1973a). Relative availability of phosphorus in nitric phosphate fertilizers in a calcareous soil. *The Indian Journal of Agricultural Sciences* 43: 781-785.

Bijay-Singh, and G.S. Sekhon. (1973b). Relative availability of neutral ammonium acetate soluble phosphorus in nitric phosphate fertilizers. *Journal of the Indian Society of Soil Science* 21: 295-299.

Bijay-Singh, and G.S. Sekhon. (1976a). Some measures of reducing leaching loss of nitrates beyond potential rooting zone II. Balanced fertilization. *Plant and Soil* 44: 391-395.

Bijay-Singh, and G.S. Sekhon. (1976b). Some measures of reducing leaching losses of nitrates beyond potential rooting zone I. Proper co-ordination of nitrogen splitting with water management. *Plant and Soil* 44: 193-200.

Bijay-Singh, and G.S. Sekhon. (1976c). Nitrate pollution of ground water from nitrogen fertilizers and animal wastes in the Punjab, India. *Agriculture and Environment* 3: 57-67.

Bijay-Singh, and G.S. Sekhon. (1977). Some measures of reducing leaching loss of nitrates beyond potential rooting zone III. Proper crop rotation. *Plant and Soil* 47: 585-591.

Bijay-Singh, H.S. Hundal, and G.S. Sekhon. (1976). Evaluation of nitric phosphates differing in water solubility of their phosphorus fraction. *Journal of Agricultural Science, Cambridge* 87: 325-330.

Bijay-Singh, and J.C. Katyal. (1987). Relative efficacy of some new urea-based nitrogen fertilizers for growing wetland rice on a permeable alluvial soil. *Journal of Agricultural Sciences, Cambridge* 109: 27-31.

Bijay-Singh, J.C. Katyal, P.K. Malhotra, and P.L.G. Vlek. (1986). Path coefficient analysis of N nutrition on yield and yield components for rice in a highly percolating soil. *Communications in Soil Science and Plant Analysis* 17: 853-867.

Bijay-Singh, K.F. Bronson, Yadvinder-Singh, T.S. Khera, and E. Pasuquin. (2001). Nitrogen-15 balance and use efficiency as affected by rice residue management in a rice-wheat system in northwest India. *Nutrient Cycling in Agroecosystems* (accepted).

Bijay-Singh, U.S. Sadana, and B.R. Arora. (1991). Nitrate pollution of ground water with increasing use of nitrogen fertilizers in Punjab, India. *Indian Journal of Environmental Health* 33: 516-518.

Bijay-Singh, and Yadvinder-Singh. (1997). Green manuring and biological N fixation: North Indian perspective. In *Plant Nutrient Needs, Supply, Efficiency and Policy Issues: 2000-2025,* eds. J.S. Kanwar and J.C. Katyal, New Delhi, India: National Academy of Agricultural Sciences, pp. 29-44.

Bijay-Singh, and Yadvinder-Singh. (1998). Integrated nutrient management for a sustainable rice-wheat system. In *Integrated Plant Nutrient Supply System for Sustainable Productivity.* Bhopal, India: Indian Institute of Soil Science, pp. 46-58.

Bijay-Singh, and Yadvinder-Singh. (1999). *Evaluation of different nitrogenous and phosphatic fertilizers in rice-wheat cropping system.* Terminal Report of the Project Funded by Rallis India Ltd., Ludhiana, India: Department of Soils, Punjab Agricultural University. 43 pp.

Bijay-Singh, Yadvinder-Singh, and C.S. Khind. (1992). Fertiliser management in soils amended with green manures. *Fertiliser News* 37(8): 41-45.

Bijay-Singh, Yadvinder-Singh, and M.S. Maskina. (1987). Rice responds to higher N application in soils testing low in organic carbon. *Progressive Farming (Punjab Agricultural University)* 23(9): 13.

Bijay-Singh, Yadvinder-Singh, C.S. Khind, and O.P. Meelu. (1991). Leaching losses of urea-N applied to permeable soils under wetland rice. *Fertiliser News* 28: 179-184.

Bijay-Singh, Yadvinder-Singh, K.F. Bronson, and C.S. Khind. (1999). Use of chlorophyll meter technology for nitrogen management in rice and wheat in the coarse textured soils of Punjab, India. In *Proceedings of the 2nd CREMNET Workshop cum Group Meeting,* Thanjavur, India: Soil and Water Management Research Institute, pp. 55-68.

Bijay-Singh, Yadvinder-Singh, M.S. Maskina, and O.P. Meelu. (1997). The value of poultry manure for wetland rice grown in rotation with wheat. *Nutrient Cycling in Agroecosystems* 47: 243-250.

Cassman, K.G. and P.L. Pingali. (1995). Intensification of irrigated rice systems: learning from the past to meet future challenges. *GeoJournal* 35: 299-305.

Chatterjee, B.N., K.I. Singh, A. Pal, and S. Maiti. (1979). Organic manures as substitute for chemical fertilizers for higher yielding rice cultivars. *The Indian Journal of Agricultural Sciences* 44: 188-192.

Chaudhary, T.N. and K.K. Katoch. (1981). Fertiliser-N management for wheat in coarse textured soils. *Fertiliser News* 26(12): 36-39.

Chhonkar, P.K. (1994). Mobilization of soil phosphorus through microbes: Indian experience. In *Phosphorus Research in India,* ed. G. Dev, Gurgaon, India: Potash and Phosphate Institute of Canada-India Programme, pp. 120-125.

Chhonkar, P.K., and K.V.B.R. Tilak. (1996). Biofertilizers for sustainable agriculture: Research gaps and future needs. In: *Plant Nutrient Needs, Supply, Efficiency and Policy Issues: 2000-2025,* eds. J.S. Kanwar and J.C. Katyal, New Delhi, India: National Academy of Agricultural Sciences, pp. 52-66.

Choudhary, O.P., B.R. Arora, and H.S. Hundal. (1997). Phosphate sorption isotherms in relation to P nutrition of wheat (*Triticum aestivum*). *Nutrient Cycling in Agroecosystems* 47: 99-106.

Dhillon, N.S., A.S. Sidhu, J.S. Brar, and G. Dev. (1987). Soil test based fertilizer requirements for varying targeted yield of cereals. *Indian Journal of Ecology* 14: 83-87.

Dobermann, A., K.G. Cassman, C.P. Mamaril, and J.E. Sheehy. (1998). Management of phosphorus, potassium, and sulfur in intensive, irrigated lowland rice. *Field Crops Research* 56: 113-138.

Dobermann, A., K.G. Cassman, P.C. Sta. Cruz, M.A.A. Adviento, and M.F. Pampolino. (1996). Fertilizer inputs, nutrient balance and soil nutrient-supplying power in intensive, irrigated rice systems. III. Phosphorus. *Nutrient Cycling in Agroecosystems* 46: 111-125.

Dobermann, A., P.C. Sta. Cruz, and K.G. Cassman. (1995). Potassium balance and soil potassium supplying power in intensive, irrigated rice ecosystems. In *Potassium in Asia-Balanced Fertilization to Increase and Sustain Agricultural Production*. Basel, Switzerland: International Potash Institute, pp. 199-234.

Driessen, P.M. (1986). Nutrient demand and fertilizer requirements. In *Modelling of Agricultural Production: Weather, Soils and Crops*, eds. H. van Keulen and J. Wolf, Wageningen, The Netherlands: Pudoc, pp. 182-202.

FAO. (1998). *Current World Fertilizer Situation and Outlook–1996/97–2002/2003*. Rome, Italy: Food and Agriculture Organization of the United Nations. 31 pp.

Faroda, A.S. (1992). A decade of agronomic research in rice-wheat system in Haryana. In: *Rice-Wheat Cropping System*, eds. R.K. Pandey, B.S. Dwivedi and A.K. Sharma, Modipuram, India: Project Directorate Cropping System Research, pp. 233-238.

Fertiliser Association of India. (1999). Fertiliser Statistics 1997-98. New Delhi, India: Fertiliser Association of India. 396 pp.

Gaur, A.C. (1990). *Phosphate Solubilizing Microorganisms as Biofertilisers*. New Delhi, India: Omega Scientific Publishers. 176 pp.

Gill, H.S. and O.P. Meelu. (1983). Studies on the utilization of P and causes for its differential response to rice-wheat rotation. *Plant and Soil* 74: 211-222.

Gill, K.S. (1992). *Research on Rice-Wheat Cropping System in the Punjab*. Ludhiana, India: Punjab Agricultural University. 56 pp.

Goswami, N.N. (1986). Soil tests as a guide to the fertilizer needs of irrigated rice. *Fertiliser News* 31(9): 26-33.

Goswami, N.N. and M.B. Kamath. (1984). Fertiliser use research on phosphorus in relation to its utilization by crops and cropping systems. *Fertiliser News* 29(2): 22-26.

Goswami, N.N. and M. Singh. (1976). Management of fertilizer phosphorus in cropping systems. *Fertiliser News* 21(9): 56-59.

Goswami, N.N., S.R. Bapat, and V.N. Pathak. (1971). Studies on the relationships between soil tests and crop responses to phosphorus under field conditions. In *Proceedings of the International Symposium on Soil Fertility Evaluation*, New Delhi, India: Indian National Science Academy, pp. 351-354.

Gupta, M.B.S. and P.N. Narula. (1973). Fertilizer value of different nitrogen sources in calcareous soil and method to increase to increase their efficiency. *Journal of the Indian Society of Soil Science* 21: 309-314.

Hegde, D.M. (1992). *Cropping System Research Highlights. Annual Report of the All India Coordinated Agronomic Research Project*. New Delhi, India: Indian Council of Agricultural Research. 185 pp.

Hegde, D.M., and A. Sarkar. (1992). Yield trends in rice-wheat system in different agro-ecological regions. In *Rice-Wheat Cropping System*, eds. R.K. Pandey, B.S. Dwivedi and A.K. Sharma, Modipuram, India: Project Directorate Cropping System Research, pp. 15-31.

Huke, R. and E. Huke. (1992). *Rice/Wheat Atlas of South Asia*, Los Banos, Philippines: International Rice Research Institute. 50 pp.

Hundal, H.S., P.S. Deol, and G.S. Sekhon. (1977). Evaluation of some sources of fertilizer phosphorus in two cycles of a paddy-wheat cropping sequence. *Journal of Agricultural Sciences, Cambridge* 88: 625-629.

Islam, A. (1995). Review of soil fertility research in Bangladesh. In *Improving Soil Management for Intensive Cropping in the Tropics and Sub-Tropics*, eds. M.S. Hussain, S.M. Imamul Huq, M. Anwar Iqbal, and T.H. Khan, Dhaka, Bangladesh: Bangladesh Agriculture Research Council, pp. 1-18.

Islam, M.S. and K.M. Hossain. (1986). Nutrient requirements for non-irrigated and irrigated wheat. In *Proceedings of the Third National Wheat Training Workshop*, Dhaka, Bangladesh: Bangladesh Agriculture Research Council, pp. 74-90.

Jaggi, T.N. and K.L. Luthra. (1983). Mussoorie phosphate rock as an economic but effective source of fertilizer phosphorus. *The Indian Journal of Agricultural Chemistry* 15(3): 41-49.

Jain, J.M. and M.C. Sakar. (1979). Transformation of inorganic phosphorus under field condition and its effect on P uptake and grain yield of wheat. In *Phosphorus in Soils, Crops, and Fertilizers*, Bulletin No. 12, New Delhi, India: Indian Society of Soil Science, pp. 460-464.

Janssen, B.H. and J. Wolf. (1988). A simple equation for calculating the residual effect of phosphorus fertilizers. *Fertilizer Research* 15: 79-87.

Jianguo, H. and L.M. Shuman. (1991). Phosphorus status and utilization in the rhizosphere of rice. *Soil Science* 152: 360-364.

Kanwar, J.S., and M.D. Joshi. (1964). *Fertilizer Responses of Major Crops on Hill Soils of the Punjab*. Research Bulletin No.1, Ludhiana, India: Punjab Agricultural University. 48 pp.

Kanwar, J.S., and M.S. Mudahar. (1986). *Fertilizer Sulphur and Food Production*. Dordrecht, The Netherlands: Martinus Nijhoff/Dr. Junk Publishers. 247 pp.

Kapur, M.L., Bijay-Singh, A.L. Bhandari, D.S. Rana, and J.S. Sodhi. (1982). Relative performance of three dwarf wheat (*Triticum aestivum* L.) varieties as affected by fertilizer nitrogen level and time of sowing. *Journal of Research (Punjab Agricultural University)* 19: 207-210.

Katyal J.C., Bijay-Singh, and E.T. Craswell. (1984). Fate and efficiency of urea supergranules applied to a high percolating rice growing soil. In *Nitrogen in Soils, Crops and Fertilizers*, Bulletin No. 13, New Delhi, India: Indian Society of Soil Science, pp. 229-237.

Katyal J.C., Bijay-Singh, P.L.G. Vlek, and R.J. Buresh. (1987). Efficient nitrogen use as affected by urea application and irrigation sequence. *Soil Science Society of America Journal* 51: 366-370.

Katyal J.C., Bijay-Singh, P.L.G. Vlek, and E.T. Craswell. (1985a). Fate and efficiency of nitrogen fertilizers applied to wetland rice II. Punjab, India. *Fertilizer Research* 6: 279-290.

Katyal J.C., Bijay-Singh, V.K. Sharma, and E.T. Craswell. (1985b). Efficiency of some modified urea fertilizers for wetland rice grown on a permeable soil. *Fertilizer Research* 8: 137-146.

Kawaguchi, K. and K. Kyuma. (1977). *Paddy Soils in Tropical Asia. Their Material Nature and Fertility.* Honululu: The University Press of Hawaii. 258 pp.

Kemmler, G. (1980). Potassium deficiency in the soils of the tropics as a constraint to food production. In *Priorities for Alleviating Soil-Related Constraints to Food Production in the Tropics.* Los Banos, Philippines: International Rice Research Institute, pp. 253-275.

Khanna, S.S. and M.L. Chaudhary. (1979). Residual and cumulative effect of phosphatic fertilizers. In *Phosphorus in Soils, Crops, and Fertilizers*, Bulletin No. 12, New Delhi, India: Indian Society of Soil Science, pp. 142-148.

Kirk, G.J.D., and M.A. Saleque. (1995). Solubilization of phosphate by rice plants growing in reduced soil. Prediction of the amount solubilized and the resultant increase in uptake. *European Journal of Soil Science* 46: 247-255.

Kirk, G.J.D., T.R. Yu, and F.A. Choudhury. (1990). Phosphorus chemistry in relation to water regime. In *Phosphorus Requirements for Sustainable Agriculture in Asia and Oceania.* Los Banos, Philippines: International Rice Research Institute, pp. 211-223.

Kolar, J.S. and H.S. Grewal. (1989). Response of rice to potassium. *International Rice Research Newsletter* 14 (3): 33.

Koyama, T. (1981). The transformations and balance of nitrogen in Japanese paddy soils. *Fertilizer Research* 2 : 261-268.

Kuldip-Singh. (1993). *Effects of Different Factors and Management Practices on Nitrate Dynamics in Soils.* M.Sc. Thesis, Ludhiana, India: Punjab Agricultural University. 121 pp.

Kuldip-Singh, M.S. Aulakh, Bijay-Singh, and J.W. Doran. (1996). Effect of soil pH on kinetics of nitrification in semi-arid subtropical soils under upland and flooded conditions. *Journal of the Indian Society of Soil Science* 44: 378-381.

Kulkarni, K.R., S.B. Hukeri, and O.P. Sharma. (1978). Fertilizer response experiments on cultivators fields in India. In *Proceedings of the India/FAO/Norway Seminar on Development of Complimentary use of Mineral Fertilizers and Organic Materials.* New Delhi, India: Ministry of Agriculture and Cooperation, pp. 27-31.

Kundu, D.K. and J.K. Ladha. (1995). Enhancing soil nitrogen use and biological nitrogen fixation in wetland rice. *Experimental Agriculture* 31: 261-277

Mahapatra, I.C., and V.C. Srivastava. (1984). Management of rainfed upland rice: Problems and prospects. *Oryza* 21: 20-37.

Maskina, M.S., Bijay Singh, Yadvinder-Singh, H.S. Baddesha, and O.P. Meelu. (1988). Fertilizer requirement of rice-wheat and maize-wheat rotations on coarse textured soils amended with farmyard manure. *Fertilizer Research* 17: 153-164.

Mathur, B.S. and S. Lal. (1989). Effect of graded levels of rock phosphate on plant nutrition in an Alfisol. *Journal of the Indian Society of Soil Science* 37: 491-494.

Meelu, O.P. and A.L. Bhandari. (1978). Fertilizer management of paddy in Northern India. *Fertiliser News* 23(3): 3-10.

Meelu, O.P., K.R. Kulkarni, and A.L. Bhandari. (1982). Crop response to nutrients under irrigated conditions. In *Review of Soil Research in India*. New Delhi, India: Indian Society of Soil Science, pp. 424-441.

Meelu O.P., R.S. Rekhi, J.S. Brar, and J.C. Katyal. (1983). Comparative efficiency of different modified urea materials in rice. In *Proceedings of the Fertiliser Association of India–Northern Region Seminar on Fertilizer Use Efficiency*, New Delhi, India: The Fertiliser Association of India, pp. 47-59.

Meelu, O.P., S. Saggar, M.S. Maskina, and R.S. Rekhi. (1987). Time and source of nitrogen application in rice and wheat. *Journal of Agricultural Sciences, Cambridge* 109: 387-391.

Meelu O.P., Yadvinder-Singh, and Bijay-Singh. (1990). Relative efficiency of ammonium chloride under different agro-climatic conditions–A review. *Fertiliser News* 35(4): 25-29.

Meelu, O.P., Yadvinder-Singh, and Bijay-Singh. (1994). *Green Manuring for Soil Productivity Improvement*. World Soil Resources Reports No. 76, Rome, Italy: Food and Agricultural Organization of the United Nations. 123 pp.

Meelu, O.P., Yadvinder-Singh, Bijay-Singh, and A.L. Bhandari. (1995). Response of potassium application in rice-wheat rotation. In: *Use of Potassium in Punjab Agriculture*, eds. G. Dev and P.S. Sidhu, Gurgaon, India: Potash and Phosphate Institute of Canada-India Programme, pp. 94-98.

Meelu, O.P., Yadvinder-Singh, M.S. Maskina, Bijay-Singh, and C.S. Khind. (1992). Balanced fertilization with NPK and organic manures in rice. In *Balance Fertiliser Use for Increasing Foodgrains Production in Northern States*, Gurgaon, India: Potash and Phosphate Institute of Canada–India Programme, pp. 63-74.

Mishra, B. and R.D. Mishra. (1992). Phosphorus management in rice-wheat system. In: *Rice-Wheat Cropping System*, eds. R.K. Pandey, B.S. Dwivedi and A.K. Sharma, Modipuram, India: Project Directorate Cropping System Research, pp. 84-92.

Mishra, B., N.P. Mishra, and R.D. Sharma. (1980). Direct and residual effects of Mussoorie rock phosphate applied in conjunction with pyrites or superphosphate in maize-wheat rotation. *Indian Journal of Agricultural Science* 50: 691-697.

Misra, R.D., D.S. Pandey, and V.K. Gupta. (1996). Crop residue management for increasing the productivity and sustainability in rice-wheat system. In *Abstract of Poster Sessions*, 2nd International Crop Science Congress, New Delhi, India: National Academy of Agricultural Sciences, pp. 42.

Modgal, S.C., Y. Singh, and P.C. Gupta. (1995). Nutrient management in rice-wheat cropping system. *Fertiliser News* 40 (5): 49-54.

Nambiar, K.K.M. and A.B. Ghosh. (1984). *Highlights of Research on Long-Term Fertilizer Experiments in India (1971-82)*. New Delhi, India: Indian Agricultural Research Institute. 189 pp.

Narang, R.S. and A.L. Bhandari. (1992). Integrated nutrient management in rice-wheat system. In *Rice-Wheat Cropping System*, eds. R.K. Pandey, B.S. Dwivedi and A.K. Sharma, Modipuram, India: Project Directorate Cropping System Research, pp. 68-83.

Nitant, H.C. and K.S. Dargan. (1974). Influence of notrogenous fertilizers on yield and nitrogen uptake of wheat in saline-sodic soil. *Journal of the Indian Society of Soil Science* 22: 121-124.

Palmer, B., M. Ismunadji, and V.T. Xuan. (1990). Phosphorus management in lowland rice-based cropping system. In *Phosphorus Requirements for Sustainable Agriculture in Asia and Oceania.* Los Banos, Philippines: International Rice Research Institute, pp. 325-331.

Panda, M., D. Panda, and B. Jena. (1981). Critical limits of available phosphorus for rice in some soils of Orissa. *Oryza* 18: 101-104.

Pandey P.C., P.S. Bisht, R.C. Gautam, and Pyare Lal. (1989). Evaluation of modified urea forms for increasing N use efficiency in irrigated transplanted rice. In *Soil Fertility and Fertilizer Use*, volume 3, eds. V. Kumar, G.C. Shrotriya and S.V. Kaore, New Delhi, India: Indian Farmers Fertilizer Cooperative, pp. 114-126.

Pathak, H. and M.C. Sarkar. (1997). Nitrogen supplementation with rice straw in an Ustochrept. *Journal of the Indian Society of Soil Science.* 45: 103-107.

PDCSR. (1990). *Annual Report.* Modipuram, India: Project Directorate Cropping System Research. 396 pp.

PDCSR. (1998). *Annual Report.* Modipuram, India: Project Directorate Cropping System Research. 382 pp.

Peng, S., F.V. Garcia, R.C. Laza, A.L. Sanico, R.M. Visperas, and K.G. Cassman. (1996). Increased nitrogen use efficiency using a chlorophyll meter in high yielding irrigated rice. *Field Crops Research* 47: 243-252.

Peng, S. and K.G. Cassman. (1998). Upper thresholds of nitrogen uptake rates and associated nitrogen fertilizer efficiencies in irrigated rice. *Agronomy Journal* 90: 178-185.

Peterson, T.A., T.M. Blackmer, D.D. Francis, and J.S. Schepers. (1993). *Using a Chlorophyll Meter to Improve N Management.* Nebguide G93-1171A. Lincoln, Nebraska, U.S.A: Cooperative Extension Service, University of Nebraska. 4 pp.

Pingali, P.L., M. Hossain, S. Pandey, and L. Price. (1995). *Economics of Nutrient Management in Asian Rice Systems: Toward Increasing Knowledge Intensity.* Los Banos, Philippines: International Rice Research Institute. 203 pp.

Ponnamperuma, F.N. (1972). The chemistry of submerged soils. *Advances in Agronomy* 24: 29-96.

Prakasa Rao, E.V.S.P. and R. Prasad. (1980). Nitrogen leaching losses from conventional and new nitrogenous fertilizers in lowland rice culture. *Plant and Soil* 57: 383-392.

Prasad, B.L. and J. Prasad. (1992). Availability and critical limits of potassium in rice and calcareous soils (Calciorthents). *Oryza* 29: 310-316.

Prasad R., S.N. Sharma, S. Singh, and M. Prasad. (1989). Relative efficiency of prilled urea and urea supergranules for rice. In *Soil Fertility and Fertilizer Use*, volume 3, eds. V. Kumar, G.C. Shrotriya and S.V. Kaore, New Delhi, India: Indian Farmers Fertilizer Cooperative, pp. 38-46.

Prasad, R., S. Singh, S.N. Sharma, and M. Prasad. (1992). Relative efficiency of N sources in wheat. *Fertiliser News* 37(6): 39-40.

Prem Narain, P.N. Soni, and A.K. Pandey. (1990). Economics of long term fertilizer use and yield sustainability. In: *Soil fertility and Fertilizer Use*, volume 4, eds. V. Kumar, G.C. Shrotriya and S.V. Kaore, New Delhi, India: Indian Farmers Fertilizer Cooperative, pp. 251-264.

Rajput, A.L. (1995). Effect of fertilizer and organic manure on rice (*Oryza sativa*) and their residual effect on wheat (*Triticum aestivum*). *Indian Journal of Agronomy* 40: 292-294.

Rajput, D.R., Om Parkash, and T.A. Singh. (1984). Movement and fertilizer use efficiency of nitrogen in wheat as affected by crop residues. In *Nitrogen in Soils, Crops and Fertilizers*, Bulletin of the Indian Society of Soil Science No. 13, New Delhi, India: Indian Society of Soil Science, pp. 268-276.

Rana D.S., Bijay-Singh, M.L. Kapur, and A.L. Bhandari. (1984). Relative efficiency of new urea based nitrogen fertilizers for rice grown in a light textured soil. *Journal of the Indian Society of Soil Science* 32: 284-287.

Rana, D.S., P.S. Deol, K.N. Sharma, Bijay-Singh, A.L. Bhandari, and J.S. Sodhi. (1985). Interaction effect of native soil fertility and fertilizer application on yield of paddy and wheat. *Journal of Research (Punjab Agricultural University)* 20: 431-436.

Randhawa, N.S., and H.L.S. Tandon. (1992). Advances in soil fertility and fertiliser use research in India. *Feriliser News* 26(3): 11-26.

Rattan, K.R., M.P. Singh, R.O. Singh, and U.S.P. Singh. (1996). Long term effect of inorganic and organic-inorganic nutrient supply system on yield trends in rice-wheat cropping system. *Journal of Applied Biology* 6: 56-58.

Reddy K.R. and W.H. Patrick, Jr. (1975). Effect of alternate aerobic and anaerobic conditions on redox potential, organic matter decomposition and nitrogen loss in a flooded soil. *Soil Biology and Biochemistry* 7: 87-91.

Reddy K.R. and W.H. Patrick, Jr. (1976). Effect of frequent changes in aerobic and anaerobic conditions on redox potential and nitrogen loss in flooded soil. *Soil Biology and Biochemistry* 8: 491-495.

Regmi, A.P. (1994). *Long-Term Effects of Organic Amendments and Mineral Fertilizers on Soil Fertility in a Rice-Wheat Cropping System in Nepal.* M.Sc. Thesis, Los Banos, Philippines: University of Philippines. 142 pp.

Rekhi, R.S. and O.P. Meelu. (1983). Effect of complementary use of mung straw and inorganic fertilizer N on the nitrogen availability and yield of rice. *Oryza* 20: 125-129.

Run-Kun, L., J. Bai Fan, and L. Cheng Kwei. (1982). P management for submerged soils. In *Proceedings of the 12th International Congress of Soil Science.* New Delhi, India: Indian Society of Soil Science, pp. 182-191.

Saggar, S., O.P. Meelu, and G. Dev. (1985). Effect of Phosphorus applied in different phases in rice-wheat rotation. *Indian Journal of Agronomy* 30: 199-206.

Salim, M. (1995). Rice crop residue use for crop production. In *Organic Recycling in Asia and the Pacific*, RAPA Bulletin volume 11, Bangkok, Thailand: Regional Office for Asia and the Pacific, Food and Agriculture Organization of the United Nations, 99 pp.

Saunders, D.A. (1990). *Report of an On-Farm Survey–Dinajpur District: Farmer's Practices and Problems, and Their Implications,* Monograph No. 6, Nashipur, Bangladesh: BARI Wheat Reserach Centre. 39 pp.

Savant, N.K., S.K. DeDatta, and E.T. Craswell. (1982). Distribution pattern of ammonium nitrogen and ^{15}N uptake by rice after deep placement of urea supergranules in wetland soil. *Soil Science Society of America Journal* 46: 567-573.

Sekhon, G.S. (1995). Characterization of K availability in paddy soils–present status and future requirements. In *Potassium in Asia–Balanced Fertilization to Increase and Sustain Agricultural Production*, Basel, Switzerland: International Potash Institute. pp 115-133.

Shankaracharya, N.B. and B.V. Mehta. (1971). Note on the losses of nitrogen by volatilization of ammonia from loamy sand soil of Anand treated with different nitrogen carriers under field conditions. *Indian Journal of Agricultural Sciences* 41: 131-133.

Sharma, A.N. and R. Prasad. (1980). Nutrient removal (NPK) in rice-wheat rotation. *Fertiliser News* 25(10): 34-36, 44.

Sharma, H.L., C.M. Singh, and S.C. Modgal. (1987). Use of organics in rice-wheat crop sequence. *Indian Journal of Agricultural Sciences* 57: 163-168.

Sharma, K.N., J.S. Brar, M.L. Kapur, O.P. Meelu, and D.S. Rana. (1978). Potassium soil test values and response of wheat, bajra and gram to fertilizer K. *Indian Journal of Agronomy* 23: 10-13.

Sidhu, A.S., H.S. Sur, and G.C. Aggarwal. (1994). Effect of methods of urea application on yield of wheat, and nitrogen fertilizer and water use efficiency. *Journal of the Indian Society of Soil Science* 42: 10-14.

Singandhupe, R.B. and R.K. Rajput. (1990). Nitrogen use efficiency in rice under varying moisture regimes, sources and levels in semi-reclaimed sodic soil. *Indian Journal of Agronomy* 35: 73-81.

Singh, B. and A.L. Bhandari. (1995). Response of cereals to applied potassium. In *Use of Potassium in Punjab Agriculture*, eds. G. Dev and P.S. Sidhu, Gurgaon, India: Potash and Phosphate Institute of Canada-India Programme, pp. 58-68.

Singh D., R.S. Antil, V. Kumar, and M. Singh. (1985). Response of paddy to nitrogen source, their application time and nitrogen inhibitors. *Journal of Research (Haryana Agricultural University)* 15: 87-93.

Singh, G.B. and B.S. Dwivedi. (1996). Integrated nutrient management for sustainability. *Indian Farming* 46(8): 9-15.

Singh, H.P. and K.C. Sharma. (1979). Response of rice and wheat to nitrogen in rice-wheat rotation. *The Indian Journal of Agricultural Sciences* 49: 674-679.

Singh, K.D. and B.M. Sharma. (1994). Soil test crop response coorelations for phosphorus. In *Phosphorus Research in India*, ed. G. Dev, Gurgaon, India: Potash and Phosphate Institute of Canada-India Programme, pp. 47-68.

Singh, K.N., and M. Singh. (1987). Effect of levels and methods of potash application on the uptake of K by dwarf wheat varieties. *Mysore Journal of Agricultural Science* 21: 18-26.

Singh, M., and V. Singh. (1992). Balanced fertilisation with NPK for increasing wheat production. In *Balance Fertiliser Use for Increasing Foodgrains Production in Northern States*, Gurgaon, India: Potash and Phosphate Institute of Canada–India Programme, pp. 75-84.

Singh, M.P., R.P. Singh, V.P. Singh, and S.C. Verma. (1990). Effect of modified urea materials on the performance of rice under varying nitrogen levels. *Indian Journal of Agronomy* 35: 384-390.

Singh, R.P., R.P. Singh, and J. Seth. (1980). Nitrogen and phosphorus management of wheat under the condition of delayed availability of fertilizers. *Indian Journal of Agronomy* 25: 433-440.

Singh, S. and R. Prasad. (1992). DCD for increasing fertilizer nitrogen efficiency in wheat. In *Annual Report 1991-92*, New Delhi, India: Division of Agronomy, Indian Agricultural Research Institute, 33 pp.

Singh,Y., D. Singh, and R.P. Tripathi. (1996). Crop residue management in rice-wheat cropping system. In *Abstract of Poster Sessions*, 2nd International Crop Science Congress, New Delhi, India: National Academy of Agricultural Sciences, pp. 43.

Skogley, E.O. and A. Dobermann. (1996). Synthetic ion exchange resins for soil and environmental studies. *Journal of Environmental Quality* 25: 13-24.

Stalin P., A. Dobermann, K.G. Cassman, T.M. Thiyagarajan, and H.F.M. Ten Berge. (1996). Nitrogen supplying capacity of lowland rice soil of southern India. *Communications in Soil Science and Plant Analysis* 27: 2851-2874.

Swarup, A., and K.N. Singh. (1989). Effect of 12 years of rice-wheat cropping sequence and fertilizer use on soil properties and crop yields in sodic soils. *Field Crops Research* 21: 277-287.

Tandon, H.L.S. (1980). Fertilizer use research on wheat in India. *Fertiliser News* 25(10): 45-78.

Tandon, H.L.S. (1987). *Phosphorus Research and Agricultural Production in India*. Fertiliser Development and Consultation Organization, New Delhi. 160 pp.

Tandon, H.L.S. and G.S. Sekhon. (1988). *Potassium Research and Agricultural Production in India*. Fertiliser Development and Consultation Organization, New Delhi. 144 pp.

Thakur, K.S. and M.N. Singh. (1987). Effect of organic wastes and N levels on transplanted rice. *Indian Journal of Agronomy* 32: 161-164.

Thind, H.S., Bhajan Singh and M.S. Gill. (1984). Relative efficiency of nitrogenous fertilizers for rice-wheat rotation. In *Nitrogen in Soils, Crops and Fertilizers*, Bulletin of the Indian Society of Soil Science 13: 181-184.

Tiwari, K.N. (1985). Changes in potassium status of alluvial soils under intensive cropping. *Fertiliser News* 30(9): 17-24.

Tiwari, K.N., A.N. Pathak, and S.P. Tiwari. (1980). Fertilizer management in cropping system for increased efficiency. *Fertiliser News* 25(3): 13-20.

Tiwari, K.N., B.S. Dwivedi, and A. Subba Rao. (1992). Potassium management in rice-wheat system. In *Rice-Wheat Cropping System*, eds. R.K. Pandey, B.S. Dwivedi and A.K. Sharma, Modipuram, India: Project Directorate Cropping System Research, pp. 93-114.

Tiwari, K.N., G. Dev, D.N. Sharma and U.V. Singh. (1998). Maximising yield of of a rice-wheat sequence in recently reclaimed saline-sodic soils. *Better Crops International* 12 (2): 9-11.

Toor, G.S. and V. Beri. (1991). Extent of fertilizer N immobilized by the application of rice straw and its availability in soil. *Bioresource Technology* 37: 189-191.

Varvel, G.E., J.S. Schepers, and D.D. Francis. (1997). Ability for in-season correction of nitrogen deficiency in corn using chlorophyll meters. *Soil Science Society of America Journal* 61: 1233-1239.

Velayutham, M., K.C.K. Reddy, and G.R.M. Sankar. (1985). All India coordinated research project on soil test-crop response correlation and its impact on agricultural production. *Fertiliser News* 30(4): 81-95.

Verma, S., and R.S. Mathur. (1990). The effects of microbial innoculation on the yield of wheat when grown in straw-amended soil. *Biological Wastes* 33: 9-16.

Verma, T.S. and R.M. Bhagat. (1992). Impact of rice straw management practices on yield, nitrogen uptake and soil properties in a wheat-rice rotation in northern India. *Fertilizer Research* 33: 97-106.

Verma, U.N. and V.C. Srivastava. (1995). Balanced fertilization with NPK for increasing wheat production. In *Balanced Fertiliser Use for Increasing Foodgrains Production in Eastern States*, ed. G.Dev, Gurgaon, India: Potash and Phosphate Institute of Canada-India Programme, pp. 22-34.

Vig, A.C., B.S. Brar, C.R. Biswas, and Milap Chand. (1986). P uptake and recovery of applied P in wheat as influenced by clay content. *Journal of Nuclear Agriculture and Biology* 15: 53-59.

Vig, A.C., and N.T. Singh. (1983). Yield and P uptake by wheat as affected by P fertilization and soil moisture regime. *Fertilizer Research* 4: 21-29.

Vlek P.L.G., B.H. Byrnes, and E.T. Craswell. (1980). Effect of urea placement on leaching losses of nitrogen from flooded rice soils. *Plant and Soil* 54: 441-449.

Walia, S.S., S.S. Brar, and D.S. Kler. (1995). Effect of management of crop residues on soil properties in rice-wheat cropping system. *Environment and Ecology* 13: 503-507.

Willet, I.R. (1986). Phosphorus dynamics in relation to redox processes in flooded soils. *Transactions of the International Congress of Soil Science* 13(6): 748-755.

Wolf, J., C.T. De Wit, B.H. Janssen, and D.J. Lathwell. (1987). Modeling long-term crop responses to fertilizer phosphorus. 1: The model. *Agronomy Journal* 79: 445-451.

Yadav, R.L. (1997). Urea-N management in relation to crop residue recycling in rice-wheat cropping system in north-western India. *Bioresource Technology* 61: 105-109.

Yadav, R.L., S.R. Singh, K. Prasad, B.S. Dwivedi, R.K. Batta, A.K. Singh, N.G. Patil, and S.K. Chaudhary. (2000). Management of irrigated ecosystem. In *Natural Resource Management for Agricultural Production in India*, eds. J.S.P. Yadav and G.B. Singh, New Delhi, India: Indian Society of Soil Science, pp.775-870.

Yadvinder-Singh, Bijay-Singh, O.P. Meelu, and C.S. Khind. (2000a). Long-term effects of organic manuring and crop residues on the productivity and sustainability of rice-wheat cropping system in Northwest India. In *Long-term Soil Fertility Experiments in Rice-Wheat Cropping Systems*, eds. I.P. Abrol, K.F. Bronson, J.M. Duxbury, and R.K. Gupta, Rice-Wheat Consortium Paper Series 6, New Delhi, India: Rice-Wheat Consortium for the Indo-Gangetic Plains, pp. 149-162.

Yadvinder-Singh, A. Dobermann, Bijay-Singh, K.F. Bronson, and C.S. Khind. (2000b). Optimal Phosphorus management strategies for wheat-rice cropping on a loamy sand. *Soil Science Society of America Journal* 64:1413-1422..

Yadvinder-Singh, Bijay-Singh and C.S. Khind. (1992). Nutrient transformations in soils amended with green manure. *Advances in Soil Science* 20: 237-309.

Yadvinder-Singh, Bijay-Singh, M.S. Maskina, and O.P. Meelu. (1988). Effect of organic manures, crop residues and green manure (*Sesbania aculeata*) on nitrogen and phosphorus transformations in a sandy loam soil at field capacity and under waterlogged conditions. *Biology and Fertility of Soils* 6: 185-187.

Yadvinder-Singh, Bijay-Singh, M.S. Maskina, and O.P. Meelu. (1995). Response of wetland rice to nitrogen from cattle manure and urea in a rice-wheat rotation. *Tropical Agriculture (Trinidad)* 72: 91-96.

Yadvinder-Singh, C.S. Khind, and Bijay-Singh. (1991). Efficient management of leguminous green manures in wetland rice. *Advances in Agronomy* 45: 135-189.

Yadvinder-Singh, O.P. Meelu, and Bijay-Singh. (1990). Relative efficacy of calcium ammonium nitrate under different agro-climatic conditions–A review. *Fertiliser News* 35(4): 41-45.

Yadvinder-Singh, and T.S. Khera. (1998). Balanced fertilization in rice-wheat cropping system. 74-87. In *Balanced Fertilization in Punjab Agriculture*, eds. M.S. Brar and S.K. Bansal, Ludhiana, India: Department of Soils, Punjab Agricultural University; Basel, Switzerland: International Potash Institute; Gurgaon, India: Potash Research Institute of India, pp. 74-87.

Yang, J.E., E.O. Skogley, S.J. Georgitis, B.E. Schaff, and A.H. Ferguson. (1991). Phytoavailability soil test; development and verification of theory. *Soil Science Society of America Journal* 55: 1358-1365.

Zia, M.S., M. Munsif, M. Aslam, and M.A. Gill. (1992). Integrated use of organic manures and inorganic fertilizers for the cultivation of lowland rice in Pakistan. *Soil Science and Plant Nutrition* 38: 331-338.

Management
of Soil Micronutrient Deficiencies
in the Rice-Wheat Cropping System

V. K. Nayyar
C. L. Arora
P. K. Kataki

SUMMARY. Within the last three decades, the rice-wheat cropping system has triggered, and with time, aggravated soil micronutrient deficiencies in the Indo-Gangetic Plains (IGP). This has largely been due to the shift from an earlier rice and wheat monoculture with low yielding, long duration indigenous varieties, to an intensive rice-wheat rotation cropping system with short duration modern high yielding varieties on the same piece of land. The problems related to micronutrient deficiency in the IGP are more due to the size of its available pools in the soil rather than its total contents and are greatly influenced by crop management, or rather its mismanagement. Deficiency of zinc is widespread in the IGP, but with the extensive use of zinc sulfate, zinc deficiency has reduced in some areas of the region. Meanwhile, the deficiency of Fe, Mn and B has increased in the IGP. Deficiency of Cu and Mo is location specific and can limit rice and wheat yields. The adoption and spread of the rice-wheat system in permeable coarse textured soils, particularly in the western IGP, not only caused iron deficiency in rice but also resulted in the emergence of manganese deficiency in wheat. In highly calcareous and acidic soils, boron is the next limiting micronutrient in

---

V. K. Nayyar and C. L. Arora are Senior Soil Scientists, Department of Soils, Punjab Agricultural University, Ludhiana, Punjab-141004, India. P. K. Kataki is Cornell University Coordinator and CIMMYT Adjunct Scientist, CIMMYT South Asia Regional Office, Lazimpat, P.O. Box 5186, Kathmandu, Nepal.

[Haworth co-indexing entry note]: "Management of Soil Micronutrient Deficiencies in the Rice-Wheat Cropping System." Nayyar, V. K., C. L. Arora, and P. K. Kataki. Co-published simultaneously in *Journal of Crop Production* (Food Products Press, an imprint of The Haworth Press, Inc.) Vol. 4, No. 1 (#7), 2001, pp. 87-131; and: *The Rice-Wheat Cropping System of South Asia: Efficient Production Management* (ed: Palit K. Kataki) Food Products Press, an imprint of The Haworth Press, Inc., 2001, pp. 87-131. Single or multiple copies of this article are available for a fee from The Haworth Document Delivery Service [1-800-342-9678, 9:00 a.m. - 5:00 p.m. (EST). E-mail address: getinfo@haworthpressinc.com].

crop production after zinc. Bumper rice and wheat harvests in the past decade, the declining use of organic manures in the region and except for the widespread use of zinc sulfate, a general lack of awareness amongst farmers on micronutrient deficiency problems has contributed to micronutrients limiting rice and wheat yields in the IGP. Approaches to alleviating micronutrient deficiencies include matching the crop removals of the micronutrients with its replenishments through their respective external carriers, supplementation through organic sources and mobilization/utilization through cultivation of micronutrient efficient crop cultivars. Identification of efficient micronutrient carriers and finding the optimum rate, mode and time of its application is important in ameliorating the micronutrient deficiencies. This article reviews the extent of micronutrient deficiency and discusses various management options available to reduce micronutrient deficiency induced crop yield reduction for rice and wheat in the Indo-Gangetic Plains. *[Article copies available for a fee from The Haworth Document Delivery Service: 1-800-342-9678. E-mail address: <getinfo@haworthpressinc.com> Website: <http://www.HaworthPress.com> © 2001 by The Haworth Press, Inc. All rights reserved.]*

KEYWORDS. Boron, copper, cropping system, deficiency, Indo-Gangetic Plains, iron, manganese, micronutrients, molybdenum, rice, South Asia, wheat, zinc

INTRODUCTION

Micronutrient deficiencies in the Indo-Gangetic Plains (IGP) started emerging with the adoption and spread of intensive agriculture in this region. Imbalance use of high analysis macronutrient (NPK) fertilizers, decreased use of organic manure, reduced recycling of crop residues, and bumper harvests in the last three decades have induced secondary and micronutrient deficiencies in the IGP. The first fall out of such an intensive cropping practice was the appearance of zinc deficiency in many parts of the IGP. Subsequently, the deficiencies of iron, manganese, boron and molybdenum have also been recorded and its severity depended on the soil conditions and the crop grown. In a three year rice-wheat cropping system experiment (Gupta and Mehla, 1993), the mean Zn, Cu, Mn and Fe uptake by rice has been reported to be 252, 199, 801, and 3431 g ha^{-1} compared to 170, 80, 281 and 1833 g ha^{-1} uptake by wheat at medium fertility level (Table 1). These uptake rates represented 19.5, 15.4, 62 and 265 g t^{-1} biomass for rice and 17, 8.3, 29 and 190 g t^{-1} biomass for wheat for Zn, Cu, Mn and Fe, respectively. The rice crop removed larger quantities of micronutrients compared to wheat.

TABLE 1. Uptake (kg ha^{-1}) of Zn, Cu, Mn and Fe by rice and wheat.

Fertility levels	Biomass (t ha^{-1})		Rice				Wheat			
	Rice	Wheat	Zn	Cu	Mn	Fe	Zn	Cu	Mn	Fe
Control*	6.5	4.1	0.12	0.09	0.35	1.68	0.07	0.03	0.11	0.70
Low	8.8	7.5	0.17	0.13	0.57	2.51	0.12	0.05	0.21	1.30
Medium	12.9	10.2	0.25	0.19	0.80	3.43	0.17	0.08	0.28	1.83
High	14.5	12.4	0.30	0.22	0.97	4.06	0.20	0.09	0.33	2.23

*Control: 0:0:0; Low: 60:30:30; Medium: 120:60:60 and High:180:90:90 of N:P_2O_5:K_2O.
Adapted from Gupta and Mehla (1993).

Increasing levels of fertility progressively increased the total removal (Table 1) of micronutrients due to increased dry matter production (Gupta and Mehla, 1993).

Rice and wheat occupies and will continue to occupy a pivotal role in the national food security of the IGP. This cropping system, though remunerative in the short run, has resulted in the over-exploitation of the natural soil resource base and this trend has been enhanced by the imbalanced use of inputs. The continuing emphasis on maximization of food grain production without appropriate management practices from a shrinking land resource base will result in still further depletion of the micronutrient reserves and will cause the deficiency of other micronutrients to limit crop production, besides aggravating the existing ones. The simplest solution to alleviating micronutrient deficiency is the application of micronutrient fertilizers to the crop. However, this solution is complicated because:

- Field crop deficiency symptoms are often not easy to identify, for example, Ca deficiency can be confused with B deficiency, Fe with Mn deficiency, leaf strip disease with Mn deficiency, virus infection and little leaf with Zn or B deficiency, brown streak disease of rice with Zn deficiency. Often micronutrient deficiency induces disease susceptibility in crops and the resulting diagnosis becomes complicated. Crop varieties also vary in its micronutrient uptake efficiency.
- The IGP spreads across the four countries of South Asia: Pakistan, India, Nepal and Bangladesh from the west to the east. Though the soils in general are coarse textured, with low organic matter content and CEC; the soil pH, rainfall duration and intensity, seasonal fluctuations in the ground water levels and temperature regimes in this region vary widely from the west to the east. Soil micronutrient status also varies tremendously between and within states, districts and fields and an extensive database on the soil micronutrient status has not yet been completed.

- Critical limits for soil and plant micronutrient status for the IGP have been advocated, but field tests to validate these levels and facilities for testing micronutrients in this region has not been adequate to cope with the required volume of analyses.

If these constraints to high productivity are not identified, monitored and alleviated, the fertilizer use efficiency of costly NPK fertilizers and other agricultural inputs will be markedly reduced. Reliable data on the consumption and production of micronutrient fertilizer is not available for the four countries of the IGP. According to recent estimates, the projected micronutrient demand (Sehgal, 1999) for the year 2025 in India is expected to increase by 5 to 10 times of the present output (Table 2). This would entail the use of fertilizer nutrient in its right proportion, their application at the most appropriate time of crop growth, coupled with efficient delivery methods. In addition, augmenting the availability of native nutrients by modifying soil environment through management practices in different crops and cropping systems will be needed.

Information on crop micronutrient requirement in the IGP generated so far has mostly been on a single crop basis. Meager information is, however, available on micronutrient management in the rice-wheat cropping system. In this chapter an attempt is being made to review the extent and management of micronutrient deficiency in the IGP for maintaining and perhaps increasing the productivity of the rice-wheat system.

SOURCES OF MICRONUTRIENTS

The primary source of micronutrients in soils is the parent material, but a significant contribution, especially to surface soils, may arise from industrial and urban pollution, agricultural sprays as well as through fertilizers and irrigation waters.

TABLE 2. Projected demand for micronutrients in India for the year 2025 ('00,000 tons).

Micronutrient	Present consumption	Projected requirement in 2025
Zinc (as sulfate)	0.58	5.60
Iron (as sulfate)	0.34	1.70
Manganese (as sulfate)	0.04	0.54
Copper (as sulfate)	0.12	1.21
Boron (as Borax)	0.22	1.26

Adapted from Sehgal (1999).

Soil Parent Material as a Source of Micronutrient

The parent rocks constitute the major source of micronturient to soils, but in highly variable amounts. In immature soils, the total micronutrient content approximates the composition of the parent material. In such soils, the total micronutrient content has even been compared to rocks that have contributed to the sediments constituting the soils parent material (Sharma et al. 1999). Under the influence of different pedogenic factors, the micronutrients get redistributed during synthesis of secondary minerals. The wide variations in the total micronutrient content of Indian soils (Table 3) reflects the difference in the composition of parent material and the pedological conditions under which soils have developed. Katyal and Sharma (1991) reported wide variations in the total and available micronutrient content in 57 benchmark soils representing eight orders. For example, flood plain alluvial soils (Fluvents) derived from silicon sandstone has strikingly lower contents of Zn, Cu, Mn and Fe than some chromusterts developed on basaltic alluvium. Similarly, soils developed from mixed shale parent material contained higher amount of total zinc than those developed on sandstone material (Yaduvanshi et al. 1988). Soils that develop over sandstone are inherently low in total micronutrients and have significant negative correlation coefficients with its total Zn, Cu, and Fe content (Katyal and Sharma, 1991; Sharma et al. 2000). In twenty benchmark profiles of the IGP, the total micronutrient content ranged between 23 and 99 mg kg^{-1} for Zn, 10 and 64 mg kg^{-1} for Cu, 169 to 627 mg kg^{-1} soil for Mn and 1.5 to 4.5 percent for Fe. Soils in the aridic moisture regime had a low reserve of total micronutrients compared to those occurring in aquic moisture regime (Sharma et al. 2000). Total micronutrient contents in the Ap horizon of soils derived from phyllite and slate alluvium were higher than those derived from aeolian alluvium under aridic moisture regime. Similarly, in the aquic/udic moisture regime, soils derived from granit-

TABLE 3. Total and available micronutrient content of Indian soils.

Micronutrient	Total content (mg kg^{-1} soil) Takkar (1982)	Available micronutrient (mg kg^{-1} soil) Singh (1999)	
		Content	Mean
Zinc	2 to 1,019	0.2 to 6.9	0.9
Copper	1.9 to 960	0.1 to 8.2	2.1
Iron	2700 to 1,91,000	0.8 to 196	19.0
Manganese	37 to 11,500	0.2 to 118	21.0
Boron	3.8 to 630	0.08 to 2.6	–
Molybdenum	0.01 to 18.1	0.07 to 7.67	–

ic alluvium had appreciably lower content of micronutrients than those developed from mixed alluvium but had a larger content than those developed from silty alluvium (Table 4).

Agricultural Chemicals and Fertilizers as Sources of Micronutrients

The major nutrient fertilizers generally contain appreciable amounts of micronutrients as contaminants. These impurities arise from the raw materials and are also introduced during the manufacturing process. Some fertilizer sources contain sufficient quantity of micronutrients to benefit crop growth (Table 5), however, undesirable impurities such as Cd, Pb, Cr, etc., may also be present (Williams and David, 1973; Arora, Nayyar and Randhawa, 1975). The quantity of micronutrients that can be added through different combinations of fertilizers to supply 120, 60 and 60 kg of N, P_2O_5 and K_2O ha^{-1} is given in Table 6 (Arora, Nayyar and Randhawa, 1975). Agricultural chemicals used to control pest and diseases may increase the concentration of some metals in surface soils appreciably when used continuously for several years particularly in horticultural crops.

Industrial Pollution as a Source of Micronutrients

Appreciable amounts of micronutrients may also arise from the use of untreated sewage and industrial effluents for irrigation purposes, particularly in the developing countries (Bansal, Nayyar and Takkar, 1992). These sources may also contribute toxic elements to soils, which not only enter the food chain through plants and is a source of concern but also affects the availability of other micronutrients to plants (Nayyar and Bansal, 1989). Many underground waters may also add considerable amount of toxic elements such as selenium, arsenic and accumulate in soils and crops with their

TABLE 4. Total micronutrient content in Ap horizon of soils of the IGP as influenced by parent material and moisture regime.

Moisture regime	Parent material	Total micronutrients (mg kg^{-1} soil)			
		Zn	Cu	Mn	Fe
Aridic	Phyllite and slate alluvium	52	25	476	29300
Aridic	Aeolian alluvium	36	15	507	254000
Aquic/Udic	Granite alluvium	42	25	210	214000
Aquic/Udic	Mixed alluvium	68	31	360	250000
Aquic/Udic	Silty alluvium	39	21	257	168000

Adapted from Sharma et al. (2000).

TABLE 5. Elemental content of fertilizers and manures.

Fertilizer	Element (parts per million)										
	Zn	Cu	Fe	Mn	B	Mo	Co	Al	Sr	Cr	Pb
Urea	4.0	0.6	36	0.5	1.0	5.3		25		6.3	
Calcium ammonium nitrate	7.6	2.8	407	24.8	9.0	56.0	6.6	281	21.8	8.5	116.3
Ammonium sulfate nitrate	54.7	1.9	490	53.8	6.5	5.0	1.8	51	18.9	3.0	41.8
Ammonium sulfate	11.3	0.8	23	3.5		6.0	23.7			4.0	
Triple superphosphate	418.0	49.3	3483	75.0	212.5	270.0	46.7	2495	419.0	392.5	237.5
Single superphosphate (powder)	165.3	15.5	4050	890.0	132.5	335.0	76.7	3063	403.0	87.5	487.5
Rock phosphate	187.0	32.0	19917	975.0	71.5	555.0	108.7	6600	822.0	183.8	962.3
Muriate of potash	10.0	3.1	110	3.5	16.3	26.0	22.4	93	93.8	12.8	117.5
Diammonium phosphate	112.3	7.2	11275	307.0	396.3	75.3	15.8	5100	83.8	80.5	195.0

Adapted from Arora et al. (1975).

TABLE 6. Estimates for addition of secondary and micro elements to soils by different combinations of fertilizers.

Combination of fertilizers to supply 120:60:60	Elements to be added to soil (g ha^{-1})										
	Zn	Cu	Fe	Mn	B	Mo	Co	Al	Sr	Cr	Pb
Ammonium sulfate: Single superphosphate:KCl	69.6	6.56	1542	336.2	51.3	131.7	44.9	1160	168.1	36.4	194.6
Ammonium sulfate nitrate:Single superphosphate:KCl	88.2	6.99	1756	358.9	54.3	130.5	31.8	1180	176.9	35.5	213.9
Calcium ammonium nitrate:Single superphosphate:KCl	66.6	7.48	1725	345.9	55.6	155.1	34.2	1290	178.6	38.2	250.4
Urea: Single superphosphate:KCl	64.0	6.27	1539	334.2	51.6	129.6	31.0	1160	168.1	35.7	194.6
Urea:Triple superphosphate:KCl	53.3	6.62	455	9.9	28.5	37.7	8.1	330	69.4	52.0	41.4
Urea:Diammonium phosphate:KCl	16.5	1.36	1489	40.5	53.5	13.5	4.3	680	27.9	13.1	37.1

Adapted from Arora et al. (1975).

continuous use. In recent years much emphasis has been given to such toxic metals but will not be reviewed here.

EXTENT OF MICRONUTRIENT DEFICIENCIES IN THE INDO-GANGETIC PLAINS

The total micronutrient contents of soils are generally of limited value as far as plant growth and responses to their application are concerned. In most cases, total contents have been insignificantly related to the plant content. In order to match the levels of micronutrients in soil with plant requirement, their available contents in soils are determined. Like total contents, the available micronutrient status of soils is also highly variable (Table 3). Soil properties exercise a considerable influence on the availability of micronutrients. The extent of micronutrient deficiency, therefore, varies not only in different states but also in different districts of the same state or in various blocks of the same district depending upon the soil characteristics and other management conditions.

During a world-wide effort to compile information on the micronutrient status of 30 countries, soil samples of the IGP of India, Nepal and Pakistan were also analyzed (Sillanpaa, 1982), and the results are summarized below:

- A majority of the sampled soils from India were coarse-textured, alkaline, low in organic matter content and cation exchange capacity (CEC), and the mean pH was among the highest of the thirty countries. Soil samples from the east, especially that of Bihar, were deficient in Zn, B and Mn. High B values were recorded from Punjab and Delhi. Many soils of the Punjab and Haryana states were low in Fe but were high in Mo. The severity and numbers of micronutrient deficiency varied from the west to the east of the Indian-IGP region, a trend perhaps influenced by the distribution of the parent material in the Indian-IGP.
- Most of the sampled soils of Nepal were coarse to medium textured, varied widely in its pH–though mostly acidic, had a low organic matter content and CEC. Though the micronutrient deficiencies of Cu, Fe and Mn were considered unlikely due to its high content in the sampled soils, the Mo content of soils and plants were low. Micronutrient deficiencies of B (considering HWS B critical level of 0.3 to 0.5 mg kg^{-1}) and Zn of Nepalese soils were considered to be the most severe in the international ratings.
- Soil samples from Pakistan were also coarse to medium textured, low in organic matter content and CEC. A larger percentage of the sampled sites of Pakistan had alkaline soils and therefore, the availability of Mn and Zn were low but that of Mo was high. Contents of Cu and Fe in

soils of Pakistan were within the normal international range, but B content varied widely–from high to very low values, perhaps influenced by the pH of the sampled soils.

Areas of Deficiency

Zinc deficiency is the most acknowledged and widespread in the four countries of the IGP and to some extent, has overshadowed the importance of diagnosis of other micronutrient deficiencies in the region. Analysis of 90,218 samples in different states of the IGP of India revealed the predominance of Zn deficiency in divergent soils (Table 7). Of these samples 51, 10, 3 and 2 per cent samples were found to be deficient in available Zn, Fe, Mn and Cu, respectively. The proportion of Zn deficient soils in different states varied from 54-60% in the sates of Bihar and Haryana to 36% in West Bengal. Coarse textured soils, alkaline soils, flood plain soils and calcareous soils, which have relatively high pH with low organic matter content are more prone to Zn deficiency (Figure 1). Similarly, fine textured calcareous black soils (vertisols) and highly leached red soils are expected to show greater incidence of Zn deficiency. Awareness amongst farmers on the menace of soil Zn deficiency has resulted in the increased use of zinc sulfate. As a result of this, the status of available Zn in soil has increased in many areas and thereby, the extent of Zn deficiency in many states has decreased depending on the rate and frequency of zinc sulfate application. A random sampling of some of the areas in Punjab revealed a marked decrease in the extent of Zn deficiency, but there has been an increase in the deficiency of Fe and Mn (Table 8).

The extent of Fe deficiency is approximately a fifth that of Zn deficiency along the IG7P and is largely influenced by the vast areas under alkaline to calcareous soil tracts. Iron deficiency seems to be only second in importance

TABLE 7. Extent of micronutrient deficiency in the soils of India.

State	No. of soil samples	Percent soil samples deficient			
		Zn	Cu	Fe	Mn
Bihar	19214	54.0	3	6	2
Haryana	21848	60.5	2	20	4
Punjab	16483	48.1	1	14 ·	2
Uttar Pradesh	26126	45.7	1	6	3
West Bengal	6547	36.0	0	0	3
Total	90218	50.6	2	10	3

Adapted from Singh (1999).

FIGURE 1. Association of soil characteristics with the occurrence of Zn deficiency in the IGP. (Texture: s = sandy; sl = silty loam; l = loam; cl = clay loam; OC = organic carbon.)

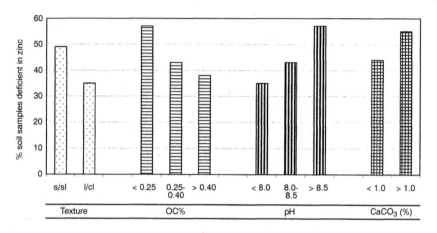

TABLE 8. Changes in the extent of soil micronutrient deficiency in Punjab soils.

District	Year	No. of soil samples	Percent soil samples deficient in			
			Zn	Cu	Fe	Mn
Ludhiana	1970	1730	56	0	0	2
	1998	200	6	0	2	23
Sangrur	1977	360	71	4	7	0
	1998	100	14	0	18	35
Jalandhar	1977	742	45	6	0	0
	1998	100	14	0	13	39

Adapted from Nayyar et al. (1999).

to Zn deficiency in the states of Punjab and Haryana of the Indian-IGP. Manganese deficiency in field crops in the IGP seems to be in localized sites where rice-wheat crop rotation is practised in coarse-textured soils. Copper deficiency is not regarded to be widespread in the IGP, but deficiency based on plant analyses seems to be higher than that of soil analysis. The critical limits used for soil Cu or plant Cu perhaps need to be re-calibrated. The incidence of boron deficiency was highest in the acid soils of West Bengal followed by the calcareous soils of Bihar (Figure 2). Borkakati and Takkar (2000) reported that alluvial soils of Assam were more deficient (44%) in B as compared to lateritic soils (34%) based on the critical level of 0.5-mg

FIGURE 2. Extent of boron deficiency in different states of India

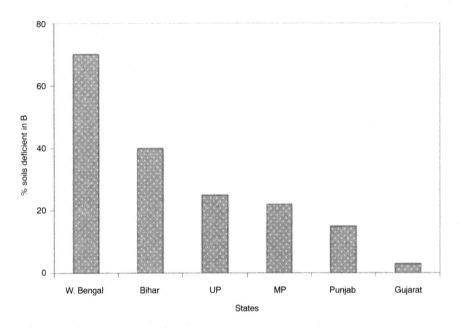

kg^{-1} HWS-B. The deficiency of B was highest under fruit crop ecosystem (56-65%) followed by field crop (42-47%), pasture (37-45%) and forest ecosystem (5-17%). In acid soils of Meghalaya under rice cultivation, 35 percent of the 136 soil samples indicated deficiency of B (Nongkynrih, Dkhar and Khathing, 1996). Similarly, in acid alluvial soils of Coochbehar, Jalpaiguri and West Dinajpur districts of Bengal, 70 out of 88 surface soil samples were deficient in available B (Mondal et al., 1991). In young alluvial soils of Bihar, the incidence of B deficiency was higher (47%) in calcareous as compared to non-calcareous soils (Sakal and Singh, 1999). Deficiency of B and response of crops to applied B has also been reported from Nepal (Srivastava et al. 1997). In Bangladesh, yield increases of 29%, 69%, 146%, 7%, 48% and 20% has been recorded to applied B over minus B plots for wheat, mustard, chickpea, potato, cauliflower and groundnut, respectively (Ahmed and Hossain, 1997). However, response of crops to applied B can also be influenced by climatic conditions (Saifuzzaman, 1995, and, Saifuzzaman and Meisner, 1995).

DIAGNOSIS OF MICRONUTRIENT DEFICIENCY

The assessment of micronutrient deficiency can be made through visual leaf symptoms and soil and plant analysis. Response of crops to the application of micronutrients not only confirms the deficiencies but also helps in determining nutrient needs.

Deficiency Symptoms

The visual recognition of the leaf symptoms is one of the important and oldest techniques of identifying nutrient disorders in crops. Visual deficiency symptoms are generally characteristic enough to permit identification of the deficiency of a micronutrient as these appear on particular plant part(s), and at specific growth stage(s) depending upon the mobility of the element within the plant, but requires good diagnostic experience. Nutrients like iron and boron, which are not readily translocated from old to young leaves under stress conditions within the plant, are called immobile nutrients and their deficiency symptoms first appear on young leaves. In view of the variable mobility of Zn, Cu, Mo and Mn under conditions of their deficiency, the location of their symptoms in different crops and crop species may vary depending upon the degree of their mobility. Confident diagnosis by this method, however, requires much experience, as the symptoms of some nutrient deficiencies are difficult to differentiate without a thorough knowledge. Another limitation of this technique is that by the time the deficiency symptoms appear, the crop has undergone a marked set back and the corrective measures taken at that time may not produce optimum yields. The diagnosis by this method should be confirmed through soil and plant analyses followed by designing experiments of crop response to the added micronutrient of interest.

Plant Analysis

Plant analysis is widely used as a means of detecting micronturient deficiency and the need for fertilizing the crops. Under conditions of hidden micronutrient deficiency, plant analyses is the most effective method for diagnosis of micronutrient disorders. This is because neither the plant exhibits visual deficiency symptoms, nor do the soil tests precisely predict such situations. However, to use plant test data as an effective aid, the critical levels or ranges for deficiency and sufficiency need to be well defined for different crops and crop species at specific growth stages and for different soil types. The critical values reported for rice and wheat in different soils of the IGP are given in Table 9.

Soil Analysis

Soil analysis is an invaluable and a convenient diagnostic tool for quick and timely assessment of micronturient availability. Several chemical extrac-

TABLE 9. Critical levels for micronutrients in the Indo-Gangetic Plains.

Micronutrient	Soil test method	Critical level in soil (mg kg^{-1})	Critical level in plant dry matter (mg kg^{-1})
Zinc	DTPA	0.45 to 1.20	15 to 20
Manganese	DTPA	2 to 3.5	20 (10 to 25)
Copper	DTPA/Amm.acetate	0.14 to 0.60	4 (3 to 6)
Iron	DTPA	3.2 to 6.9	50 (25 to 80)
Boron	Hot Water Soluble	0.5	11 to 18
Molybdenum	Amm. acetate	0.2 to 0.5	0.8 to 1.8

Compiled from different sources.

tion procedures have been used to measure the plant available micronutrients. Most of the researchers used DTPA soil test method as proposed by Lindsay and Norvell (1978) for determining the available content of Zn, Cu, Fe, and Mn particularly in alkaline calcareous soils. Analysis of B is commonly conducted by Hot Water Soluble (HWS) extraction method of Berger and Troug (1944) and of Mo by acid ammonium oxalate method of Grigg (1953). The common soil tests and critical levels of micronutrients advocated in the IGP are given in Table 9. However most of the critical levels reported have not been adequately field tested, but nevertheless, provide a guideline. The results of numerous investigations revealed that for proper management of micronutrient deficiencies, critical limits specific to soil and crops should be used.

MANAGEMENT OF MICRONUTRIENT DEFICIENCIES

Once the micronutrient limiting crop growth is detected, it becomes imperative to use the appropriate fertilizer materials and techniques to ameliorate the deficiency. In view of the large-scale deficiency of zinc in different soils of the IGP, most of the studies in the past were confined to zinc. In the last two decades, due to the increasing incidences of other micronutrient deficiencies under specific situations, focus has shifted to their deficiency management. The studies related to rice-wheat system are enumerated below.

Zinc

Visual symptoms of zinc deficiency in rice plants first appear on older leaves as light yellow spots scattered in the interveinal areas. These spots later turn yellowish-brown imparting pale yellowish-brown color to the affected leaves. With time, leaves develop brown rusty spots, which later co-alesce to form a continuous chlorotic area and under severe deficiency condi-tions, the entire leaf becomes rusty (Plate 1), and is termed the "Khaira" disease in the IGP. Under acute deficiency, the affected leaves wither and die (Plate 2). In wheat, Zn deficiency symptoms is manifested as the develop-ment of a light yellowish white tissue between the midrib and margins in the middle or lower half of the affected leaf. Later, the affected leaves develop minute reddish brown spots which coalesce to form reddish brown lesions (Plate 3) leading to necrosis and limping of the leaf (Plate 4).

PLATE 1. "Khaira" disease in rice due to zinc deficiency. Leaves form brown rusty spots, which later coalesce to form continuous chlorotic areas and entire leaf becomes rusty.

PLATE 2. Severe and prolonged zinc deficiency in rice. Leaves wither and die.

Carriers

Of the several inorganic Zn carriers, $ZnSO_4.7H_2O$ (subsequently referred to as $ZnSO_4$ throughout the text) has been rated as the most efficient for correcting Zn deficiency in rice and wheat in different soil situations. Water insoluble ZnO and Zn-frits were distinctly inferior to water soluble $ZnSO_4$ in coarse textured and sodic soils for rice and wheat (Gupta, Potalia and Katyal, 1986; Nayyar and Takkar, 1980; Takkar and Nayyar, 1986), but were as effective in fine textured soils (Rathore et al., 1995; Takkar and Nayyar, 1986). In a loamy soil, the direct as well as the residual effect of Zn-frits applied at the rate of 10 kg Zn ha^{-1} to rice in a rice-wheat rotation was at par

PLATE 3. Zinc deficiency in wheat induces light yellowish white tissue formation between the mid-rib and margins of the middle and lower half of the affected leaves. Leaves develop minute reddish brown spots, which later coalesce to form reddish brown lesions.

with $ZnSO_4$. The ZnO, however, proved significantly inferior to zinc sulfate (Chhibba, Nayyar and Takkar, 1989). In a calcareous soil of Bihar, the residual effect of ZnO and Zn-frits on the grain yield of rice and wheat was comparable with that of soluble $ZnSO_4$ applied direct to barley (Sakal et al., 1985b). The greater fixation of Zn from the soluble source in fine textured soils as compared to coarse textured soils may be responsible for their equal efficiency in fine textured soils. However, both ZnO and Zn-frits were inferior to soluble $ZnSO_4$ in increasing the grain yield of rice in alkali soils due to their low supply of Zn in the early growth stages of severely Zn-deficient rice

PLATE 4. Acute zinc deficiency in wheat leads to necrosis and limping of affected leaves.

crop (Nayyar and Takkar, 1980). Efforts have also been made to develop either a mixture of $ZnSO_4$ and superphosphate or a compound of zincated urea for uniform distribution of zinc through these materials which also reduces the cost of application, thereby increasing its efficiency many fold. Field results indicate that the yield of rice and wheat with zincated superphosphate (2.5% Zn) were either lower or comparable with that of zinc sulfate (Nayyar et al., 1990). In a silt loam calcareous soil of Bihar, the residual effect of zincated superphosphate on rice when applied to maize was at par with that of zinc sulfate (Sakal, Singh and Singh, 1985a). Zinc chelates have also been evaluated and proved to be superior to zinc sulfate at equivalent rate of Zn application for rice on Zn-deficient loamy sand soil of Punjab (Bansal and Nayyar, 1989) but their high costs made them economically less efficient (Table 10). However, foliar sprays of Zn-EDTA produced higher yields than $ZnSO_4$ at equivalent Zn concentration in the spray solution (Bansal and Nayyar, 1992).

Traditionally, farmers have frequently used farmyard manure (FYM) as a source of nutrients. Application of FYM increased the availability of Zn by mobilizing the native and applied Zn through chelation. Continuous application of 10 t FYM ha^{-1} yr^{-1} for 20 years not only maintained the initial DTPA-extractable Zn level in soil but also significantly increased its content over no FYM plots (Table 11). Six year application of 12, 5 and 2.5 t ha^{-1} $year^{-1}$ of farmyard, poultry and piggary manure, respectively to maize only in a maize-wheat crop rotation on a Zn-deficient loamy sand soil of Punjab, was as effective as 11 kg Zn ha^{-1} for the rotation (Nayyar et al. 1990).

TABLE 10. Effect of sources and rates of zinc application on yield and zinc uptake by rice.

Zinc source	Zinc rates	Grain yield (t ha^{-1})	Zn uptake (g ha^{-1})
ZnSO$_4$·7H$_2$O	2.8	6.10	225
	5.6	6.65	283
	11.2	7.10	351
Zn-EDTA	2.8	6.25	240
	5.6	6.90	304
	11.2	7.60	378
Control		5.30	123
LSD (0.05)		0.54	43

Adapted from Nayyar et al. (1994).

TABLE 11. Effect of continuous application of farmyard manure on DTPA-extractable Zn (μg g^{-1} soil) in surface (0-15 cm) soil.

Treatment	Year				
	1971	1976	1981	1986	1991
Control	1.08	0.87	0.70	0.63	0.49
NPK	1.10	0.83	0.63	0.58	0.46
NPK+FYM	1.12	1.08	1.20	1.26	1.36

NPK = 150 kg N, 32.7 kg P and 62.2 kg K ha^{-1} for maize and 150 kg N, 32.7 kg P and 31.1 kg K ha^{-1} for wheat; FYM = 10 t ha^{-1} to every maize crop.
Adapted from Biswas and Benbi (1997).

Experiments with integrated use of organic manure and zinc sulfate have revealed encouraging results. In a black soil of Madhya Pradesh, 5 t FYM enriched with 5 kg Zn ha^{-1} gave as much grain yield of rice as 10 kg Zn ha^{-1} applied alone (Table 12) with a similar residual effect on the subsequent wheat crop (Rathore et al. 1995). In a calcareous soil of Bihar, application of 10 t ha^{-1} of compost was as effective as 12.5 kg ha^{-1} of ZnSO$_4$ to rice. The mixing of 12.5 kg ZnSO$_4$ with compost further improved the crop yields and were at par with 25 kg ZnSO$_4$ ha^{-1} (Singh, Sakal and Singh, 1982). In a similar soil, increasing rates of biogas slurry (BGS) from 0.625 t to 10 t ha^{-1} increased the grain yield as well as Zn uptake by both the rice and wheat crops for two consecutive years both in the presence and absence of Zn. The Zn use efficiency, calculated from cumulative Zn uptake by all the four crops, was also increased with BGS (Singh et al. 1998).

Rate and Frequency of Zinc Application

The rate and frequency of zinc application should be based on the soil type, crop variety grown and the method of application as discussed in this

TABLE 12. Effect of zinc enriched organic manures on the grain yield of rice and wheat in a black soil.

Treatment	Grain yield (t ha^{-1})		Zn uptake (g ha^{-1})	
	Rice	Wheat	Rice	Wheat
Control	2.25	2.86	107	137
10 kg Zn/ha	3.47	3.46	250	444
10 t FYM/ha	3.06	3.43	185	603
5 t FYM + 2.5 kg Zn/ha	2.91	3.13	104	187
5 t FYM + 5 kg Zn/ha	3.46	3.39	139	384
LSD (p = 0.5)	0.68	NS		

Adapted from Rathore et al. (1995).

section. The quantity of Zn required for alleviating zinc deficiency varied with the severity of its deficiency and/or the soil type. In highly sodic (pH > 10) and flood plain soils, the optimum yield of rice was obtained with 22 kg Zn ha^{-1} (Takkar and Nayyar, 1981a, 1986). This was in comparison to 11 kg ha^{-1} in moderately alkali soils (pH 9.4-9.7; Takkar and Singh, 1978) or 5.5 to 11 kg ha^{-1} in sandy to loam alkaline soils (Chhibba, Nayyar and Takkar, 1989; Nayyar et al. 1990). In a majority of the cases, Zn deficiency in wheat and rice was corrected by an application of 11 kg Zn ha^{-1} (Bansal et al. 1980; Takkar, Chhibba and Mehta, 1989). The rate of Zn application also varied with the soil texture. The best rate of Zn application for wheat on coarse textured loamy sand soil was 11 kg ha^{-1}. This was in contrast to 5.5 kg Zn ha^{-1} required for obtaining optimum yields in fine textured loamy soils (Nayyar et al. 1990). This difference may be due to the greater diffusivity of Zn in fine as compared to coarse textured soils (Singh, Sinha and Randhawa, 1980). The best rate of Zn for wheat in mixed red and black soils was 5 kg Zn ha^{-1} as compared to 10 kg Zn ha^{-1} in alluvial and black soils (Rathore et al. 1995). The optimum rate of Zn application is also governed by the native available Zn status of soils. In calcareous alkaline soils (Typic Ustochrepts and Ustipsamments) of Punjab, the optimum rates of Zn for wheat was 11.0 and 5.5 kg Zn ha^{-1} on low (< 0.60 mg kg^{-1} DTPA extractable Zn) and medium (0.6-1.2 mg kg^{-1} DTPA-Zn) Zn status soils, respectively (Bansal et al. 1980). The ideal Zn application rate also differs with the crop variety. Four of the five varieties tested at Jabalpur center of "All India Coordinated Research Project on Micronutrients" produced optimum yields at 11 kg Zn ha^{-1}. In calcareous soils of Bihar, Ratna, UPR-338 and Sita varieties of rice yielded optimum at 5 kg Zn ha^{-1} and Rajendra Dhan and Pusa S-21 at 10 kg Zn ha^{-1} (Takkar, Chhibba and Mehta, 1989).

Crops require only small quantities of Zn for their normal growth. A

normal crop of paddy (7.0 t ha^{-1}) removes about 250 g of Zn. Soil application rates are much higher, i.e., 50-60 kg zinc sulfate per hectare and therefore a considerable amount of Zn remains in the soil which can be utilized by
the subsequent crops. For sustaining high productivity and increasing the
efficiency of applied Zn fertilizers, it is essential to work out the optimum Zn
requirement as well as the frequency of its application for a cropping system.

In Ustochrepts of Ludhiana 11.2 kg Zn ha^{-1} to first rice crop followed by
its repeat application to fourth crop proved most efficient in terms of grain
yield response (Unpublished data). Similarly, in a calcareous soil (calcior-
thent) of Bihar, soil application of 10 kg Zn ha^{-1} as $ZnSO_4$ to first crop of
rice and 10 kg Zn ha^{-1} at four crops interval was found to be the most ideal
dose and frequency of Zn application. This was followed by 10 kg Zn ha^{-1}
to first rice crop and 5 kg Zn ha^{-1} at three crops interval with corresponding
cumulative grain yield response of 96 and 86 q ha^{-1}, respectively, in five
cycles of crop rotation (Sakal, Nayyar and Singh, 1997).

In moderately alkaline soils (pH 9.4-9.7), 11.2 kg Zn ha^{-1} and application
of 50 percent of the gypsum requirement, produced the optimum yields of
rice and wheat (Takkar and Singh, 1978). However, double the quantity of
Zn, i.e., 22.4 kg Zn ha^{-1} was required for obtaining optimum yield on a
highly sodic soil (pH 10.4) and the fifth crop in the rotation needed a repeat
application of Zn (Nayyar et al. 1990). In a gypsum amended sodic soil
(Aquic Natrustalp) of Karnal (pH 9.2), continuous application of 2.25 kg Zn
ha^{-1} produced as much yield as a single initial application of 18 kg Zn ha^{-1}
after the seventh crops (Singh and Abrol, 1985).

Mode and Method of Zinc Application

Soil application and foliar sprays are the most common Zn application
methods. Results of experiments conducted in different states (Nayyar et al.
1990; Gupta et al. 1994; Sakal et al. 1993, 1996; Table 13) revealed the
superiority of soil application of Zn over its foliar application. The lower
efficiency of the foliar mode is primarily due to delayed cure of the deficiency as well as the low concentration of Zn in the spray solution. Coating and
soaking of seeds in Zn solution, dipping roots of rice seedlings in ZnO
suspension and transplanting rice seedlings from Zn enriched nurseries have
also been tested and proved either inferior to or just at par with that of soil
application of Zn (Yoshida et al. 1970; Nayyar et al. 1990).

Application of zinc sulfate to soil through broadcast and mix method was
most efficient compared either to its drill application, band placement or its
top dressing at the optimum dose (Takkar, Mann and Randhawa, 1974; Gupta
et al. 1994). At lower rates, Zn drilled below the seed had an edge over the
broadcast method but the optimum yield levels were not achieved.

TABLE 13. Response of rice and wheat to soil and foliar application of zinc.

Treatments	Grain yield response (t ha^{-1})	
	Wheat (Takkar & Bansal, 1987)	Rice (Nayyar et al., 1994)
Soil application (Kg Zn ha^{-1})		
2.8	0.20	0.80
5.6	0.60	1.35
11.2	0.77	1.80
Foliar application (% ZnSO$_4$ sol.)		
0.5	0.60	0.90*
1.0	0.50	–
Control yield	1.45	5.30
LSD (p = 0.05)	0.36	0.54

* 0.2% unneutralized zinc sulphate solution.

Time of Zinc Application

Irrespective of the rate, mode and method of application, the efficiency of Zn use is also determined by the appropriate time of application of Zn fertilizer. Soil application of Zn prior to seeding or transplanting gives the best results in terms of efficient correction of Zn deficiency. Delayed application of Zn particularly in rice causes appreciable yield decline (Takkar, Mann and Randhawa, 1974; Sadana and Takkar, 1983; Gupta et al. 1994). Sakal et al. (1993) applied 25 kg ZnSO$_4$ ha^{-1} as a single dose at transplanting comparing it to its half dose application at transplanting and the remaining half at tillering stage and recording similar rice yields for both the treatments. Delaying the half dose of Zn from tillering to panicle initiation stage significantly lowered the grain yield indicating that Zn requirements of the rice crop must be met at or prior to the tillering stage.

Zinc Efficient Cultivars

Crop varieties differ in their zinc requirement and their tolerance to Zn stress, and numerous studies have been carried out to screen Zn-efficient varieties of different crops in the IGP (Takkar 1993; Sakal et al. 1984; Rathore et al. 1986). Cultivation of Zn-efficient crop varieties provides an alternative to combating Zn deficiency, but the agronomic means of solving Zn deficiency via its soil or foliar application in the IGP is more common.

Iron

Iron deficiency limits crop production in highly calcareous and saline-sodic soils. Iron chlorosis in rice grown on permeable coarse textured soils of Punjab is due to the inherent low Fe content of soils and also due to the inadequate degree of reduction of iron oxides due to the difficulty of impounding water for longer periods in sandy soils (Takkar and Nayyar, 1979).

Deficiency of Fe in almost all the crops is visually exhibited as interveinal chlorosis of the new or younger leaves. Soon after, the leaf veins loose its green color, the entire leaf turns yellow. Under acute deficiency conditions, affected leaves become bleached and the newly emerging leaves also look white or bleached (Plates 5 and 6). In some crops, reddish spots/lesions may also develop at a later stage. With time, the affected leaves turn papery, become necrotic and die.

Carrier, Mode, Rate and Time of Application of Iron

A number of inorganic Fe sources are available to combat its deficiency, of which, ferrous sulfate heptahydrate ($FeSO_4.7H_2O$) is the most commonly used source in the region. Chelated products such as Fe-EDTA, Fe-EDDHA and multi-micronutrient mixtures containing variable amounts of Fe are also available. Organic manure also contains variable amounts of Fe depending upon the source and storage conditions. Total Fe content in poultry, piggery and FYM was reported to be 1075, 1600 and 1465 μg g^{-1}, respectively (Arora, Nayyar and Randhawa, 1975). Sewage sludge, municipal wastes, poultry manure, FYM and pressmud have been reported to contain 7882, 5375, 3400, 2528 and 3910 μg Fe g^{-1}, respectively (Sakal et al. 1996). In a calcareous soil of Bihar, 5 t poultry manure or 1 t pyrite or 10 t compost ha^{-1} produced as much rice grain yield as 50 kg $FeSO_4$ ha^{-1} (Sakal et al. 1996). The response of rice to $FeSO_4$ or pyrite improved appreciably when applied in combination with compost (Table 14). However, application of 10 t ha^{-1} of different organic wastes significantly increased the rice yield over soil application of 10-20 kg Fe ha^{-1} as ferrous sulfate and enriching the organic wastes with Fe further increased yields (Prasad, Prasad and Prasad, 1989). Singh et al. (1992b) have also recorded the highest yield of rice when 25 kg Fe ha^{-1} was applied to a sandy loam soil of Varanasi as $FeSO_4$ in combination with compost @ 10 t ha^{-1}. The yield of the following wheat crop was also increased by the residual effect of ferrous sulfate and compost.

Both soil and foliar applications of ferrous sulfate have been found to be effective for the wheat crop (Table 15). Since ferrous sulfate applied to soil is susceptible to transformations into unavailable forms, the rates of application to soils are very high and, therefore, uneconomical as compared to foliar application. Even the application of 200 kg $FeSO_4.7H_2O$ ha^{-1} was found to

PLATE 5. Deficiency of Fe in rice induces interveinal chlorosis of younger leaves. Leaves loose their green color and appear "bleached."

be inferior to three foliar applications with 2% unneutralized ferrous sulfate solution in mending the Fe deficiency in rice grown on coarse textured soils of Punjab (Nayyar and Takkar, 1989). In calcareous soils of Bihar, variable results have been reported. In one experiment, two foliar sprays of 1% $FeSO_4$ solution produced significantly higher grain yield of rice as compared to the soil application of 50 kg $FeSO_4$ ha^{-1} while in another experiment both the modes were equally efficient (Sakal et al. 1996). The rates for soil application to both rice and wheat have been reported to vary from 10 kg Fe ha^{-1} to 20 kg Fe ha^{-1} depending upon the soil environment and deficiency status.

PLATE 6. Deficiency of Fe in rice induces interveinal chlorosis of younger leaves. Leaves loose their green color and appear "bleached."

TABLE 14. Effect of iron carriers on yield and iron uptake by rice.

Treatment	Rice yield (t ha^{-1})		Fe uptake (kg ha^{-1})
	Grain	Straw	
Control	3.40	5.93	1.25
50 kg FeSO$_4$ ha^{-1}	4.27	7.27	1.25
10 t compost ha^{-1}	4.00	6.98	1.19
1 t pyrite ha^{-1}	3.67	6.53	1.53
50 kg FeSO$_4$ + 10 t compost ha^{-1}	4.80	7.83	1.73
1 t pyrite + 10 t compost ha^{-1}	4.47	7.40	2.07
LSD (p = 0.05)	0.84	1.16	0.28

Adapted from Sakal et al. (1996).

Soil Puddling and Submergence as a Means to Correct Fe Deficiency

Generally, Fe chlorosis in rice occurs in upland or highly percolating coarse textured soils because of less mobilization of Fe^{2+} ion as the desired degree of reduction does not occur. The puddling of such soils has been shown to markedly reduce the extent of Fe-chlorosis in rice thereby increasing the grain yield significantly (Table 16). Addition of FYM under such conditions further decreased the severity of Fe deficiency in rice. Maji and Bandyopadhyay (1992) have also reported higher availability of Fe with submergence in all the thirteen soils of the coastal areas of West Bengal. Similarly, the findings of Srivastava and Srivastava (1994) have shown 3.9-fold increase in the available Fe content when sodic soils were subject to waterlogging in comparison to field capacity conditions.

Use of Organic and Green Manure

Nayyar and Takkar (1989) evaluated the efficiency of green manure in comparison to soil and foliar application of ferrous sulfate to rice for three

TABLE 15. Grain yield of wheat as influenced by soil and foliar application of iron.

Treatment	Grain yield (t ha^{-1})	
	Expt. I	Expt. II
Control	3.8	3.8
Soil application (kg Fe ha $^{-1}$)		
10	4.1	4.4
20	4.7	4.7
Foliar application (FeSO$_4$ sol.; %)		
0.5	–	4.5
2.0	4.4	–
LSD (0.05)	0.5	0.3

Adapted from Gupta et al. (1994).

TABLE 16. Rice grain yield as influenced by soil and foliar mode of iron application.

Treatments	Mode of application	Rice grain yield (t ha^{-1})	
		Puddled	Unpuddled
Control		5.63	1.50
FeSO$_4$ 100 kg ha^{-1}	Soil	6.54	1.57
FeSO$_4$ + FYM 10 t ha^{-1}	Soil	8.14	2.78
FeSO$_4$, 2% solution	Foliar	8.46	4.47
FeSO$_4$, 3% solution	Foliar	8.76	5.51
LSD (0.05)		0.60	0.33

Adapted from Nayyar and Takkar (1989).

years. The results have shown that the combination of green manure and the foliar spray of 1% ferrous sulfate solution produced the highest grain yield followed by green manuring or foliar sprays. Soil application of Fe proved significantly inferior to green manuring and foliar application (Figure 3). The DTPA-extractable soil Fe also increased significantly with green manuring. In a greenhouse experiment, the effect of green manuring the soil with *Sesbania aculeata* in terms of increase in the iron uptake by rice grain was equivalent to an application of 10 mg Fe kg^{-1} soil (Chahal et al. 1998). A marked increase in water soluble and DTPA-extractable soil Fe over control with the addition of *Sesbania* under submerged conditions has been recorded (Khind et al. 1987; Singh et al. 1992a; Nayyar and Chhibba, 2000).

Iron Efficient Cultivars

Among the micronutrients, Fe deficiency is the most difficult to correct. Therefore, identification and/or evolving Fe-efficient varieties appear to be the best alternative to fertilization of crops with Fe in alkaline soils where the deficiency is generally encountered. The differential susceptibility of crop varieties to Fe deficiency may be attributed to the nature and amount of

FIGURE 3. Rice grain yield (three-year mean) as affected by green manuring (GM), Fe application methods, and its combinations.

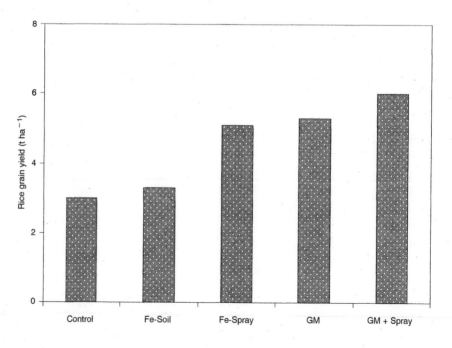

reductant present in the root exudates. Singh et al. (1985) categorized rice genotypes into four classes viz. highly susceptible, susceptible, moderately tolerant and tolerant based on the degree of iron chlorosis. They explained the variability amongst the genotypes on the basis of shoot yield, chlorophyll content, orthophenanthroline reactive Fe and total Fe content in leaves in addition to the extent of Fe chlorosis. All these parameters increased with the increase in tolerance or decrease in Fe chlorosis except that total Fe was generally higher in chlorotic as compared to green leaves.

Manganese

The adoption of rice-wheat rotation in non-conventional rice soils not only caused Fe deficiency in rice but also over several years resulted in the appearance of Mn deficiency in wheat. The deficiency has been increasing since its recognition in Punjab during 1979-80 (Takkar and Nayyar, 1981b) on highly permeable coarse textured alkaline soils. The response of wheat to Mn application has also been reported in calcareous soils of Bihar and coarse textured soils of Haryana. The increasing occurrence of this deficiency and the failure of the wheat crop at many sites without Mn application necessitated studies to evolve techniques for the efficient management of this problem.

Starting at the lower and middle leaves of the affected plants, symptoms of manganese deficiency appear as small chlorotic spots in the interveinal area at the basal part of the leaf and gradually extends towards the tip of the leaf. With time, these spots turn light greyish-yellow to greyish-brown (Plate 7). Under acute deficiency, these spots enlarge and coalesce to form streaks of pinkish brown to buff color in the interveinal areas (Plate 8). At heading stage of crop growth, these symptoms are very conspicuous on the flag leaf (Plate 9). Ears emerge with great difficulty and they look sickle shaped and deformed (Plate 10). Grain development is also inhibited in some varieties due to Mn stress, most of the grains developing to half seed stage (Kaur, Arora and Takkar, 1988a).

Carriers, Mode, Rate and Time of Application

Manganese sulfate, MnO_2, Mn-frits, Mn-EDTA as well as some multi-micronutrient mixtures have been evaluated for their efficiency to correct Mn deficiency in wheat. At equivalent yield levels, manganese sulfate was four times more efficient than Mn-frits. Manganese oxide proved to be the least effective as the plants could not reach the reproductive stage up to 40 mg kg^{-1} soil application of Mn (Kang, 1985). Several micronutrient mixtures have been tested. The concentration of Mn in the multi-micronutrient mix-

PLATE 7, 8, 9 and 10. Manganese deficiency in wheat induces small chlorotic interveinal spots, which turn grayish-yellow to grayish-brown in color (Plate 7). Under acute Mn deficiency, these spots enlarge, coalesce and form interveinal streaks of pinkish brown to buff color (Plate 8). These symptoms are conspicuous on the flag leaf at the heading stage of crop growth (Plate 9). Acute Mn-deficiency induces deformed, sickle-shaped earheads (Plate 10).

Plate 7

tures is critical for its effectiveness. In field experiments on Mn-deficient soils, foliar sprays of 0.5 and 1% $MnSO_4.H_2O$ solution significantly gave higher wheat grain yield than that of other materials tested at similar concentration of the product. This was because of the very low content of Mn in the micronutrient products compared to manganese sulfate (Sadana, Nayyar and Kaur, 1989). When the product was tried at equivalent concentration of Mn in the spray solution, wheat grain yield obtained with $MnSO_4$ was at par

Plate 8

with that of the multi-micronutrient product (Table 17, Nayyar, Bansal and Singh, 1998). However, Mn-EDTA was found to be four times more efficient than $MnSO_4.H_2O$ as three sprays of 0.25% Mn-EDTA produced as much wheat grain yield as the same number of sprays of 1.0% $MnSO_4.H_2O$ solution (Table 18).

Both soil and foliar application of Mn significantly increased wheat grain yield (Nayyar, Sadana and Takkar, 1985; Takkar, Nayyar and Sadana, 1986; Sadana, Nayyar and Takkar, 1991). But the rates of soil application of Mn (40 to 75 kg ha^{-1}) was not economical compared to three foliar sprays of 0.5% to 1% $MnSO_4$ solution (7.5 to 15 kg ha^{-1}). In a severely Mn-deficient field, three sprays of 0.5% $MnSO_4$ solution produced significantly higher wheat grain yield (3.7 t ha^{-1}) as compared to 2.5 t ha^{-1} obtained with 20 kg Mn broadcast and mixed in a hectare field (Takkar, Nayyar and Sadana, 1986). Band placement of $MnSO_4.H_2O$ or its delayed application at the time of first

Plate 9

irrigation to wheat did not improve its efficiency (Sadana, Nayyar and Takkar, 1991; Nayyar, Sadana and Takkar, 1985). Treating seed with $MnSO_4.H_2O$ or Teprosyn-Mn was significantly inferior to foliar application of Mn (Table 18). The low efficiency of soil applied Mn is due to its reversion to higher oxides in alkaline soils. Only 1 mg and 8.4 mg Mn kg^{-1} soil remained in the water soluble and exchangeable forms, respectively, of the 40 mg Mn kg^{-1} soil added to a Mn deficient soil after 30 days incubation at field capacity (Kang, 1985). Time of foliar application of Mn is very critical for managing Mn deficiency as the deficiency symptoms in wheat appear just after the first irrigation to the crop (about four weeks after seeding). Foliar sprays initiated 26 days after sowing, i.e., 2 days before first irrigation to wheat proved significantly superior to the same number of sprays started 3 days after the crop was first irrigated (Table 19, Takkar, Nayyar and Sadana, 1986; Nayyar et al. 1990; Sadana, Nayyar and Takkar, 1991).

Plate 10

TABLE 17. Effect of rates and sources of Mn on the yield and Mn uptake by wheat.

Source of Mn	Mn Solution conc. (%)	Yield (t ha⁻¹) Grain	Yield (t ha⁻¹) Straw	Mn uptake (g ha⁻¹)
MnSO₄·H₂O	0.06	3.54	7.65	145
	0.12	3.94	8.60	186
	0.16	4.18	9.10	230
Micronutrient mixture (4% Zn and 12% Mn)	0.06	3.50	8.20	132
	0.12	3.72	8.80	182
Control		2.23	6.45	79
LSD (p = 0.05)		0.47	0.95	36

Adapted from Nayyar et al. (1998).

Manganese Efficient Cultivars

While screening a large number of wheat genotypes, some genotypes did not show any visible symptoms of Mn deficiency under stress conditions. But these genotypes with a hidden Mn deficiency registered poor yield, high straw/grain ratio, lower Mn translocation index (MTI; percentage of total Mn uptake (tops) that was translocated to grain) and Mn use efficiency (MUE, ratio of grain yield produced to total Mn uptake) compared to check genotype

TABLE 18. Effect of mode, source and rate of Mn application on grain and straw yield of wheat.

Mode/Source	· Rate	Yield (t ha^{-1})	
		Grain	Straw
	SEED		
$MnSO_4 \cdot H_2O$	75 g kg-1 seed	2.90	5.25
Teprosyn-Mn	5 ml kg-1 seed	2.50	4.20
	FOLIAR		
$MnSO_4 \cdot H_2O$	0.5%	3.85	7.03
	1.0%	4.00	7.25
Mn-EDTA	0.10%	3.73	6.83
	0.25%	4.05	7.10
	0.50%	2.25	7.33
Control yield		0.69	4.15
LSD (0.05)		0.69	0.94

Adapted from Nayyar and Bansal (2000).

TABLE 19. Effect of time of manganese application on Mn content and wheat yield.

Treatment	Yield (t ha^{-1})		Mn content ($\mu g\ g^{-1}$)	
	Grain	Straw	Grain	Straw
	Foliar Application (0.5 % $MnSO_4.H_2O$ solution)			
1[*] + 2[**]	4.62	8.28	19.7	17.3
0 + 3	4.30	7.82	18.3	15.8
	Foliar Application (1 % $MnSO_4.H_2O$ solution)			
1 + 2	4.92	8.52	21.1	18.0
0 + 3	4.52	8.06	19.9	16.9
Control	2.73	5.05	12.9	11.4
LSD (p = 0.05)	0.30	0.33	3.0	1.8

Number of sprays before (*) and after (**) first irrigation to wheat.
Adapted from Sadana et al. (1991).

HD 2009. On the other hand, genotypes that exhibited slight to moderate symptoms of Mn deficiency had higher harvest indices (> 30.0), higher relative grain yield (89-155%), higher MTI and MUE compared to the control genotype (Kaur, Nayyar and Takkar, 1988b). It was stressed that economic yield along with high harvest index and MUE under Mn stress conditions should form a better criterion for screening wheat genotypes.

In subsequent field experiments, it was observed that all the tested cultivars though responded to foliar sprays of 1% solution of $MnSO_4.H_2O$, they differed in the magnitude of their response (Bansal, Nayyar and Takkar, 1991; Bansal and Nayyar, 1998). Durum varieties of wheat (PBW 34, PDW 215) are more sensitive to Mn deficiency than the aestivums (Figure 4). Cultivars HD 2329 and WH 542 were rated as Mn-efficient considering the percent grain yield increase and harvest index in response to Mn application. Efficient cultivars have more fertile tillers, longer spikes, more grains per spike, higher harvest index and higher photosynthetic oxygen evolution and chloroplast Mn than the less efficient cultivars (Kaur and Takkar, 1987; Kaur, Takkar and Nayyar, 1989a; Kaur et al. 1989b).

Copper

The incidence of Cu deficiency in the IGP of India is less than 3 percent. Its deficiency has been reported to be higher (31 to 40%) in the states of Kerala and Gujarat (Singh, 1999). The visual symptoms of Cu deficiency in rice and wheat have not been recorded under field condition in such soils.

FIGURE 4. Percent yield response of durum (PBW 34, PBW 215) and aestivum (PBW 343, HD 2329) wheat to Mn application over control treatments.

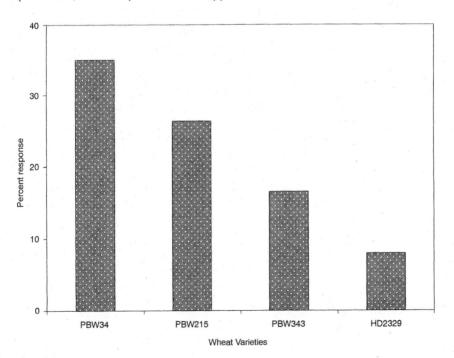

However, symptoms of Cu deficiency have been developed and characterized under sand culture (Sharma et al. 1996; Karim and Vlamis, 1962).

The leaves of Cu deficient plants first appear bluish-green and later turn yellowish-white near the tips on either side along the middle rib. Thereafter, dark brown necrotic lesions appear at the tips and then spread downwards along the midrib. The newly emerging leaves fail to unroll and they maintain a needle like appearance usually of the entire leaf or occasionally of the top-half leaf, leaving the basal end to develop normally. Effective tillering is severely depressed resulting in considerable loss in grain yield. The panicles are small with numerous small sterile grains.

In wheat, the youngest leaves are pale green and they become rolled and twisted. The tips of the young leaves become severely discolored or bleached which later spreads along the leaf margins towards the base of the leaves. In case of severe deficiency, the discolored or bleached part of the lamina turns necrotic and papery and the leaf apices curl into spirals confined largely to the apical one-third part of the leaves. Ears and stems may become noticeably darker on maturation. Plants may fail to head under severe Cu deficiency and seed set is considerably impaired with many florets being sterile under moderate deficiency.

Source, Rate and Method of Copper Application

In view of the limited deficiency of Cu, very few experiments have been conducted in the IGP to study the response of crops and of the management of its deficiency particularly of rice and wheat to Cu application. Takkar, Chhibba and Mehta (1989) summarized the results of 110 experiments on rice and 34 on wheat in soils of Bihar and reported an average response of 0.46 and 0.38 t ha^{-1}, respectively, to 25 kg CuSO$_4$ ha^{-1}. In about 60 percent of the experimental soils, wheat responded significantly to Cu application. Earlier, Singh et al. (1979) summarized the results of 124 field experiments conducted under AICARP, according to which, significant response of rice (up to 1.13 t ha^{-1}) to foliar application of Cu was observed in one year at Varanasi and Nandyal and two years at Chiplima. Similarly, results of 87 experiments on wheat have shown response to the foliar application of Cu to the order of 0.25 to 0.98 t ha^{-1}.

The most common and convenient carrier of Cu for both soil and foliar application is copper sulfate (CuSO$_4$.5H$_2$O). Copper sulfate is frequently applied as fungicide and germicide in vegetables and fruit trees for controlling diseases because Cu^{2+} ion is toxic to lower forms of life. Thus, when copper sulfate is applied to foliage as a routine for disease control, the deficiency of copper is taken care of. However, for long-term correction of Cu deficiency, soil application may be more beneficial compared to its foliar application. In a loamy sand soil of Ludhiana, the increase in wheat grain yield over control with soil application of 5 and 10 kg Cu ha^{-1} was 0.31 and 0.49 t ha^{-1}, respectively, compared to 0.11 and 0.09 t ha^{-1} with foliar application of 0.2 and 0.4% CuSO$_4$ solution (Lal, Grewal and Randhawa, 1971). The response of wheat to foliar application of Cu has also been reported by Grewal, Randhawa and Bhumbla (1967) in a coarse textured soil of Punjab. The field experiments conducted by Sharma, Swami and Singh (1992) on clay loam soil at Banswara showed 0.91 t ha^{-1} increase in wheat grain yield over a control yield of 3.80 t ha^{-1} by an application of 20 kg CuSO$_4$ ha^{-1}. Agrawal and Gupta (1994) reported that the straw and grain yields of rice grown on a sandy loam soil (Inceptisol) at Varanasi increased with 10 and 20 kg of Cu as copper sulfate ha^{-1} while the higher rate of 40 kg Cu ha^{-1} decreased grain yield. In acid Hill soil of Uttar Pradesh, soil application of 10 kg CuSO$_4$ ha^{-1} has been shown to be significantly superior to foliar sprays of 0.2% CuSO$_4$ solution in increasing the grain yield of soybean (Dwivedi, Dwivedi and Pal, 1990). The above observations suggest that though the deficiency of Cu is not exhibited by the soil and plant analysis, the crops do have some kind of hidden hunger of Cu so that they respond to its foliar or soil application. Therefore, the critical levels used to evaluate the soil and plant test data need to be re-established to match with the crop responses.

The Lucknow Center of AICRP on micronutrients examined eight varieties of wheat for their relative tolerance to copper deficiency (Takkar, Chhibba and Mehta, 1989). Only three varieties, viz., UP 270, UP 262 and UP 2001 were rated as moderately susceptible to Cu stress while WL 711, UP 2202, UP 368, UP 2203 and UP 115 were highly susceptible.

Boron

Boron deficiency in soils and crops is also becoming an important limiting factor in the IGP. Its deficiency is a common phenomenon on highly calcareous soils, highly leached coarse textured soils, acid soils and in soils poor in organic matter. Boron is considered as the next limiting micronutrient after Zn in calcareous soils of Bihar and significant responses of crops to its application have been recorded (Sakal and Singh, 1995).

Cereals have a low boron requirement. In rice, longitudinal growth of plants is inhibited and the plants remain stunted, bushy and have a succulent appearance. Plants develop small white necrotic streaks on the youngest unfolded leaf and the one next to it. The streaks later intensify and coalesce to form larger white irregular patches. The young emerging leaves generally fail to unroll and remain partially enclosed within the subtending leaf (Sharma et al. 1996). The plants sprout new side shoots on which the symptoms appear. Seed formation is drastically inhibited even by moderate deficiency.

In case of wheat, deficient plants have short stems and yellowish-green leaves. The splitting of the newer leaves close to the midrib may also occur. In case of severe B deficiency, an increase in tillering occurs and the stalks and/or roots may sprout from the lower nodes of stem and the new shoots have water soaked appearance and are paler than the older parts of the plant. Later, newer growth becomes necrotic and shoots whither. The development of the inflorescence and setting of grains is restricted. Reduction in grain yield due to B deficiency may take place even without exhibiting any symptoms on foliage.

Sources, Rate and Method of Boron Application

Borax ($Na_2B_4O_7.10H_2O$, 11% B) and sodium tetraborate ($Na_2B_4O_7.5H_2O$, 14% B or $Na_2B_4O_7$, 20% B) have been the most popular sources for soil application and boric acid (H_3BO_3, 17% B) for foliar sprays. Colemanite ($Ca_2B_6O_{11}.5H_2O$, 10% B) and B-frits (2-6% B) are expected to be more useful fertilizers on sandy soils prone to B leaching because of their less solubility and or slow release of boron. Long duration field crops and fruit trees may also be benefited from this property of colemanite and B-frits.

Most of the studies relating to the rate, mode and time of application of B have been carried out on oilseed and pulse crops due to their higher B requirement as compared to cereals. Based on the quadratic relation between rate of B application and yield, Singh, Sakal and Sinha (1988) reported an optimum level of B to be 2.08 kg ha^{-1} for chickpea and 1.66 kg ha^{-1} for maize. Similarly in blackgram, the optimum rate of B has been reported to be 1.96 kg B ha^{-1} with a response of 0.23 t ha^{-1} (Sakal and Singh, 1999). However, at a higher rate of B application (beyond 2.5 kg ha^{-1}) the crop yield started declining and at 4 kg B ha^{-1}, visual symptoms of B toxicity in crops and a marked decline in yield was recorded (Sakal and Singh, 1999). Therefore, the range between optimum and toxic levels of B is quite narrow and the application of B should be made only after confirmation of its deficiency through soil and/or plant tests. Its indiscriminate use may result in toxicity to crop plants. The rate of B application for obtaining optimum yields also varies with available B status and soil texture. In seven field experiments on calcareous soils of Bihar with variable available B content (0.15-0.65 mg kg^{-1} soil), the optimum rate of B application was 3 kg B ha^{-1} for chickpea grown on soil having an available B status of less than 0.35 mg kg^{-1} and the grain yield response ranged from 0.19 to 0.89 t ha^{-1}. However, for other locations with higher soil B status, the optimum rate of B application was 2 kg B ha^{-1} and grain yield response was 0.09 to 0.59 t ha^{-1}. Also, the optimum rate of B application for different crops in calcareous heavy textured soils ranged between 1.5 and 2.0 kg B ha^{-1} and in light textured soils, it varied between 1.0 and 1.5 kg ha^{-1} (Sakal et al. 1996).

In a three year field study on a silty loam acid soil of North Bengal, significant increase in mean grain yield of wheat of 1.89 t ha^{-1} was recorded with 20 kg borax ha^{-1} but yields significantly declined with 30 kg borax ha^{-1} (Mitra and Jana, 1991). Application of half dose B to soil and the remaining half as a foliar spray recorded significantly higher grain yield than its full dose application either to soil or as foliar application. Roy and Pradhan (1994) also reported similar results from the Tarai region of West Bengal. In the acid and hill soils of Uttar Pradesh, soil application of 20 kg sodium tetraborate ha^{-1} and two foliar sprays of its 0.20 percent solution proved equally effective in increasing the grain yield of soybean. But the residual effect of soil application on the following wheat crop significantly out yielded the direct effect of foliar sprays. Addition of the same rate of B after liming the soil produced significantly higher yields of soybean and wheat as compared to its application without lime (Dwivedi, Dwivedi and Pal, 1990). In calcareous soils of Bihar, three foliar sprays of 0.25% boric acid solution proved markedly inferior to soil application of B for both rice and wheat (Sakal et al. 1998). Crop varieties also differ in their susceptibility to B stress. Sakal et al. (1996) have classified varieties of different crops in terms of their tolerance to B stress based on their percent response to B application. In case of wheat, Sonalika and UP 262 were reported to be more tolerant as compared to HD 2285 and HD 2329.

Molybdenum

The early symptoms of Mo deficiency are similar to that of nitrogen deficiency. In wheat, young leaves become chlorotic along the apex and apical margins, which later intensifies and extends to the apical half, turning dry and necrotic. Severely affected leaves turn light brown. Deficiency of Mo also reduces the yield of rice, mustard, chickpea, greengram, blackgram, pigeonpea, cotton, sugarbeet and cauliflower.

Sodium molybdate, ammonium molybdate and molybdenum trioxide are the common sources of Mo, but sodium molybdate is most commonly used. In view of the limited deficiency of Mo, very little work has been done on the management of its deficiency.

In acid hill soil of Uttar Pradesh, the increase in soybean yield with application of 0.5 kg ha^{-1} of sodium molybdate to a limed soil far exceeded the yield obtained with the same rate of application to an unlimed soil. The residual effect of this treatment was also apparent on wheat. Two foliar sprays of 0.05 percent solution of sodium molybdate proved significantly inferior to its soil application (Dwivedi, Dwivedi and Pal, 1990).

CONCLUSION:
FUTURE MICRONUTRIENT RESEARCH NEEDS IN THE IGP

The relatively minor problem of soil micronutrient deficiency at the beginning of the green revolution three decades ago, is a factor to reckon with in present times in sustaining the productivity of rice and wheat in the IGP. The monoculture of rice and wheat of the past has given rise to an exhausting soil micronutrient feeder rice-wheat cropping system of the present day. Proper management of the rice-wheat cropping system is a key to minimizing micronutrient deficiency induced crop yield reduction. Except for the acknowledgement of zinc deficiency by farmers, existence of other micronutrient deficiencies is not common knowledge and under a "perceived" optimum management condition, poor yields of rice and wheat is often described by farmers as being due to "sickly" soils.

The critical micronutrient values for both soils and plants in the IGP have not been extensively field tested; its field validation will be of immense help to researchers and extension specialists. Crop response to micronutrients and soil micronutrient analysis in the IGP has been scattered. An eco-regional soil micronutrient status-analysis within the IGP and synthesis of this information using GPS and GIS will facilitate delineating regions of specific deficiencies. This will help formulate on-farm diagnostic and adaptive research, spreading of awareness amongst farmers, and extrapolating results to similar sub-eco-regions within the IGP where intervention programs on micronutrients can be undertaken.

More information on the transformation and availability of micronutrients for different soils and the effect of manipulating the soil physical environment and its moisture regimes on plant available micronutrients, need to be generated. Past practices and present research have indicated that continuous use of FYM or of other organic sources arrests the depletion of available micronutrient pools from soils. Development of integrated micronutrient technology using available organic materials is needed not only to increase micronutrient use efficiency but also to decrease the pressure on the use of costly inorganic micronutrient carriers. Field experiments have proved the superiority of zinc sulfate as a zinc carrier. Increasing costs coupled with a shortfall in supply of zinc sulfate has necessitated investigations on evaluating sparingly soluble zinc sources or ores combined with zinc mobilizers. The residual availability of various sources of micronutrients for a cropping system needs to be worked out. Micronutrients like boron whose deficiency and toxicity limits for crops are very narrow require careful investigations including its field testing for rates and frequency of application. Finally, the identification and/or breeding for micronutrient efficient crop cultivars having either low micronutrient requirement for potential yields or capable of mining micronutrients from the less available pools should be given priority.

REFERENCES

Agrawal, H.P. and M.L. Gupta. (1994). Effect of copper and zinc on copper nutrition of rice. *Annals of Agricultural Research* 15(2):162-166.

Ahmed, S. and M.B. Hossain. (1997). The problem of boron deficiency in crop production in Bangladesh. In *Boron in Soils and Plants*, eds. R.W. Bell and B. Rerkasem. Kluwer Academic Publishers, pp 1-5.

Arora, C.L.,V.K. Nayyar and N.S. Randhawa. (1975). Note on secondary and micro-element contents of fertilizers and manures. *Indian Journal of Agricultural Sciences* 45:80-85.

Bansal, R.L. and V.K. Nayyar. (1989). Effect of zinc fertilizers on rice grown on Typic Ustochrepts. *IRRI Newsletter* 14:24-25.

Bansal, R.L. and V.K. Nayyar. (1992). Response to different Zn carriers of rice grown on Ustifluvents in India. *IRRI Newsletter* 17(3):15-16.

Bansal, R.L. and V.K. Nayyar. (1998). Screening of wheat (*Triticum aestivum*) varieties tolerant to manganese deficiency stress. *Indian Journal of Agricultural Sciences* 29:107-112.

Bansal, R.L., V.K. Nayyar and P.N. Takkar. (1991). Field screening of wheat cultivars for manganese efficiency. *Field Crops Research* 29:107-112.

Bansal, R.L., V.K. Nayyar and P.N. Takkar. (1992). Accumulation and bioavailability of Zn, Cu, Mn and Fe in soils polluted with industrial waste water. *Journal of Indian Society of Soil Science* 40:796-99.

Bansal, R.L., P.N. Takkar, N.S. Sahota and M.S. Mann. (1980). Evaluation of soil procedures for predicting zinc availability to wheat under calcareous alkaline field conditions. *Field Crops Research* 3: 43-51.

Berger, K.C. and E. Truog. (1944). Boron tests and determination for soils and plants. *Soil Science* 57:25-36.

Biswas, C.R. and D.K. Benbi. (1997). Dynamics of physical, chemical and biological properties of soil and yield sustainability in a long-term fertilizer experiment. *Research Bulletin*, Department of Soils, Punjab Agricultural University 3: 1-66.

Borkakati, K. and P.N. Takkar. (2000). Forms of boron in acid alluvial and lateritic soils in relation to ecosystem and rainfall distribution. In *International Conference on Managing Resources for Sustainable Agricultural Production in the 21st Century*. Vol 2: 127-128.

Chahal, D.S., S.S. Khehra, G.S. Dhaliwal, R. Arora, N.S. Randhawa and A.K. Dhawan. 1998. Rice yield and iron availability as influenced by green manuring and ferrous sulfate application under different moisture regimes. *Ecological Agriculture and Sustainable Development: Volume 1*. Proceedings of an International Conference on Ecological Agriculture: Towards Sustainable Development, Chandigarh, India, 15-17 Nov. 1998, 652-658.

Chhibba, I.M., V. K. Nayyar and P.N. Takkar. (1989). Direct and residual effect of some zinc carriers in rice-wheat rotation. *Journal of Indian Society of Soil Science* 37: 585-587.

Dwivedi, G.K., M. Dwivedi and S.S. Pal. (1990). Modes of application of micronutrients in acid soil in soybean-wheat crop sequence. *Journal of Indian Society of Soil Science* 38: 458-463.

Grewal, J.S., N.S. Randhawa and D.R. Bhumbla. (1967). Effect of micronutrients on yield and their contents in wheat. *Journal of Research,* Punjab Agricultural University 4: 315-322.

Grigg, J.L. 1953. Determination of available molybdenum of soils. *New Zealand Journal of Science and Technology* 34: 404-414.

Gupta, V.K., S.P. Gupta, Ram Kala, B.S. Potalia and R.D. Kaushik. (1994). *25 years of micronutrient research in soils and crops of Haryana,* Department of Soil Science, CCS Haryana Agricultural University, Hissar, pp. 1-99.

Gupta, V.K. and D.S. Mehla. (1993). Depletion of micronutrients from soil and their uptake in rice-wheat rotation. *Journal of Indian Society of Soil Science* 4: 704-706.

Gupta, V.K., B.S. Potalia and J.C. Katyal. (1986). Response of wheat to different zinc carriers in a loamy sand (Typic torripsamment) soil. *Journal of Indian Society of Soil Science* 34: 631-632.

Kang, N.S. (1985). *Availability of manganese from different manganese carriers to wheat.* M.Sc. Thesis, Punjab Agricultural University, Ludhiana, Punjab, India.

Karim, A.Q.M.B. and J. Vlamis. 1962. Micronutrient deficiency symptoms of rice grown in nutrient culture solutions. *Plant and Soil* 16: 347-360.

Katyal, J.C. and B.D. Sharma. (1991). DTPA-extractable and total Zn, Cu, Mn and Fe in Indian soils and their association with some soil properties. *Geoderma* 49: 165-179.

Kaur, N.P., C.L. Arora and P.N. Takkar. (1988a). Abnormalities in growth and development of grains of wheat, *Triticum aestivum* and *Triticale cultivars* in relation to manganese deficiency. *Indian Journal of Experimental Biology* 26: 147-148.

Kaur, N.P., V.K. Nayyar and P.N. Takkar (1988b). Screening of wheat germplasm tolerant to manganese deficiency stress. In *International symposium on management in soils and plants* ed. M.J., Webb, R.O., Nable, R.D., Graham, R.J., Hannam, Adelaide. pp 119-120.

Kaur, N.P. and P.N. Takkar. (1987). Effect of Mn stress on the growth and yield parameters in wheat and triticale cultivars. *Wheat Information Service* 64: 24-33.

Kaur, N.P., P.N. Takkar and V.K. Nayyar. (1989a). Some physiological studies on differential susceptibility of wheat and triticale to manganese deficiency. *Annals of Biology* 5: 115-122.

Kaur, N.P., P.N. Takkar, C.L. Arora and V.K. Nayyar. (1989b). Relative Mn efficiency of wheat cultivar HD 2009. *Indian Journal of Plant Physiology* 32: 306-310.

Khind, C.S., A. Jugsujinda, C.W. Lindare and W.H. Patrick, Jr. (1987). Effect of Sesbania straw in a flooded soil on soil pH, redox potential and water soluble nutrients. *IRRI Newsletter* 12(3): 42-43.

Lal, C., J.S. Grewal and N.S. Randhawa. (1971). Response of maize and wheat crops to soil and spray application of copper in the Ludhiana soils. *Journal of Research* Punjab Agricultural University 8: 52-56.

Lindsay, W.L. and W.A. Norvell. (1978). Development of DTPA soil test for zinc, iron, manganese and copper. *Journal of the Soil Science Society of America,* 42: 421-428.

Maji, B. and B.K. Bandyopadhyay. (1992). Effect of submergence on Fe, Mn and Cu contents in some coastal rice soils of West Bengal. *Journal of Indian Society of Soil Science,* 40: 390-392.

Mitra, A.K. and P.K. Jana. (1991). Effect of doses and method of boron application on wheat in acid *Terai* soils of North Bengal. *Indian Journal of Agronomy* 36: 72-74.

Mondal, A.K., S. Pal, B. Mandal and L.N. Mandal. (1991). Available boron and molybdenum content in alluvial acidic soils of north Bengal. *Indian Journal of Agricultural Sciences,* 61: 502-504.

Nayyar, V.K. and R.L. Bansal. (1989). *Chemical pools of zinc, lead and cadmium in a polluted soil.* Proceedings of ILZiC Silver Jubliee Conference, pp. 56.1-56.6.

Nayyar, V.K. and R.L. Bansal. (2000). Response of wheat to rate and mode of manganese application in a coarse textured Mn-deficient soil. *International Conference on Managing Resources for Sustainable Agricultural Production in the 21st Century* Vol. 2: 136-138.

Nayyar, V.K., R.L. Bansal and S.P. Singh. (1998). *Annual Report of AICRP of Micro- and Secondary nutrients and Pollutant Elements in Soils and Plants (1997-1998).* Punjab Agricultural University, Ludhiana, India.

Nayyar, V.K., R.L. Bansal and S.P. Singh. (1999). *Annual Report of AICRP of Micro- and Secondary nutrients and Pollutant Elements in Soils and Plants (1998-1999).* Punjab Agricultural University, Ludhiana, India.

Nayyar, V.K., R.L. Bansal, S.P. Singh and M.P.S. Khurana. (1994). *Salient Achivements of AICRP on Micro- and Secondary-Nutrients and Pollutant Elements in Soils and Plants (1984-94).* Department of Soils, Punjab Agricultural University, Ludhiana, India, pp. 1-133.

Nayyar, V.K. and I.M. Chhibba. (2000). Effect of green manuring on micronutrient availability in rice-wheat cropping system of northwest India. In *Long-Term Soil Fertility Experiments in Rice-Wheat Cropping System.* Rice-Wheat Consortium Paper Series 6: 68-72.

Nayyar, V.K., U.S. Sadana and P.N. Takkar. (1985). Methods and rates of application of Mn and its critical levels for wheat following rice on coarse textured soils. *Fertilizer Research* 8: 173-178.

Nayyar, V.K. and P.N. Takkar. (1980). Evaluation of various zinc sources for rice grown on alkali soil. *Zeitschrift fur Plfanzenernahrung und Bodenkunde* 143: 489-493.

Nayyar, V.K. and P.N. Takkar. (1989). Combating iron deficiency in rice-grown in sandy soils of Punjab. *International Symposium on Managing Sandy Soils,* CAZRI, Jodhpur, India. Vol. 1: 379-384.

Nayyar, V.K., P.N. Takkar, R.L. Bansal, S.P. Singh, N.P. Kaur and U.S. Sadana. (1990). *Micronutrients in Soils and Crops of Punjab.* Research Bulletin, Department of Soils, Punjab Agricultural University, Ludhiana, pp. 1-146+xiv

Nongkynrih, P., P.S. Dkhar and D.T. Khathing. (1996). Micronutrient elements in acid alfisols of Meghalaya under rice cultivation. *Journal of Indian Society of Soil Science* 44: 455-457.

Prasad, B., J. Prasad and R. Prasad. (1989). Influence of iron-enriched organic wastes on crop yield and iron nutrition of crops in calareous soil. *Indian Journal of Agricultural Sciences* 59: 359-364.

Rathore, G.S., S.B. Dubey, R.S. Khamparia and B.L. Sharma. (1986). *Annual Progress Report of All India Coordinated Scheme of Micronutrient and Secondary*

Nutrients and Pollutant Elements in Soils and Plants. Department of Soil Science, JNKVV, Jabalpur, Madhya Pradesh, India.

Rathore, G.S., R.S. Khamparia, G.P. Gupta, S.B. Dubey, B.L. Sharma and V.S. Tomar. (1995). *Twenty Five Years of Micronutrient Research in Soils and Crops of Madhya Pradesh. Research Bulletin.* Department of Soil Science and Agricultural Chemistry, JNKVV, Jabalpur, Madhya Pradesh, India, pp. 1-101+vii.

Roy, S.K. and A.C. Pradhan. (1994). Effect of method and time of boron application on wheat (*Triticum aestivum*) production in tarai region of West Bengal. *Indian Journal of Agronomy* 39: 643-645.

Sadana, U.S., V.K. Nayyar, and N.P. Kaur. (1989). Response of wheat grown on manganese deficient soils to foliar application of different micronutrient formulations. *Micronutrient News* II(9): 1-2.

Sadana, U.S., V.K. Nayyar, and P.N. Takkar. (1991). Response of wheat grown on manganese-deficient soil to methods and rates of manganese sulphate application. *Fertilizer News* 36(3): 55-57.

Sadana, U.S. and P.N. Takkar. (1983). Methods of zinc application to rice on sodic soil. *IRRI Newsletter* 8: 21-22.

Saifuzzaman, M. (1995). Influcence of seeding date, genotyope and boron on sterility of wheat in Bangladesh. In *Sterility in Wheat in Subtropical Asia: Extent, Causes and Solutions,* eds. H.M. Rawson and K.D. Subedi. Proceedings of a workshop 18-21 September 1995, Lumle Agricultural Research Centre, Pokhara, Nepal. ACIAR Proceedings No. 72, 46-50.

Saifuzzaman, M. and C. Meisner. (1995). Wheat sterility in Bangladesh: An overview of the problem, research and possible solutions. In *Sterility in Wheat in Subtropical Asia: Extent, Causes and Solutions,* eds. H.M. Rawson and K.D. Subedi. Proceedings of a workshop 18-21 September 1995, Lumle Agricultural Research Centre, Pokhara, Nepal. ACIAR Proceedings No. 72, 46-50.

Sakal, R., V.K. Nayyar and M.V. Singh. (1997). Micronutrient status under rice-wheat cropping system for sustainable soil productivity. In S*ustainable Soil Productivity Under Rice-Wheat System.* ISSS Bulletin No. 18, 39-47.

Sakal, R. and A.P. Singh. (1995). Boron Research and Agricultural Production. In *Micronutrient Research and Agricultural Production* ed. H.L.S. Tandon. Fertilizer Development and Consultation Organization, New Delhi, India, pp. 1-31.

Sakal, R. and A.P. Singh. (1999). Available zinc and boron status of Bihar soils and response of oilseeds and pulses to zinc and boron application. In *National Symposium on Zinc Fertilizer Industry Whither To,* eds. Ramendra Singh and Abhay Kumar. Session III.

Sakal, R., A.P. Singh and B.P. Singh. (1985a). Use of zincated superphosphate to control zinc deficiency in calcareous soils. *Journal of Indian Society of Soil Science* 33: 443-446.

Sakal, R., A.P. Singh, R.B. Sinha and N.S. Bhogal. (1996). *Twenty Five Years of Research on Micronutrients in Soils and Crops of Bihar (1967-1992).* Department of Soil Science, RAU, Pusa, Bihar, India, pp. 1-207.

Sakal, R., A.P. Singh, R.B. Sinha and N.S. Bhogal. (1998). *Annual Report of AICRP on Micro- and Secondary-Nutrients and Pollutant Elements in Soils and Plants.* RAU, Pusa, Bihar, India.

Sakal, R., B.P. Singh, A.P. Singh and R.B. Sinha. (1985b). A comparative study of the direct and residual effect of zinc carriers on the response of crops in calcareous soil. *Journal of Indian Society of Soil Science* 33: 836-840.

Sakal, R., B.P. Singh, R.B. Sinha and A.P. Singh. (1984). Response of some wheat germplasm to zinc application in calcareous soil. *Annals of Agricultural Research* 5:137-142.

Sakal, R., R.B. Sinha, A.P. Singh and N.S. Bhogal. (1993). Evaluation of methods and time of zinc application to rice. *Journal of Indian Society of Soil Science* 41: 195-196.

Sehgal, V. (1999). *Indian Agriculture 1999.* Indian Economic Data Research Centre, B-713, Panchvati, Vikaspuri, New Delhi-110018, pp. 600.

Sharma, B.D., H.S. Jassal, J.S. Sawhney and P.S. Sidhu. (1999). Micronutrient distribution in different physiographic units of Siwalik Hills of semiarid tract of Punjab, India. *Arid Soil Research and Rehabilitation* 13:189-200.

Sharma, B.D., S.S. Mukhopadhyay, P.S. Sidhu and J.C. Katyal. (2000). Pedospheric attributes in distribution of total and DTPA-extractable Zn, Cu, Mn and Fe in Indo-Gangetic Plains. *Geoderma* 96: 131-151.

Sharma, C.P., C. Chatterjee, P.N. Sharma, N. Nautiyal, N. Khurana and P. Sinha. (1996). *Deficiency Symptoms and Critical Concentration of Micronutrients in Crop Plants.* ed. C.P. Sharma, Lucknow University, Bull. No. 1, 1-110.

Sharma, S.K., B.N. Swami and R.K. Singh. (1992). Response of wheat (*Triticum aestivum*) to micronutrients at two fertility levels in black soil. *Indian Journal of Agronomy* 37: 255-257.

Sillanpaa, M. (1982). *Micronutrients and the Nutrient Status of Soil: A Global Study.* FAO, Soils Bulletin 48:1-444.

Singh, A.P., R. Sakal and B.P. Singh. (1982). Effect of zinc enriched compost and other methods of zinc application on zinc nutrition of rice in calcareous soils. *Journal of Indian Society of Soil Science* 30:572-573.

Singh, A.P., R. Sakal and R.B. Sinha. (1988). Boron nutrition of chickpea and winter maize grown on calcareous soils of north Bihar. *Indian Fertilizer Scene Annual* 1: 26-30.

Singh, A.P., R. Sakal, R.B. Sinha and N.S. Bhogal. (1998). Use efficiency of applied zinc alone and mixed with biogas slurry in rice-wheat cropping system. *Journal of Indian Society of Soil Science* 46:75-80.

Singh, B.P., R.A. Singh, M.K. Sinha and B.N. Singh. 1985. Evaluation of techniques for screening Fe-efficient genotypes of rice in calcareous soil. *Journal of Agricultural Science, Cambridge* 105: 193-197.

Singh, Bijay, Yadvinder Singh, U.S. Sadana and O.P. Meelu. (1992a). Effect of green manure, wheat straw and organic manures on DTPA-extractable Fe, Mn, Zn and Cu in a calcareous sandy loam soil at field capacity and waterlogged conditions. *Journal of Indian Society of Soil Science* 40: 114-118.

Singh, D., C.R. Leelawathi, K.S. Krishanan and S. Sarup. (1979). *Monograph on Crop Responses to Micronutrients.* Indian Agricultural Statistics Research Institute (ICAR) New Delhi.

Singh, K., C. Deo, J.S. Bohra, J.P. Singh, R.N. Singh and K. Singh. (1992b). Effect of iron carriers and compost application on rice yield and their residual effect on succeeding wheat crop in Entisols. *Annals of Agricultural Research* 13:181-183.

Singh, M.V. (1999). Micronutrient deficiency delineation and soil fertility mapping. In *National Symposium on "Zinc Fertilizer Industry-Whither To,"* eds. Ramendra Singh and Abhay Kumar. Session II.

Singh, M.V. and I.P. Abrol. (1985). Direct and residual effect of fertilizer zinc application on yield and chemical composition of rice-wheat crops in an alkali soil. *Fertilizer Research* 8: 179-191.

Singh, S.P., M.K. Sinha and N.S. Randhawa. (1980). Diffusion of ^{65}Zn as influenced by rates of applied zinc in soils of divergent texture. *Journal of Indian Society of Soil Science* 28: 290-294.

Srivastava, S.P., C.R. Yadav, T.J. Rego, C. Johansen and N.P. Saxena. (1997). Diagnosis and alleviation of boron deficiency causing flower and pod abortion in chickpea (*Cicer arietinum* L.) in Nepal. In *Boron in Soils and Plants*, eds. R.W. Bell and B. Rerkasem. Proceedings of the International Symposium on Boron in Soils and Plants held at Chiang Mai, Thailand, 7-11 September, 1997, pp. 95-99.

Srivastava, A.K. and O.P. Srivastava. (1994). Response to iron and manganese application as affected by their availability in a Typic Natraqualf. *Arid Soil Research and Rehabilitation* 8: 301-305.

Takkar, P.N. (1982). *Micronutrients-Forms, Contents, Distribution in Profile, Indices of Availability and Soil Test Methods.* Reviews of Soil Research in India. Part I, 12th International Congress of Soil Science, New Delhi, pp. 361-91.

Takkar, P.N. (1993). Requirement and response of crop cultivars to micronutrients in India-a review. In *Genetic Aspects of Plant Mineral Nutrition*, eds. P.J. Randall, E. Delhaize, R.A. Richards and R. Munns, CSRIO, Canberra, Australia, Kluwer Academic Publisher, pp. 341-348.

Takkar, P.N. and R.L. Bansal. (1987). Evaluation of rates, methods and sources of zinc application to wheat. *Acta Agronomica Hungarica* 36: 277-283.

Takkar, P.N., I.M. Chhibba and S.K. Mehta. (1989). *Twenty Years of Coordinated Research on Micronutrients in Soil and Plants*-Bulletin I, IISS, Bhopal. pp. 1-314.

Takkar, P.N., M.S. Mann and Randhawa, N.S. (1974). Methods and time of zinc application. *Journal of Research*, Punjab Agricultural University 11: 278-283.

Takkar, P.N. and V.K. Nayyar. (1979). Iron deficiency affects rice yield in Punjab. *Indian Farming* 29: 9-12.

Takkar, P.N. and V.K. Nayyar. (1981a). Effect of gypsum and zinc on rice mutrition on sodic soil. *Experimental Agriculture* 17: 49-55.

Takkar, P.N. and V.K. Nayyar. (1981b). Preliminary field observation of manganese deficiency in wheat and berseem. *Fertilizer News* 26: 22-23.

Takkar, P.N. and V.K. Nayyar. (1986). *Integrated Approach to Combat Micronutrient Deficiency.* Proceedings of FAI Seminar on Growth and Modernisation of Fertilizer Industry PS III/2.1-2.16.

Takkar, P.N., V.K. Nayyar and U.S. Sadana. (1986). Response of wheat on coarse textured soils to mode and time of manganese application. *Experimental Agriculture* 22:19-152.

Takkar, P.N. and T. Singh. (1978). Zinc nutrition of rice (*Oryza sativa*) as influenced by rates of gypsum and zinc fertilization of alkali soils. *Agronomy Journal* 70: 447-450.

Williams, C.H. and D.J. David. (1973). The effect of superphosphate on the cadmium content of soils and plants. *Australian Journal of Soil Research* 11: 43-56.

Yaduvanshi, H.S., K.L. Sharma, K. Singh and T.S. Verma. (1988). Distribution of chemical forms of zinc in soils of varying orders under different climatic zones. *Journal of Indian Society of Soil Science* 36: 427-432.

Yoshida, S., G.W. Mclean, M. Shafi and K.E. Muller. (1970). Effects of different methods of zinc application on growth and yields of rice in a calcareous soil, West Pakistan. *Soil Science and Plant Nutrition*, 16:147-149.

Sterility in Wheat
and Response of Field Crops
to Applied Boron
in the Indo-Gangetic Plains

P. K. Kataki
S. P. Srivastava
M. Saifuzzaman
H. K. Upreti

SUMMARY. Results of field experiments across the Indo-Gangetic Plains (IGP) region indicates that soil B deficiency induces sterility in wheat and results in poor crop yields of legumes and cereals. The deficiency of soil B and the response of crops to applied B generally increases from the northwestern to the eastern end of the IGP, this trend being influenced by the distribution of the soil parent material and the variation in the climatic conditions within the IGP. An earlier FAO study rated Nepal as the lowest of thirty countries in its soil B status and showed that a higher soil B deficiency problem exists in the eastern half of the Indian-IGP, therefore crop response to applied B is more likely in these areas. Few studies have made qualitative and quantitative assessment of sterility in wheat in the IGP and the positive response of other

---

P. K. Kataki is Cornell University Coordinator and CIMMYT Adjunct Scientist. S. P. Srivastava is Senior Soil Scientist and H. K. Upreti is Senior Scientist (Agronomy) with the Nepal Agricultural Research Council. M. Saifuzzaman is Senior Scientific Officer, Bangladesh Agricultural Research Institute.

Address correspondence to: P. K. Kataki, CIMMYT South Asia Regional Office, Lazimpat, P.O. Box 5186, Kathmandu, Nepal (E-mail: pkataki@vsnl.com).

[Haworth co-indexing entry note]: "Sterility in Wheat and Response of Field Crops to Applied Boron in the Indo-Gangetic Plains." Kataki, P. K. et al. Co-published simultaneously in *Journal of Crop Production* (Food Products Press, an imprint of The Haworth Press, Inc.) Vol. 4, No. 1 (#7), 2001, pp. 133-165; and: *The Rice-Wheat Cropping System of South Asia: Efficient Production Management* (ed: Palit K. Kataki) Food Products Press, an imprint of The Haworth Press, Inc., 2001, pp. 133-165. Single or multiple copies of this article are available for a fee from The Haworth Document Delivery Service [1-800-342-9678, 9:00 a.m. - 5:00 p.m. (EST). E-mail address: getinfo@haworthpressinc.com].

133

crops to B application also suggest B deficiency related sterility problems in these crops. Micronutrient research in the IGP often rates Zn
followed by Fe and Mn deficiencies in some instances, as its major soil
micronutrient deficiency problems. However, the deficiency of B perhaps is as important if not more, than Fe and Mn deficiency. The
awareness of soil B deficiency is not as widespread as its occurrence in
the IGP region. This article reviews and discusses the sterility problems
in wheat and the response of several field crops to applied B to highlight the growing importance of soil B deficiency in the IGP. *[Article
copies available for a fee from The Haworth Document Delivery Service:
1-800-342-9678. E-mail address: <getinfo@haworthpressinc.com> Website:
<http://www.HaworthPress.com> © 2001 by The Haworth Press, Inc. All rights
reserved.]*

KEYWORDS. Bangladesh, boron, cropping system, India, Indo-Gangetic Plains, Nepal, Pakistan, rice, sterility, and wheat

INTRODUCTION

Most of the results from the Indo-Gangetic Plains (IGP) indicate sterility
in wheat resulting from soil boron deficiency. Morphologically, the wheat
crop shows very little B deficiency symptoms early in the crop growth stage,
especially under "hidden" deficiency conditions. Open wheat florets (Plate 1)
or "gaping glumes" for a longer period than normal at post-anthesis stage of
crop growth, and transparent ear heads (Plate 2) when viewed against the
sunlight during the grain filling period, are typical symptoms of sterile wheat
plants in the field. At harvest, a field of sterile wheat crop is dark-green in
color due to the absence of an appropriate sink for the photosynthates to be
deposited. This is in contrast to a non-sterile wheat field with a golden-yellow
color and grain filled spikes. This difference in the color of a wheat crop at
maturity could perhaps be used as a tool to survey the severity and the extent
of wheat sterility. The ear head of a wheat plant, which protrudes the top of
the crop canopy, remains transparent at grain filling stage due to it being
sterile, and is therefore easier to detect compared to other field crops. Application of B to wheat has in most cases reduced its sterility and increased
yield in the IGP. Other field crops including rice, cotton, and various legumes
also respond positively to applied B, suggesting sterility problems in these
crops. Floret sterility in field crops other than wheat is often difficult to detect
visually because of drooping ear heads within the crop canopy during grain
filling stage (e.g., several rice varieties) or pods being hidden between the
leaves (e.g., legumes) and therefore sterility not being as apparent as in
wheat. In chickpea, boron deficiency leads to excessive flower and pod drop

PLATE 1. Two sterile and transparent wheat spikes with open or "gaping glumes" for a longer period than normal (the two central wheat spikes in this picture) compared to a fertile, compact, grain filled wheat spike (to the right of the two central sterile spikes in this photograph).

leaving severed peduncles on stems (Plate 3) compared to seed bearing pods with the application of B (Plate 4).

Even the most perfect wheat heads are partially sterile because of its sequential nature of development and grain filling, therefore, Rawson and Bagga (1979), and Rawson (1996), assumed that in all wheat spikelets, terminal florets with lemmas less than approximately 3 mm should be discounted from the sterility index. Soil Cu deficiency can also induce sterility in wheat. But in the IGP most of the evidence till date indicate limited and scattered soil Cu deficiency sites and higher frequency of responses of field crops to applied B. Therefore, soil B deficiency is the primary reason for inducing wheat sterility in the IGP. However, several factors, which reduce crop transpiration, is thought to influence, modify and sometimes, to nullify the positive effect of B application to crops in the IGP, especially to wheat. These factors include: low radiation due to cloudy conditions, fog or mists, high or low humidity, high or low temperature, drought or water shortages, water logging or water excess, and high soil pH (Bell and Dell, 1995; Dawson and Wardlaw, 1989; Misra et al., 1992; Saini and Aspinall, 1981; Saifuzzaman, 1995; Sthapit et al., 1989; Willey and Holliday, 1971).

PLATE 2. A field of sterile but otherwise vigorously growing wheat crop. The crop is at the grain filling stage. Note the number of transparent wheat spikes indicating ear-head sterility. The wheat variety shown here was the newly released wheat variety BL 1135, grown in a farmer's field in Sipaghat village of Kavre district, Nepal, during the 1998-99 wheat season.

Wheat varieties differ in their susceptibility to sterility due to soil B deficiency (Figure 1) and with the application of B, sterility in susceptible varieties can be reduced significantly with a corresponding increase in grain yield (see also Mehrotra, Srivastava and Mishra, 1980; Rerkasem 1993, Rerkasem and Jamjod, 1997). Wheat anther concentrations of B can be six times those in the bulk ear (Rerkasem, 1995). B is essential for the cell wall development of generative organs during its growth (Matoh et al., 1992; Matoh, 1997). In wheat, male gametogenesis is very sensitive to low B supply (Rerkasem and Loneragan, 1994) and therefore is associated with male sterility due to poorly developed anthers and pollen resulting in fertilization failure (Rerkasem and Jamjod, 1997). Reviewing the recent research results of the importance of B in plant physiology, Blevins and Lukaszewski (1998) discusses the important structural role of B in cell walls, its role in plant membrane function and its involvement in the metabolic activities of plants.

This chapter reviews the extent of wheat sterility, and compiles information on field crop response to applied B from the Indo-Gangetic Plains region. Crop response to applied B from the states outside the IGP but within

PLATE 3. Soil B deficiency leads to excessive flower drop and hence poor seed yield in chickpea (note the severed flower stalks). This photograph was taken in a farmer's field at Rampur, Chitwan district, Nepal, during the 1999-2000 winter season.

South Asia are also being reviewed to highlight the extent of B deficiency in this region. Management of B deficiency, and the source, rate and method of boron application is also discussed in the previous chapter of this publicatoin entitled "Management of Soil Micronutrient Deficiencies in the Rice-Wheat Cropping System."

EXTENT OF B DEFICIENCY
AND HENCE WHEAT STERILITY IN THE IGP

Deficiency of soil B induces floret sterility in wheat. Research results from the IGP often report the response of field crops to applied B in terms of increase in crop yield over minus B plots, which is a good, simple but an indirect assessment of sterility. Few studies have quantified sterility in these experiments, and others assess sterility qualitatively. Compilation of the results from these experiments indicates that sterility and crop response to applied B is more frequent in the eastern half of the IGP. This would include the eastern Terai and the eastern mid-hills of Nepal, the states of Bihar, West

PLATE 4. Excessive flower drop in chickpea due to B deficiency (plant to the right of the photograph) can be reduced by application of B (plant to the left of the photograph), thereby increasing number of pods per plant and hence seed yield. This photograph was taken in a farmer's field demonstration on chickpea response to B application in Rampur, Chitwan district, Nepal, during 1999-2000 winter season.

Bengal, Orissa, Assam and Meghalaya of India, and the north and northwestern districts of Bangladesh. This trend is probably influenced by the general trend of the soil B status in the region. However, few studies also indicate crop response to applied B in the western IGP, especially in certain areas of Pakistan.

India

The average B content of soils of India were close to that of the international mean of a FAO soil micronutrient status study of thirty countries (Sillanpaa, 1982). However, this study recorded a high degree of variation between soil samples from different parts of northern and eastern India. High soil B values were recorded in only 10% of the samples, and low B values were twice as frequent. Considering 0.3 to 0.5 mg l^{-1} as the critical limit for soil B status, half of the sampled Indian soils were rated as having hidden to severe B deficiency. The highest B values were of the soils from the north-

FIGURE 1. Response of wheat varieties to B application and differences between varieties in sterility occurrence and grain yield. The data has been sorted in an ascending order of sterility for the minus B treatment. Note the decrease in sterility percentage and a corresponding increase in the grain yield with the application of B for the sterility susceptible wheat varieties. Thirty wheat varieties were screened for sterility occurrence with and without soil applied B at sowing time, during the 1998-99 wheat season in Sipaghat village of Kavre district, Nepal (Kataki and Upreti, unpublished data).

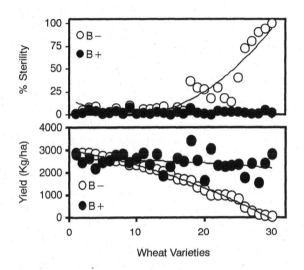

western part of the Indian IGP, i.e., the states of Punjab and Delhi. Four-fifths of the samples analyzed from the eastern state of Bihar were low to severely deficient in soil B status (Sillanpaa, 1982).

The above results have been reconfirmed by the soil B analyses conducted within the country. Of the total number of soil samples analyzed, only 0 to 6% of the soils of the northwestern parts of the Indian-IGP states of Punjab, Gujarat, Rajasthan and Delhi had B deficient soils (Tandon, 1995; Gajbhiye, Banerjee and Goswami, 1980; Takkar, Chibba and Mehta, 1989; Singh and Randhawa, 1977). The soils in these states belonged to the major soil group/order of Alfisols, Entisols, Inceptisols, and Vertisols in Gujarat; the salt affected soils, Ustocrepts and Haplustalfs of Punjab; non-saline to slightly saline soils of Delhi; and desert soils, grey brown soils, mixed red and black, medium black and red yellow soils of Rajasthan. This is in contrast to the results of soil B analyses from the eastern states of Bihar, West Bengal, Orissa, Assam and Meghalaya, where a large percentage of the analyzed soils were deficient in B (Borkakati and Takkar, 2000; Nongkynrih, Dkhar and Khathing, 1996; Mondal et al. 1991; Sakal et al., 1984, Sarengi, 1992). The red-yellow catenary soils, calcareous soils, recent and old alluvial soils of Bihar had, respectively, 40%, 48%, 41% and 32% soils deficient in B. The red lateritic and alluvial soils of West Bengal had, respectively, 64% and 84% soils deficient in soil B (Nandi et al., 1992; Ali and Basu, 1992). Similarly, 69% of the red and lateritic soils of Orissa, 17% and 43% old alluvial and acid soils of Assam, 50% of the Alfisols of Meghalaya were deficient in soil B status (Sarengi 1992; Gogoi 1992; Barthakur 1992). Though a few studies indicate pockets of higher soil B deficiency in northwestern part of the Indian-IGP (Bansal, Nayyar and Takkar, 1991), we can conclude that the severity of B deficiency to crops increases from the northwestern to the eastern end of the India-IGP. This trend of the soil B status in the Indian-IGP, which encompasses a major portion of the total IGP acreage, is also reflected in the soil B status and crop response to applied B of the neighboring countries. The eastern Terai region of Nepal and a few districts of the rice-wheat growing areas of northern and northwestern parts of Bangladesh which extends from the north-east and eastern end of the Indian IGP, crop response to applied B is likely to be expected due to deficiency of soil B.

Nepal

There are fewer numbers of state or region wise soil B status analyses and a larger number of crop responses to applied B studies that have been conducted in Nepal. The crop response to applied B is discussed in subsequent sections. Information on secondary and micornutrient status of soils of Nepal are few as compared to major nutrients (Khatri-Chhetri, 1982). The FAO study on micronutrient status of thirty countries rates Nepal the lowest in terms of its soil B status of all the countries (Sillanpaa, 1982). With 50% of its soil samples being low in soil B status in this study and the rest of the soil samples being on the lower end of the "normal" B range, B deficiency in large areas of Nepal were expected to limit crop production.

In the Chitwan district of Nepal, soil B deficiency was shown to be the most severe micronutrient deficiency limiting corn yield (Khatri-Chhetri, 1982). Deficiency of B was followed by that of Zn, Mg, S and Cu under adequate supply of N, P and K nutrients (Khatri-Chhetri, 1982). Highly significant yield increases of cauliflower and a statistically non-significant yield increase of up to 29% in mustard and maize were reported at Rampur (Khatri-Chhetri and Karki, 1979; Khatri-Chhetri, Karki and Prasad, 1979). A survey of the Chitwan valley showed the soils to be slightly alkaline to very acidic, most of the soils being loam to sandy loam, a few soils being sandy clay to silty clay loam in texture (Khatri-Chhetri, 1982).

Unlike the rest of the IGP, an additional factor that needs to be considered in Nepal is that of altitude along the mid-hill range. The mid-hill region may experience low temperatures during the critical anthesis period of crop growth stage during the cooler winter months, which may also induce sterility and therefore complicate the interpretation of B studies as is evident from the following example.

Field experiments on wheat sterility were conducted in four locations during 1987-88 wheat season (Sthapit, 1988). Of the four locations, two locations were in the temperate mid-hill region of Nepal (Lumle, 1675 msl and Pakhribas, 1700 msl) and the other two locations were in the subtropical "Terai" flat plains of Khairenitar (400 msl) and Tarahara (100 msl). Treatments included three separate wheat genotype sets for the altitudes, two sowing dates (normal and late), and two levels of B application (0 and 8 kg ha^{-1}). In the mid-hill locations, the effect of B for both the sowing dates was negligible, but late sowing significantly increased sterility percentage irrespective of B application, indicating cold injury at anthesis stage in these higher altitudes. On the contrary, in the "Terai" plains, irrespective of the sowing dates, B application significantly increased yield and reduced percent sterility (Sthapit, 1988). However, during the 1984-85 wheat sterility survey season in the western mid-hill region of Nepal, sterility percentage of 17.8% to 90.1% was recorded (Subedi et al., 1985b). It is noteworthy that the field location of conducting an experiment on sterility is of prime importance. Experiments conducted close to a farmers habitat (where run-off water with cattle dung is common), in fields which had earlier received high doses of manure because of the previous crop grown, or where large quantities of crop residues are routinely recycled into the soil, the sterility incidences will be low and hence response of crops to applied B will be negligible.

Bangladesh

The available soil B status in different soil types of Bangladesh ranges between 0.1 to 1.9 mg kg^{-1} of soil for different soil types (Ahmed, 1987; Ahmed and Hossain, 1997), with the sandy loam soils ranging from 0.1 to 0.3 mg kg^{-1} of soil. It is estimated that about 1 million hectares of cultivable land in Bangladesh is deficient in soil B status and therefore limits crop production. Though B deficiency has been reported from different parts of Bangladesh, the deficiency is more severe in the north, northwestern, and northeastern parts of the country (Figure 2). The mean soil B status of the central part of Bangladesh is of medium value but varies from medium to low and very low status (Figure 2).

Farmer's field surveys to quantify the extent of wheat sterility have been conducted in Bangladesh. Between March 30 and April 5 1998, wheat sterility survey was conducted in Rangpur-Dinajpur districts of Bangladesh covering the regions of Nashipur, Kannai, Suvra, Purba-Mollikpur, Sadarpur, Chandanbari, Boda, Kachubari, Telipara, Salander, Mustafi, Gugundi, Tista river basin of Kurigram, Taragaon and Kaunia. Similarly during 1999, wheat

FIGURE 2. Schematic map showing the boron status of soils (hot water soluble B) in Bangladesh.

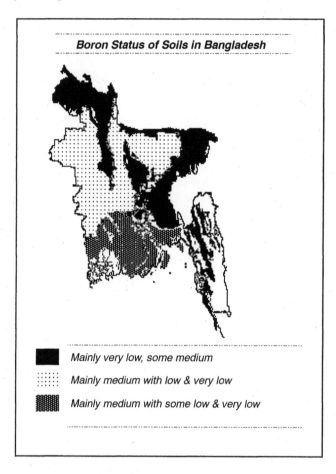

sterility survey was conducted during the 2nd week of March in Dinajpur, Thakurgaon, Panchagar, Rangpur, Kurigram, Lalmonirhat and Gazipur. Results from these surveys indicate high variability in the occurrence of wheat sterility within and between locations and the incidence of wheat sterility varied from 0% to 100% severity (Figure 3). The survey results also indicate the variation of a wheat variety to sterility when grown in different locations (Figure 4). The relation between HWS soil B and the severity of sterility for different wheat varieties grown in several locations was also not consistent. Hot water soluble (HWS) soil B analyses of the surveyed fields of Bangladesh of 1998 and 1999 did not correlate with the incidence of wheat sterility

FIGURE 3. Variations within and between locations in the occurrence of wheat sterility in Bangladesh (Saifuzzaman, unpublished data). This survey was conducted in 1998 wheat season to quantify the extent of wheat sterility in northwestern Bangladesh. Surveyed locations were–1: Dinajpur, 2: Thakurgaon, 3: Panchagar, 4: Rangpur, 5: Kurigram, and 6: Lalmonirhat.

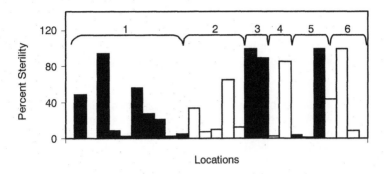

Locations

FIGURE 4. A wheat variety may differ in its sterility occurrence between locations (Saifuzzaman, unpublished data). This survey was conducted during the 1999 wheat season to quantify the extent of sterility in northwestern Bangladesh. The wheat varieties grown by farmers in the surveyed areas during this season were–1: BAW 897, 2: "Imported" variety (name unknown), 3: Inquilab, 4: Kanchan, 5: Raj, 6: Sonalika, 7: "Unknown," 8: UP262.

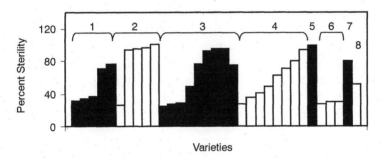

Varieties

(Figure 5). Wheat sterility of 0% to 20% is a common feature for many wheat varieties, as the terminal floret of a spikelet, and the basal and terminal spikelets of a spike often remains empty. This is due to the sequential nature of grain filling in wheat ascending from basal towards the terminal florets within a spikelet and from the mid-spikelets towards both ends of a spike. The HWS soil B content also varied greatly for a sterility range of 0% to 20% (Figure 5). Therefore, HWS soil B status is not a very good indicator of the probability of sterility occurrence for a location. However, as discussed in latter sections of this review, results from the region shows that in most cases, application of B reduced sterility and increased the yield of different crops. This indicates that in addition to the analyses of soil B status, other soil factors especially soil texture and pH, and the climatic factors in a region should be taken into account.

In central Bangladesh, foggy climatic conditions during anthesis period are suspected to induce sterility in wheat. In a controlled field experiment conducted in central Bangladesh (Figure 6), to mimic foggy days, the wheat crop was shaded during the day with "muslin" cloth for 10 days starting at

FIGURE 5. Correlation between HWS B and percent sterility in wheat from the 1998 and 1999 wheat season sterility survey (Figures 3 and 4) in Bangladesh (Saifuzzaman, unpublished data). Soil samples were collected for HWS B analyses from the same field location where the extent of sterility occurrence was quantified during both the wheat seasons.

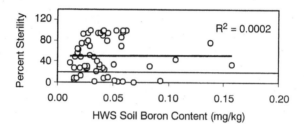

FIGURE 6. Effect of shading for 10 days with "muslin" cloth starting at different days after sowing (DAS) of a wheat crop and for the different methods of B application on the grains spike^{-1} and grain yield (kg ha^{-1}) of wheat (Saifuzzaman, unpublished data).

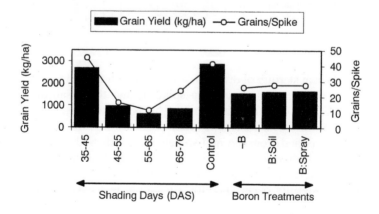

different periods after wheat sowing and B application treatments were superimposed. Shading between 55 to 65 days after sowing (DAS) significantly increased sterility percentage, and decreased grain yield and grain spike^{-1}. The mean effect of B treatment showed a marginal non-significant but higher grain yield for the plus B treatments (Figure 6). Nevertheless, since the interactions between the soil B status, environmental conditions (soil moisture status, temperature, foggy days, sunshine hour's, etc.), and varietal susceptibility to sterility are not fully understood under actual field conditions, the exact correlation's have been confusing and therefore needs more systematic research.

Pakistan

The mean B status of the soils of Pakistan was considered to be amongst the highest of the FAO study of thirty countries surveyed for soil micronutrient status including the rice-wheat growing areas of Sindh (Sillanpaa, 1982). However, soils with low B status in Pakistan were mostly from non-irrigated areas and considering critical level of 0.3 to 0.5 mg l^{-1} soil B status, B deficiency in several areas were to be expected (Sillanpaa, 1982). Approximately 45% of the sampled fields of the cotton growing areas of Multan district in Pakistan were found to be deficient in soil B status (Rashid, 1996) and application of B in these areas increased the yield of cotton. Not much information is available on soil B status of the rice-wheat growing areas of Pakistan, though occasional reports of response to B application have been made.

Indo-Gangetic Plains

The geographical distribution of micronutrients in soil is dependent on the composition of the parent materials from which it is derived. However, soils vary less in its micronutrient content compared to its parent rocks; and soil forming processes may influence micronutrient distribution within the soil profile (Sharma et al., 2000). As described earlier, analytical results of soil B status of the IGP indicate that its deficiency increases from the northwestern to the eastern end of the IGP. The northwestern region of the IGP receives less rainfall and has higher monthly temperatures compared to the eastern IGP. During the monsoon season, precipitation in the eastern IGP is one of the highest in the world. Leaching of B is likely to occur on a regular basis more in the eastern than in the western IGP, adding to its soil B deficiency problems. The distribution and the composition of the parent material, and the climatic differences between the western and the eastern IGP are therefore the major determinants of soil B status of this region.

QUALITATIVE AND QUANTITATIVE ASSESSMENT OF STERILITY

An indirect, but a simple and accurate assessment of sterility is grain yield. However, qualitative or quantitative assessment of sterility is more useful in establishing a relationship between B, sterility and crop yield. Moreover, crop yield is a manifestation of several factors. To study the effect of B on sterility, other factors of crop growth requirement are kept at an optimum level. In wheat, the vegetative growth of the plant in the IGP rarely imparts external B deficiency symptoms and therefore remains "hidden" in the field. Ear head sterility in cereals, especially in wheat can be assessed in the following ways:

- *Yield* is a simple but effective estimate of sterility in wheat and in other crops. Under optimum conditions of all other factors required for crop production (except for soil B status), crop yield differences between "plus" and "minus" B plots is a reliable estimate of sterility.
- *Visual or qualitative* assessment of sterility in wheat can be done during early grain filling period or at crop maturity. A scale of 0 to 10 could be used to visually score wheat earhead transparency, a score of 10 being 100% sterile earheads and that of 0 being a fertile grain filled earhead. Sterile earheads will have "open glumes" and will remain transparent during early grain filling period (Plate 1). At maturity, sterile earheads remain greenish-black compared to golden-yellow color of non-sterile earheads. This difference in the color of the earhead is perhaps due to the lack of a sink for the photosynthates to be deposited in sterile earheads. Differences in the color of earheads at maturity can also be used for a quick survey of sterile wheat fields.
- *Seed Set Index (SSI)* is an accurate quantitative measure of sterility. SSI can be calculated for an entire wheat spike or for the two basal florets (Basal Seed Set Index or BSSI) of the ten central wheat spikelets (Figure 7).

$$SSI = [(a - b)/a] * 100$$

$$BSSI = [(20 - c)/20] * 100$$

Where: a = number of competent florets per spike; b = number of grains per spike; c = number of grains per 2 basal florets (F1 and F2) of ten central spikelets of a wheat spike (Figure 7). The BSSI therefore calculates the seed set percent for the 20 mid-basal florets of a wheat spike. It is recommended that for wheat sterility studies, 20 earheads be collected per plot and SSI or BSSI be calculated for each earhead. The correlation between SSI and BSSI is very high and both measures of seed set correlates well with grain yield and therefore are good indices of sterility.

- *Empty earheads* per unit row length is a good semi-quantitative measure of sterility. When screening large number of wheat varieties for sterility, a meter row length can be demarcated for each varietal entry, within which, the number of empty earheads can be counted per varietal entry at maturity. The "empty" feeling of the spikes when hand held, is due to the lack of seeds and is therefore sterile. This measure is a quicker option for sterility assessment compared to the SSI or the BSSI, which are tedious methods, but requires an experienced evaluator.
- For legumes, quantitative measures of sterility are: increased flower abortion and flower drop, and reduced number of pods per plant, number of seeds per pod, and grain yield. In chickpea, flower drop due to B deficiency leaves severed peduncles (Plates 3 and 4). The number of severed peduncles can then be counted.

FIGURE 7. A schematic diagram showing parts of a wheat spike counted for the quantification of sterility by the Seed Set Index (SSI) and Basal Seed Set Index (BSSI) method. See text for details.

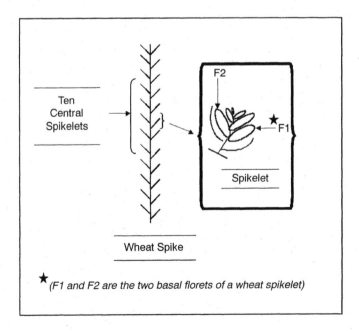

RESPONSE OF FIELD CROPS
TO APPLIED BORON IN THE IGP

The requirement for B by legumes is several times higher than that of cereals. Field crop response to applied B has been reported from several parts of the IGP. The response of field crops to applied B has been reported mostly in terms of increased yield for different crops, but the quantitative assessment of sterility has occasionally been done only for wheat. There is scattered and limited research conducted on B on the rice-wheat cropping system of the Indo-Gangetic Plains. Therefore, crop response to applied B from states outside the IGP region but within South Asia are also being discussed in this review section to highlight the growing importance of B as a limiting micronutrient for crop production in this region.

It is being assumed that instances where increase in crop yield to applied B has been reported, it is due to the decrease in the sterility of crops, increase in the seed setting, and hence greater yield. Garg, Sharma and Kona (1979), studied the effect of 0, 1 and 2.5 ppm applied B on rice variety *Jaya* in a sand culture with nutrient solution. They reported that application of B considerably improved the size, fertility, and germination of pollen grains. A concentration of 2.5 ppm B considerably increased the germination of pollens by 42.8%, size of the pollens by 18.7%, grains per panicle of rice by 18.3%, and grain per plant by 126.8% compared to the control treatment, but a higher concentration of B was inhibitory. Table 1 summarizes the results of selected B experiments on several crops from the IGP.

COUNTRY: INDIA

Crop Response to Boron in Coarse to Medium Textured Calcareous Soil

State–Bihar; Crop–Chickpea, Blackgram and Pigeonpea

More than 40% of the calcareous soils of north Bihar are deficient in available B (Sakal et al., 1984). The growth and yield of the two important pulse crop of Bihar, i.e., chickpea and pigeonpea are adversely affected in the absence of applied B.

Application of B @ 2 and 3 kg ha^{-1} significantly increased the grain yield (Table 1) of chickpea variety DG 82-13 (Sakal, Sinha and Singh, 1988). This experiment was conducted as a multi-location trial in the Samastipur and Muzaffarpur districts of Bihar, which had predominantly calcareous soils, sandy loam to silt loam in texture, with an available soil B range of 0.15 to 0.65 ppm. Chickpea grain yield was higher with 3 and then 2 kg B ha^{-1}

TABLE 1. Selected references of field crop resonse to B application in the Indo-Gangetic Plains.

State	Experiment	Soil Type	Crop	Genotype	Yield (kg ha^{-1}) B+	B–	Remarks	Reference
				India				
Bihar	Field	Coarse, calcareous	Chickpea	–	2030	1300		Sakal et al. 1988.
	Field	Coarse, calcareous	Chickpea	DHG 82-10	2480	1700		Singh et al. 1991.
			Pigeonpea	DA-11	3000	2500		
	Field	Coarse, calcareous	Blackgram	–	925	672		Sakal, Sinha and Singh, 1988.
			Chickpea	DG 82-13	2333	1433		
	Field	Coarse, calcareous	Chickpea	Pusa-256	1500	1250		Sakal et al. 1998.
			Lentil	Pant 639	1300	1250		
	Field	Coarse, calcareous	Maize	M-12	2240	1830		Sakal, Singh and Singh, 1989.
Maharashtra	Field	Coarse, calcareous	Groundnut	–	2136	1380	Dry pod	Malewar, Syed and Indikar, 1992.
	Pot	Fine, calcareous	Wheat	AKW-381	17.10	14.77	G pot^{-1}	Wankhade et al. 1996.
			Sorghum	SPH-388	23.73	19.76		
			Gram	Chaffa	28.66	21.86		
	Field	Coarse, calcareous	Groundnut	Phule Pragati	1920	1757	Dry pod	Wani, Patil and Patil, 1988.
	Field	Fine, calcareous	Groundnut	–	1495	1269	Dry pod	Shinde, Rote and Kale, 1990.
	Field	Fine, neutral	Groundnut	–	2057	1639	Dry pod	Agasimani, Babaled and Patel, 1993.
Uttar Pradesh	Field	Fine, acidic	Potato	–	20,000	11,000		Dwivedi & Dwivedi 1992.
	Field	Fine, acidic	Soybean	Bragg	1500	800		
			Wheat	–	1550	750		
	Pot	Sand culture	Rice	Jaya	31.3	13.8		

TABLE 1 (continued)

State	Experiment	Soil Type	Crop	Genotype	Yield (kg ha^{-1}) B+	B−	Remarks	Reference
West Bengal	Field	Fine, acidic	Rice	Pankaj	5000	4300	Fiber	Sarkar & Chakraborty 1980.
			Wheat	Sonalika	3216	2550		
			Jute	JRO 524	1038	906		
Gujarat	Pot	Fine, calcareous	Groundnut	GG-1	67.3	64.6	G pot^{-1}	Patel & Golakiya 1986.
Tamil Nadu	Field	Fine	Groundnut	CO2	1571	1343	Dry pod	Sudarsan & Ramaswami 1993.
Haryana	Pot	Coarse, calcareous	Chickpea	–	10.4	10.9	G pot^{-1}	Gupta et al. 1993.
			Lentil	–	6.4	7.7		
Nepal								
Mid-Hills	Field	Medium, Neutral	Wheat	–	2079	1931		Karki, 1985.
	Field	Coarse, Neutral	Wheat	–	1414	1518		Karki, 1985.
	Field	Medium	Wheat	RR-21	1890	1710		Tiwari, 1985.
Inner Terai	Field	Coarse, acidic	Chickpea	Kalika	298	0		Srivastava et al. 1997.
Bangladesh								
	Field	Typic Fluvavaquent	Wheat	–	2300	1200		Islam & Talukdar, 1991.
	Field	Typic Fluvavaquent	Wheat	–	4900	3900		Islam & Talukdar, 1991.
	Field	Typic Fluvavaquent	Wheat	–	3400	2800		Islam & Talukdar, 1991.
	Field	Typic Fluvavaquent	Wheat	–	2200	2000		Islam & Talukdar, 1991.

Field	Typic Fluvavaquent	Wheat	–	4100	3500	Islam & Talukdar, 1991.
Field	Typic Fluvavaquent	Wheat	–	2500	1600	Islam & Talukdar, 1991.
Field	Typic Fluvavaquent	Wheat	–	3100	2400	Islam & Talukdar, 1991.
Field	Typic Fluvavaquent	Mustard	–	1670	990	Islam & Talukdar, 1991.
Field	Typic Fluvavaquent	Chickpea	–	1180	480	Islam & Talukdar, 1991.
Field	Typic Fluvavaquent	Potato	–	29000	27000	Islam & Talukdar, 1991.
Field	Typic Fluvavaquent	Cauliflower	–	20000	13500	Islam & Talukdar, 1991.
Field	Typic Fluvavaquent	Groundnut	–	1800	1500	Islam & Talukdar, 1991.
Field	Typic Haplaquept	Cauliflower	–	20400	13400	Islam & Talukdar, 1991.
Field	Typic Haplaquept	Cabbage	–	67000	60000	Islam & Talukdar, 1991.
Field	Typic Haplaquept	Tomato	–	75000	69000	Islam & Talukdar, 1991.
Field	Typic Haplaquept	Broccoli	–	18300	16000	Islam & Talukdar, 1991.
Field	Typic Haplaquept	Potato	–	26000	24500	Islam & Talukdar, 1991.
Field	Typic Haplaquept	Groundnut	–	2000	1800	Islam & Talukdar, 1991.
Field	Typic Fluvaquent	Chickpea	–	1213	11	BARI 1993.

when grown in soils with less than 0.20 ppm available B compared to soils with 0.65 ppm available boron. The recommended optimum B application rate from this study was 3 kg and 2 kg B ha^{-1} for soils with < 0.35 ppm and > 0.35 ppm B, respectively.

Application of 2 and 2.5 kg B ha^{-1} increased the grain yield (Table 1) of blackgram and chickpea when grown in coarse textured highly calcareous soils of Samastipur, Bihar (Sakal, Sinha and Singh, 1988). The magnitude of yield response to applied B was higher for chickpea (69%) compared to that of blackgram (38%). Farmers frequently report sparse fruiting as a major problem in pulse crops, especially chickpea, in Bihar. Since the calcareous soils of Bihar spread over a large area, chickpea cultivation by farmers has gradually shifted from the coarse to finer textured areas.

Singh et al. (1991) reported that a sterility susceptible variety of chickpea (DHG 82-10) was very responsive to B application @ 1.5 kg ha^{-1} in field trials at Dholi, Bihar (Table 1). Similarly, pigeonpea varieties TT-6 and DA-11 were susceptible to B deficiency and responded to B application @ 1.5 kg ha^{-1} (Table 1). B application increased the mean B uptake of chickpea grain from 44 to 69.5 g and in the straw from 44.5 to 68.4 g ha^{-1} compared to the control treatments. The soil of the chickpea experimental site was sandy loam with a pH of 8.4, EC of 0.23 dS m^{-1}, and free CaCO$_3$ of 34.7%. The soils of the pigeonpea experimental site were also sandy loam with a pH of 8.6, EC 0.35 dS m^{-1}, and free CaCO$_3$ of 35.7%.

State-Bihar; Crop-Maize

Maize varieties Suwan White and M-12 also gave the highest grain yield when applied with 1.5 kg B ha^{-1} and grown in the sandy loam calcareous soil of Samastipur, Bihar (Sakal, Singh and Singh, 1989). The percent increase of B concentration of top leaves of Suwan White and M-12 maize varieties when applied with 2.5 kg B ha^{-1} was 189% and 158%, respectively, compared to the control treatment. The soil at the experimental site had a pH of 8.2, EC 0.29 dS m^{-1}, free CaCO$_3$ of 39%, and available B content of 0.38 ppm and Zn of 0.43 ppm.

State-Maharashtra; Crop-Groundnut

Application of boronated single super phosphate significantly improved groundnut crop performance compared to single super phosphate in Rahuri, Maharashtra (Wani, Patil and Patil, 1988). Application of boronated single super phosphate increased nodule number by 19.3%, dry matter yield by 3%, creeper yield by 2.9%, dry pod yield by 9.2%, shelling by 4.2%, 1000 kernel weight by 0.5%, protein content by 3.3%, oil content by 5.2%, and oil yield by 8.5% of groundnut variety Phule Pragati (Table 1). The soil of this experimental site was sandy loam, low in available B, contained 10.2% $CaCO_3$, had a pH of 8.5 and a EC of 0.34 mmhos cm^{-1}.

State-Haryana; Crop-Chickpea and Lentil

Gupta, Ramkala and Gupta (1993) reported an adverse affect of applied B on chickpea and lentil yield from a pot experiment conducted on Typic Ustipsamment loamy sand soil. The grain and straw yields of both chickpea and lentil decreased with increasing levels of B application, though the tissue B concentration increased. The available soil B content was 0.4 mg kg^{-1} soil with a pH of 8.3, EC of 0.14 dSm^{-1}, and treated with seven levels of B (0, 0.25, 0.5, 1, 2, 4 and 6 mg kg^{-1} soil as H_3BO_3). The magnitude of the adverse affect of B was more pronounced on the growth and yield of lentil compared to chickpea. Therefore, lentil is more sensitive to applied B than chickpea when the available soil B status is sufficient for crop growth.

Crop Response to Boron in Fine Textured Calcareous Soil

State-Maharastra; Crop-Groundnut, Gram, Sorghum and Wheat

In a pot experiment, Golakiya (1989) reported a critical range of calcium-boron ratio of 218 to 224 to be a favorable range for producing an optimum groundnut pod yield on calcareous clay soils of Junagadh, Maharastra. The soil had a pH of 7.4, EC of 0.18 mmhos cm^{-1}, $CaCO_3$ of 3.5% with exchangeable Ca of 4.15 meq 100g^{-1} and a HWS B of 0.66 µg g^{-1}. Application of boronated single super phosphate to groundnut which was an equivalent of 5 kg borax ha^{-1} increased dry pod yield by 29%, haulm yield by 9.5% when compared to the application of only single super phosphate (Patil et al., 1987). This experiment was conducted for three consecutive years between 1987 to 1990 in Maharashtra on a calcareous clay loam soil having an EC of 0.15 dS m^{-1}, $CaCO_3$ of 6.7%, and HWS B of 0.305 ppm. Application of boronated single super phosphate also increased groundnut dry pod yield (variety: Phule Pragati) by 28% and 13.7% when compared to diammonium phosphate (DAP) and DAP + borax, respectively.

Similar results have also been reported for gram, sorghum and wheat. A pot experiment on silty clay soil (pH 7.5, HWS B of 0.36 ppm) when applied with 5 kg borax ha^{-1} increased grain yield in the order of gram > sorghum > wheat (Wankhade et al., 1996). The uptake of B in this experiment was 374%, 341.6%, and 287% in the grain of gram, sorghum, and wheat, respectively, compared to the control treatment. Similarly, uptake of B by the straw of gram, sorghum and wheat was 275.2%, 185% and 260%, respectively, compared to the control treatment.

State-Maharashtra; Crop-Groundnut-Cotton Cropping Sequence

In a field experiment at Dhule, Shinde, Rote and Kale (1990) reported that the yearly soil application of B @ 5 kg borax ha^{-1} to a sandy clay loam soil (pH 7.8, EC 0.21 mmhos cm^{-1}, CaCO$_3$ 9.1%, HWS B 0.325 ppm), significantly increased the dry pod yield, boron uptake by groundnut plants and the available soil B content by 9.7%, 6% and 9.1%, respectively, compared to B application once in two years, and by 2.7%, 31.3% and 35.9%, respectively, compared to B application once in three years. Continuous application of B for six years did not show any toxic affect on the crops, but the available B content of the soil increased by 0.5 ppm. Residual affect of B to groundnut was not observed in the first and second year cycle of the crop. However, there was significant residual affect during the third year cycle of the groundnut crop when B was applied every year. Irrespective of the frequency of B application, B uptake by wheat increased in subsequent years. The percentage increase of the groundnut pod yield was 17.8% when 5 kg borax ha^{-1} was applied, compared to the control treatment (Table 1).

State-Maharashtra; Crop-Groundnut-Wheat Cropping Sequence

Malewar, Syed and Indlkar (1992) reported significant residual effect of 0.18% in boronated single super phosphate applied to cotton, which increased the dry pod and foliage yield of the subsequent groundnut crop. The uptake of B by cotton and groundnut was also increased in this experiment compared to treatments that received only single super phosphate or diammonium phosphate. The uptake of B by cotton (49 ppm) and groundnut (98.9 ppm) was the highest with the addition of boronated single super phosphate followed by borax + single super phosphate. This experiment was conducted in Parbhani, Maharastra, and the soil contained available HWS B of 0.35 ppm and P of 9.6 kg ha^{-1}.

State-Gujarat; Crop-Groundnut

Application of increasing levels of B (0, 0.5, 1, and 2 ppm) increased flower number and pegs pot^{-1} of groundnut, and mitigated the adverse effect of increasing levels of CaCO$_3$ (0, 10, 20 and 30%; Golakiya, 1992). This pot experiment was conducted in Junagadh, Gujarat, and the soil had a pH of 7.4, EC of 0.16 mmhos cm^{-1}, CaCO$_3$ of 3.5%, exchangeable Ca of 41.5 meq 100 g^{-1}, and a HWS B of 0.66 ppm. The flower count increased from 192 to 206 pot^{-1} with increasing levels of applied B from 0 to 2 ppm. At 30% CaCO$_3$ level, flower counts increased from 116 to 192 flowers pot^{-1} when B levels increased from 0 to 2 ppm because of higher pollen viability and flower fertility, and lower flower drop. Similarly, the total number of pegs and B uptake by plants increased and the adverse affect of $CaCO_3$ decreased with increasing levels of B application.

Application of B @ 2 ppm produced the highest pod yield of groundnut and significantly increased the uptake of N, P, K, Fe, Cu, and B, but decreased the uptake of Ca, Mn and Zn (Patel and Golakiya, 1986). Application of B @ 2 ppm without $CaCO_3$ increased pod uptake of B by 43% and total plant uptake of B by 23%, but the application of $CaCO_3$ for the same B treatment drastically reduced B uptake. This experiment was done on a vertisol, which had 12%, 29.4%, 58.6% of sand, silt and clay, respectively; and had a pH of 7.4, EC of 0.18 dS m^{-1}, $CaCO_3$ of 3.5% and HWS B of 0.66 ppm.

Crop Response to B in Fine Textured Acid Soil

State–Uttar Pradesh; Crop–Potato

The tuber yield, tuber uptake of Fe, Mn, Cu, Zn and B, starch content of potato variety Kufri Jyoti increased significantly with the application of borax @ 20 kg ha^{-1} compared to the control treatment (Dwivedi and Dwivedi, 1992). The borax applied and the control treatment tuber uptake of Fe was 139.6 and 18.2, of Mn was 23.3 and 18.2, of Cu was 4.5 and 1.5, of Zn was 15.2 and 0.42 and of B was 14 and 9.4 ppm, respectively. This experiment was conducted in Ranichauri (1900 msl), Uttar Pradesh state, on an acid Inceptisol. The soil was silty clay loam with pH 5.4 and DTPA extractable Fe of 2.87, Mn of 18.2, Cu of 1.5, Zn of 0.41, and HWS B of 0.08 ppm.

State–Uttar Pradesh; Crop–Soybean-Wheat Cropping Sequence

A field experiment in an acid silty clay loam soil (Inceptisol) of Ranichauri, Uttar Pradesh state, gave the second highest soybean (variety: Bragg) and wheat yield with 20 kg borax ha^{-1} (Dwivedi, Dwivedi and Pal, 1990). The highest yield was obtained with the application of Mo. The soil of the experimental site had a pH of 5.2, DTPA extractable Fe of 28.5, Mn of 20.4, Cu of 1.2, Zn of 3.9, HWS B of 0.08 and ammonium oxalate extractable Mo of 0.07 ppm. Of all the micronutrients applied, the concentration of B was the highest in soybean (28.4 ppm) irrespective of the mode of application followed by wheat (23.8 ppm) for the soil + lime application method.

Crop Response to B in Fine Textured Nearly Neutral Soil

State–Maharashtra; Crop–Groundnut

Significant higher pod yield was recorded for groundnut with application of 2.5 kg B ha^{-1} (Agasimani, Babaled and Patel, 1993). Further increase in

B application decreased pod yield. The soil of this experimental site had a pH of 6.8, Ca of 4900, S of 100 and HWS B of 16 B kg ha^{-1} soil.

State-West Bengal; Crop-Rice, Wheat and Jute

Application of 0.5 kg B ha^{-1} increased grain and straw yield of rice (variety: Pankaj), wheat (variety: Sonalika), and jute (variety: JR0524) fibre yield compared to the control grown in a clayey loam soil of pH 6.6 (Sarkar and Chakraborty, 1980). The response to B was higher when applied in combination with other micronutrients. Rice grain yield increase was 11% and 17.9%, respectively, for the B alone treatment and B applied in combination with other micronutrients, i.e., Mn, Zn, Cu, and Mo @ 5, 4, 1, 0.1 kg ha^{-1}, respectively. For wheat, percent grain yield increase was 26.1% and 45.7% for the B alone treatment and for B applied in combination with other micronutrients. Similarly, jute fiber yield increase was 14.5% and 50.3%, respectively, for the B alone and B applied in combination with other micronutrients.

State-Tamil Nadu; Crop-Groundnut-Blackgram Cropping Sequence

Sundersan and Ramamoorthy (1993) studied the soil and foliar application of B either alone, or in combination with Zn on groundnut (variety: CO2) in a Vertic ustrapepts sandy clay loam soil of Coimbatore, Tamil Nadu state. A significant increase of 53.3%, 16.9%, 14.7%, 34.2%, 37.4% was recorded in number of nodules, dry pod yield, shelling percentage, kernel yield and oil yield with the application of 5 kg borax ha^{-1}, compared to the nontreated control treatment. Soil application of 2.5 kg borax and 20 kg ZnSO$_4$ ha^{-1} in combination with 0.5% ZnSO$_4$ and 0.25% foliar spray of borax at 40 and 50 days after sowing, respectively, increased groundnut nodule number by 80%, dry pod yield by 32.5%, kernel yield by 47.7%, oil yield by 55%. The residual effect of B on the subsequent crop of blackgram was not significant.

COUNTRY-NEPAL

As discussed earlier, interpretation of crop response, or rather the lack of response to applied B in Nepal should be carefully reported due to the additional factor of altitude and B × environment interaction in this country. Unlike the rest of the IGP, the rice-wheat cropping system in Nepal is practiced in the flat plains called the "Terai" and the "Inner Terai" region and the higher altitudes of the mid-hill region of Nepal. In the mid-hill region, the altitude varies, and the anthesis event of the winter crop growth period, e.g., wheat, may coincide with the cold air temperature, leading to wheat sterility (Subedi et al., 1985b). Under such situations, application of B to the crop may not yield the desired results. The interaction of B and climate is perhaps of lesser importance than soil B status in the "Terai" and "Inner Terai" of Nepal. Nepal has been rated as having the lowest soil B content of the 30 country FAO survey of soil micornutrient status and crop response to applied B can therefore be expected (Sillanpaa, 1982).

Midhills

Subedi et al. (1985a) reported wheat sterility prone areas in the western hills of Nepal at Rashing Patan (420 msl). They conducted an experiment at this location where the soil was silty loam, with a neutral pH, and had a low soil B content. The wheat genotypes SW 41 and BL-1022 responded to B application of 1 kg B ha^{-1} with a decrease in sterility and an increase in grain yield.

Karki (1985) reported variable location specific (location: Kailleritar and Baireni of Rhodic Ustrochrepts and Ochric Fluvaquents soil, respectively) response to soil applied B @ 2.5 kg ha^{-1} to wheat. The soil at Kalleritar was silty loam with a pH of 6.7, and at Baireni was sandy loam with a pH of 6.1. The uptake of B in the B treated plots was found to be lower than that of the minus-B treated control plots. Mengel and Kirkby (1979), and Sillanpaa (1982) have published similar results. The lower concentration of B in the crops was possibly due to dilution phenomenon (Sillanpaa, 1982) because of enhanced crop growth.

Inner Terai

Deficiency of B in the inner Terai of Nepal is more severe, and crop response to applied B therefore is more pronounced. Srivastava et al. (1996) have shown B deficiency to be the dominant factor in inducing flower abortion and failure in pod and seed set in chickpea (variety: Kalika) at Rampur in Chitwan district of Inner Terai. They reported high flower abortion and pod sterility for treatments supplied with all nutrients except boron in chickpea. Soil application of 0.5 kg B ha^{-1} significantly reduced flower abortion, enhanced pod formation and increased grain yield in chickpea compared to the control treatment (Srivastava et al., 1997). The soils in these experimental sites are sandy loam with a pH of 6.3 and HWS B of 0.2 ppm.

The exotic accessions of bold seeded lentil show typical symptoms of B deficiency symptoms at Rampur during vegetative stage of crop growth. They further reported a ten-fold grain yield increase of bold seeded lentil variety ILL-4605 to soil applied B @ 0.5 kg B ha^{-1}, and further increase in

B application rates did not increase grain yield significantly. Sunhemp (*Crotolaria juncea*), a legume planted for green manuring to improve soil productivity on-station at Rampur, showed typical symptoms of B deficiency. The plants were stunted, had a rosette of apical leaves with white tips.

The problem of B deficiency limiting crop production is not only confined to the Rampur area of "Inner Terai" of Nepal but is also widespread in farmer's fields receiving 10 to 15 tones of compost ha^{-1} year^{-1} (Srivastava et al., unpublished data). Farmer's field trials conducted in Shukranagar and Rajahar villages of Chitwan and Nawalparasi districts in 1997-98 and 1998-99 showed varying degrees of grain yield increase of chickpea and mustard to soil applied B, and this was related to the severity of soil B deficiency.

Terai

In Nepal, wheat sterility was first reported from the eastern Terai in 1965 in farmer's field (Misra et al., 1992). However, not much research has been done in the Terai of Nepal on boron, crop response to applied boron, or on sterility. However, farmers and extension workers have reported sterility in wheat in farmer's field in different years. The spread and extent of the problems associated with B deficiency in the Terai of Nepal follows the trend of the Indian-IGP region in that there are greater number of reports of B-deficiency like symptoms of different crops from the mid to the eastern end of the Terai of Nepal. The symptoms of B deficiency are observed in chickpea at the Agricultural Research Station (RARS), Parwanipur, in groundnut, mustard, sunflower, chickpea and pigeonpea in the Oilseeds Research Program (ORP), Nawalpur, and in chickpea and maize further east at RARS, Tarahara. "Little leaves" and "flower abortion" in chickpea, "poor pod set" in pigeon pea, "dwarf plants" and "poorly developed siliqua" of mustard at ORP, Nawalpur have been reported. The "little leaf" and "flower abortion" of chickpea at Tarahara and the "white line" in the flag leaf and "small cob size" in maize at RARS, Tarahara, were observed in different years.

COUNTRY–BANGLADESH

Soil B deficiency in Bangladesh is widespread (Figure 2). Crop response to applied B has been recorded in several crop species. Sterility of wheat and its positive response to B application can be expected in much of the wheat growing areas of Bangladesh. However, response to added B has been inconsistent in experiments in the central region of Bangladesh.

Ahmed and Hossain (1997) have recorded wheat yield increase of 10% to

92% with soil applied B in wheat and 14% to 52% in different vegetables from several field trails in B deficient northern zones of Bangladesh. In an acidic soil of pH 5.2, the highest grain yield of wheat was obtained with the application of 1 kg B ha^{-1} (Ahmed and Alam, 1994).

Similar results have been reported for other crops. Ahmed and Hossain (1997), recorded a 14% grain yield increase of moongbean (*Vigna radiata*) with the application of 2.5 kg B ha^{-1} in the flood plains of Mymensingh (Typic Fluvaquent) which had a soil B content of 0.4 ppm. Wheat grain yield increased by almost 100% with the application of B (Islam and Talukdar, 1991), and similar yield increases in mustard, chickpea, potato, cauliflower, and groundnut yield at Rangpur have been reported (Table 1). In the Tista flood plain of northern Bangladesh (Rangpur and Dinajpur), there are increasing reports of B deficiency limiting yields of lentil and chickpea (BARI, 1993). On station studies indicate that chickpea yield of more than 1 ton ha^{-1} can be obtained by applying 0.5 kg B ha^{-1}.

COUNTRY: PAKISTAN

There are few reports of B related crop response studies from Pakistan. As discussed earlier, soils with a low B status was mostly from the non-irrigated areas of the country. Significant increases in all yield components and grain yield of wheat with the application of B has been reported. Approximately 45% of the sampled fields of the cotton growing areas of Multan district in Pakistan were found to be deficient in soil B status (Rashid, 1996) and application of B in these areas increased the yield of cotton.

CONCLUSION AND FUTURE RESEARCH NEEDS

Acknowledgement and the awareness of soil B deficiency are not as widespread as its occurrence in the IGP region. B deficiency and crop response to applied B is more frequent in the eastern half, compared to the western and northwestern half of the IGP. Its deficiency affects cereals, legumes and vegetables by reducing crop yields most likely due to the failure in floret fertilization and hence sterility (e.g., in wheat), and by premature flower drop (e.g., in chickpea).

The correlation between soil B status and sterility in wheat is not high. But the analyses of soil B status for the entire IGP and its synthesis using GPS and GIS will form a basic data set for research on boron. The critical limits for B sufficiency and toxicity are narrow, and the critical B limits for soil and plants need to be validated for different regions within the IGP. Field experiments designed to understand the relationship between climatic conditions, soil B status and wheat sterility is needed to clarify the confusion in its relationship. Crop varieties, especially that of wheat, differ in their susceptibility to sterility. The crop-breeding program in the region dose not tests potential varieties (to be released for cultivation) for sterility due to soil B deficiency. Testing for sterility should therefore be included as part of a breeding program. This is especially important for varieties released for the eastern half of the IGP. In recent years, high yielding wheat varieties released in Nepal have been completely sterile when grown in certain areas of the coutnry. It has been noted by several researchers and farmers in the region that wheat sterility do not occur when the crop was cultivated in fields which received high doses of Farm Yard Manure during the previous crop season, or in fields that were close to the farmers habitat or cattle shed from which it received run-off water. Therefore, organic nutrient sources can be used to manage soil B deficiency problems. However, the nutrient content of these organic sources and its management need to be properly documented. Information on the management of B deficiency using chemical and organic B sources as part of a cropping system should be compiled and awareness amongst farmers on B deficiency and its control should be created in the region.

REFERENCES

Agasimani, C.A., H.B. Babaled, and P.L. Patel. (1993). Response of groundnut to sulphur and boron. *Journal of Maharashtra Agricultural University*, 18(2): 318-319.

Ahmed, S. (1987). *Micronutrient studies under irrigated and non-irrigated conditions (BINA)*, Annual Report, 1986-87.

Ahmed, S., and M.B. Alam. (1994). Effect of zinc and boron on yield and nutrient content of wheat. *Progressive Agriculture*, 5, 55-59.

Ahmed, S., and M.B. Hossain. (1997). The problem of boron deficiency in crop production in Bangladesh. In: *Boron in soils and plants*, eds: R.W. Bell and B. Rerkasem. Proceedings of the International Symposium on Boron in Soils and Plants, Chiang Mai, Thailand, 7-11 September 1997, pp. 1-5.

Ali, M.H. and A.K. Basu. (1992). *Assessment of response to micronutrient application to important crops and its economics*. Proceedings IBFEP-HFC Workshop on Micronutrients, Bhubaneswar, India, pp. 162-171.

BARI (Bangladesh Agricultural Research institute). (1993). *Annual Report of Pulse Breeding 1992/93*, Ishurdi, Pabna, Pulses Research Center.

Bansal, R.L. V.K. Nayyar and P.N. Takkar. (1991). Availability of B, Co and S in Ustochrepts. *Journal of Indian Society of Soil Science*, 39, 181-182.

Barthakur, H.P. (1992). *Available micronutrients and crop response to their application in acid soils of Assam*. Proceedings IBFEP-HFC Workshop on Micronutrients, Bhubaneswar, India, pp. 172-179.

Bell, R.W. and B. Dell. (1995). Environmental factors affecting boron deficiency. In: *Causes of sterility in wheat*, eds. R.W. Bell and B. Rerkasem. Proceedings of a workshop held at Ciang Mai, Thailand, 25-28 July 1994, pp. 95-106.

Blevins, D.G. and K.M. Lukaszewski. (1998). Boron in plant structure and function. *Annual Review of Plant Physiology and Plant Molecular Biology*, 49:481-500.

Borkakati, K. and P.N. Takkar. (2000). Forms of boron in acid alluvial and lateritic soils in relation to ecosystem and rainfall distribution. In *International Conference on Managing Resources for Sustainable Agricultural Production in the 21st Century*, Vol 2: 127-128.

Dawson, I.A., and I.F. Wardlaw. (1989). The tolerance of wheat to high temperature during reproductive growth. III Booting to anthesis. *Australian Journal of Agricultural Research* 40, 965-80.

Dwivedi, G.K., M. Dwivedi, and S.S. Pal. (1990). Modes of application of micronutrients in acid soil in soybean-wheat crop sequence. *Journal of Indian Society of Soil Science*, 38: 458-463.

Dwivedi, G.K., M. Dwivedi. (1992). Efficacy of different modes of application of copper, zinc and boron to potato. *Annals of Agriculture Research*, 13(1): 1-6.

Gajbhiye, K.S. N.K. Banerjee, and N.N. Goswami. (1980). Correlation study of water soluble boron with ECe and silt plus clay in non-saline and low saline soils. *Journal of Indian Society of Soil Science*, 28, 251-253.

Garg, O.K., A.N. Sharma, and G.R.S.S. Kona. (1979). Effect of boron on the pollen viability and the yield of rice plants (*Oryza sativa* L. variety: Jaya). *Plant and Soil*, 52: 591-594.

Gogoi, D.N. (1992). *Available boron content of some alluvial soils of Assam*. Proceedings IBFEP-HFC Workshop on Micronutrients, Bhubaneswar, India, pp. 124-127.

Golakiya, B.A. (1989). In search of compromisation between calcium-boron antagonism in the groundnut crop. *Journal of Maharashtra Agriculture University*, 14(1): 123.

Golakiya, B.A. (1992). Calcium-boron interaction in the soils and the groundnut. *Madras Agriculture Journal*, 79(12): 665-670.

Gupta, V.K., Ramkala, and S.P. Gupta. (1993). Boron tolerance of chickpea and lentil on Typic Ustipsamment. *Journal of Indian Society of Soil Science*, 41(4): 797-798.

Islam, M.S. and K.H. Talukdar. (1997). *Research findings on the effect of boron in cereals and vegetables*. Annual Report, 1987-1991. Bangladesh Agriculture Research Institute, Joydebpur.

Karki, K.B. (1985). Effect of fertilizers and boron on yield and boron uptake in wheat. In: *Sterility in wheat in subtropical Asia* eds: H.M. Rawson and K.D. Subedi. Extent, Causes and Solutions. Proceedings of a Workshop, LARC, Pokhara, Nepal, 18-21 Sept. 39-43.

Khatri-Chhetri, T.B. (1982). *Assessment of soil test procedures for available boron and zinc in the soils of the Chitwan valley, Nepal*, Ph.D. Thesis, University of Wisconsin-Madison, pp. 299.

Khatri-Chhetri, T.B. and A.B. Karki. (1979). Effect of borax on the yield of cauliflower. *Journal of Institute of Agriculture and Animal Sciences*, 1:1-10.

Khatri-Chhetri, T.B., A.B. Karki and R.C. Prasad. (1979). Boron fertilization of maize under Rampur conditions. *Journal of Institute of Agriculture and Animal Sciences*, 1:21-23.

Malewar, G.U., I.I. Syed, and B.S. Indlkar. (1992). Effect of phosphorus with and without boron on the yield and uptake of boron and phosphorus in cotton-groundnut cropping sequence. *Annals of Agriculture Research*, 13(3): 269-270.

Matoh, T. (1997). Boron in plant cell walls. In *Boron in soils and plants: reviews*, eds: B. Dell, P.H. Brown, R.W. Bell, *International Symposium on Boron in Soils and Plants*, Chiang Mai, Thailand, 7-11 September 1997, pp. 59-70.

Matoh, T., K. Ishigaki, M. Mizutanni,W. Matsunaga and K. Takabe. (1992). Boron nutrition of cultured tobacco BY-2 cells.I. requirement for and intracellular localisation of boron and selection of cells that tolerate low levels of boron. *Plant and Cell Physiology*, 33, 1135-1141.

Mehrotra, O.N., R.D.L. Srivastava, and P.H. Mishra. (1980). Some observations on the relative tolerance of wheat genotypes to boron. *Indian Agriculture*, 24(3&4): 223-238.

Mengel, K., and E.A. Kirby. (1979). *Principles of plant nutrition*. International Potash Institute, Berne, Switzerland.

Misra, R., R.C. Munankarmi, S.P. Pandey, and P.R. Hobbs. (1992). Sterility work in wheat at Tarahara in the Eastern Tarai of Nepal. In *Boron deficiency in wheat*, eds: C.E. Mann and B. Rerkasem, pp. 65-71. Wheat Special Report No. 11, Mexico, DF, CIMMYT.

Mondal, A.K., S. Pal, B. Mandal and L.N. Mandal. (1991). Available boron and molybdenum content in alluvial acidic soils of north Bengal. *Indian Journal of Agricultural Sciences*, 61: 502-504.

Nandi, T. A.K. Karakn, M. Hossain and M.H. Ali. (1992). *Boron and molybdenum status in red and lateritic soils of Bankura, Midnapore and Purulia districts of West Bengal*. Proceedings IBFEP-HFC Workshop on Micronutrients, Bhubaneswar, India, pp. 157-160.

Nongkynrih, P., P.S. Dkhar and D.T. Khathing. (1996). Micronutrient elements in acid alfisols of Meghalaya under rice cultivation. *Journal of Indian Society of Soil Science*, 44: 455-457.

Patel, M.S. and B.A. Golakiya. (1986). Effect of calcium carbonate and boron application on yield and nutrient uptake by groundnut. *Journal of Indian Society of Soil Science*, 34: 815-820.

Patil, G.D., M.D. Patil, N.D. Patil, and R.N. Sule. (1987). Effects of boronated superphosphate, single superphosphate and borax on yield and quality of groundnut. *Journal of Maharashtra Agricultural University*, 12(7): 168-170.

Ramamoorthy, K., and S. Sudersan. (1992). Supply of zinc and boron on yield and seed quality in groundnut (*Arachis hypogea* L.). *Annals of Plant Physiology*, 6(1): 33-38.

Rashid, A. (1996). *Nutrient indexing of cotton in multan district and boron and zinc nutrition of cotton*. Micronutrient Project Annual Report 1994-95, Land Resources research Institute, National Agricultural Research Center, Pakistan Agricultural Research Council, Islamabad, pp. 76.

Rawson, H.M. (1996). Parameters Likely to be Associated with Sterility. In *Sterility in wheat in subtropical Asia: extent, causes and solutions* eds. Rawson, H.M., and

K.D. Subedi. Proceedings of a workshop held from 18 to 21 September 1995 at Lumle Agricultural Research Centre, Pokhara, Nepal. ACIAR Proceedings No. 72, 13-31.

Rawson, H.M., and A.K. Bagga. (1979). Influence of temperature between floral initation and flag leaf emergence on grain number in wheat. *Australian Journal of Plant Physiology*, 6, 319-400.

Rerkasem, B. 1993. *Screening wheat for tolerance to boron deficiency.* Publication of Agronomy Department, Chiang Mai University, Chiang Mai, Thailand 50002, August 1993, pp 23.

Rerkasem, B. (1995). Boron and Plant Reproductive Development. In *Causes of sterility in wheat*, eds. R.W. Bell and B. Rerkasem. Proceedings of a Workshop held at Ciang Mai, Thailand, 25-28 July 1994, pp. 112-117.

Rerkasem, B. and S. Jamjod. (1997). Genotypic variation in plant response to low boron and implications for plant breeding. In *Boron in soils and plants: reviews*, eds. B. Dell, P.H. Brown and R.W. Bell. International Symposium on Boron in Soils and Plants, Chiang Mai, Thailand, 7-11 September 1997, Kluwer Academic publishers, pp. 169-180.

Rerkasem, B. and J. Loneragan. (1994). Boron deficiency in two wheat genotypes in a warm, subtropical region. *Agronomy Journal*, 86, 887-90.

Saifuzzaman, M. (1995). Sterility in wheat in Bangladesh, In: *Causes of sterility in wheat*, eds. R.W. Bell and B. Rerkasem, Proceedings of a Workshop held at Ciang Mai, Thailand, 25-28 July 1994, pp. 57-71.

Saini, H.S. and D. Aspinall. (1981). Effect of water deficit on sporongenesis in wheat (*Triticum aestivum* L.). *Annals of Botany*, 48, 623-633.

Sakal, R., A.P. Singh and R.B. Singh. (1989). Differential susceptibility of maize varieties to boron deficiency in a calcareous soil. *Journal of Indian Society of Soil Science*, 37: 582-584.

Sakal, R., A.P. Singh, B.P. Singh and R.B. Sinha. (1984). Available boron of some calcerous soils of Muzaffarpur district in relation to certain soil characteristics. *Journal of Research*, (R.A.U.) 5: 45-49.

Sakal, R., R.B. Sinha and A.P. Singh. (1988). Effect of boron application on black-gram and chickpea production in calcareous soil. *Fertilizer News*, 38(2): 51-55.

Sakal, R., R.B. Sinha, A.P. Singh and N.S. Bhogal. (1988). Response of some Rabi pulses to boron, zinc and sulphur application in farmer's field. *Fertilizer News*, 43(11):37-40.

Sakal, R., A.P. Singh, R.B. Sinha and N.S. Bhogal. (1998). *Annual Report of AICRP on Micro- and Secondary-Nutrients and Pollutant Elements in Soils and Plants*. RAU, Pusa, Bihar, India.

Sarengi, S.K. (1992). *Distribution of available boron in the northern plateau region of Orissa*. Proceedings IBFEP-HFC Workshop on Micronutrients, Bhubaneswar, India, 172-179.

Sarkar, A.K. and S. Chakraborty. (1980). Effect of micronutrient on yield of major field crops grown in the gangetic delta region. *Indian Agriculture*, 24(3&4): 259-266.

Sharma, B.D., S.S. Mukhopadhyay, P.S. Sidhu and J.C. Katyal. (2000). Pedospheric attributes in distribution of total and DTPA-extractable Zn, Cu, Mn and Fe in Indo-Gangetic Plains. *Geoderma*, 96: 131-151.

Shinde, B.N., B.P. Rote and S.P. Kale. (1990). Effect of soil application of boron on yield of groundnut and its residual effects on wheat. *Journal of Maharashtra Agricultural University*, 15(2): 195-198.

Sillanpaa, M. (1982). *Micronutrient and the nutrient status of soil, a global study.* FAO Soil Bulletin, 48: 242-250.

Singh, J. and N.S. Randhawa. (1977). Boron fractionation and mineral composition of saline-alkali soils. *Journal of Indian Society of Soil Science*, 23: 277-230.

Singh, A.P., R. Sakal, R.B. Sinha and N.S. Bhogal. (1991). Relative response of selected chickpea and pigeonpea cultivars to boron application. *Annals of Agriculture Research*, 12(1): 20-25.

Srivastava, S.P., C.R. Yadav, T.J. Rego, C. Johansen, N.P. Saxena and A. Ramakrishna. (1996). Diagnosis of boron deficiency as a cause of flower abortion and failure of pod set in chickpea in Nepal. *International Chickpea and Pigeon Newsletter*, 3: 29-30.

Srivastava, S.P., C.R. Yadav, T.J. Rego, C. Johansen and N.P. Saxena. (1997). Diagnosis and alleviation of boron deficiency causing flower and pod abortion in chickpea (*Cicer arietinum* L.) in Nepal. In *Boron in soils and plants*, eds. R.W. Bell, and B. Rerkasem. Proceedings of the Workshop, Chiang Mai, Thailand, 7-11 Sept, 95-99.

Sthapit, B.R. (1988). *Studies on wheat sterility problem in the Hills, Tar and Tarai of Nepal.* Technical Report No.16/88. Lumle Agricultural Research Centre, Pokhara, Kaski, Nepal.

Sthapit, B.R., K.D. Subedi, T.P. Tiwari, S.L. Chaudhury, K.D. Joshi, B.K. Dhital and S.N. Jaisi. (1989). *Studies on causes of wheat sterility in the Hills, Tar, and Terai of Nepal.* Lumle Agricultrue Centre Technical paper No. 5/89, Nepal, LARC.

Subedi, K.D., C.B. Bidhathoki, M. Subedi and G.C., Y.D. (1985a). Effect of sowing time and boron on sterility of four genotypes of wheat (*Triticum aestivum* L.) in the western hills of Nepal. In *Sterility in wheat in subtropical Asia: extent, causes and solutions*, eds. H.M. Rawson and K.D. Subedi. Proceedings of a Workshop, LARC, Pokhara, Nepal, 18-21 Sept.

Subedi, K.D., C.B. Bidhathoki, M. Subedi, and B.R. Sthapit. (1985b). Overview of wheat sterility problem and research findings to date in the western hills of Nepal. In *Sterility in wheat in subtropical Asia: extent, causes and solutions*, eds. H.M. Rawson and K.D. Subedi. Proceedings of a Workshop, LARC, Pokhara, Nepal, 18-21 Sept.

Sundersan, S. and P.P. Ramaswami. (1993). Micronutrients in groundnut-blackgram cropping system. *Fertilizer News*, 38(2): 51-55.

Takkar, P.N., M. Chibba and S.K. Mehta (1989). *Twenty years of Coordinated Research on micronutrients in soils and plants (1967-87).* Bulletin-I. ISSS, Bhopal, India, pp. 314.

Tandon H.L.S. (1995). *Micronutrients in soils, crops and fertilisers–a source book-cum-directory.* Fertiliser Development and Consultation Organization, New Delhi, pp. 138.

Tanaka, H. (1966). Response of *Lemma pausicostata* to boron as affected by light intensity. *Plant and Soil*, 25: 425-439.

Tiwari, T.P. (1985). Review of research into wheat sterility. In *Sterility in wheat in subtropical Asia: extent, causes and solutions*, eds. H.M. Rawson and K.D. Subedi. Proceedings of a Workshop, LARC, Pokhara, Nepal, 18-21 Sept., pp. 112-118.

Wani, B.B., G.D. Patil and M.D. Patil. (1988). Effects of levels of SSP and Boronoted Superphosphate on the yield and quality of groundnut. *Journal of Maharashtra Agricultural University*, 13(3): 302-304.

Wankhade, S.G., R.C. Dakhore, D.B. Wanjari, N.R. Patil, Potdukhe and R.W. Ingle. (1996). Response of crops to micronutrients. *Indian Journal of Agricultural Research*, 30(3): 164-168.

Wiley, R.W. and R. Holliday. (1971). Plant population, shading and thinning studies in wheat. *Journal of Agricultural Science*, Cambridge, 77: 453-461.

Management
of Insect, Disease, and Nematode Pests
# of Rice and Wheat
in the Indo-Gangetic Plains

M. Sehgal
M. D. Jeswani
N. Kalra

SUMMARY. The dynamics and severity of pest attack in the Indo-Gangetic Plains (IGP) has shifted with the adoption and spread of the rice-wheat crop rotation in the last three decades. Pest attack on the rice and wheat crop in the IGP has been severe in certain years, and pest management in the region has largely been based on chemical control means. Though Integrated Pest Management (IPM) modules have been developed for rice and wheat in India, its adoption and spread has been limited to certain areas of the IGP. In general, pest infestation is higher during the hot and humid summer rice-growing season compared to the cooler wheat-growing season. This article reviews the insect, pathogen, and nematode pest spectrum and their control methods for the rice and wheat crop in the IGP. The changing dynamics of these pests, IPM modules developed in India and successful IPM strategies are also discussed in this article. *[Article copies available for a fee from The Haworth Document Delivery Service: 1-800-342-9678. E-mail address: <getinfo@haworthpressinc.com> Website: <http://www.HaworthPress.com> © 2001 by The Haworth Press, Inc. All rights reserved.]*

M. Sehgal and M. D. Jeswani are Senior Scientists, National Center for Integrated Pest Management (NCIPM), LBS Building, IARI; and N. Kalra is Senior Scientist, CASS, IARI, New Delhi–110012, India.

[Haworth co-indexing entry note]: "Management of Insect, Disease, and Nematode Pests of Rice and Wheat in the Indo-Gangetic Plains." Sehgal, M., M. D. Jeswani, and N. Kalra. Co-published simultaneously in *Journal of Crop Production* (Food Products Press, an imprint of The Haworth Press, Inc.) Vol. 4, No. 1 (#7), 2001, pp. 167-226; and: *The Rice-Wheat Cropping System of South Asia: Efficient Production Management* (ed: Palit K. Kataki) Food Products Press, an imprint of The Haworth Press, Inc., 2001, pp. 167-226. Single or multiple copies of this article are available for a fee from The Haworth Document Delivery Service [1-800-342-9678, 9:00 a.m. - 5:00 p.m. (EST). E-mail address: getinfo@haworthpressinc. com].

167

KEYWORDS. IPM, nematode, pathogen, pest, rice, wheat

INTRODUCTION

Biotic stresses, due to the infestation of insects, pathogens, nematodes and weeds may cause on an average, 20 to 30% rice and wheat yield reduction. Development of pest resistance to agro-chemicals, secondary pest outbreaks, and emergence of new pest problems with a changing cropping pattern coupled with favorable climate conditions has contributed to crop losses. Unregulated and excessive pesticide use is quite alarming in the IGP (Dhaliwal and Arora, 1996) due to its uneven distribution and usage on selective crops and the low reliance on alternative chemicals like biopesticides and botanicals. An estimated 17 percent of total pesticide use in India is on the rice crop alone while its total cropped area is about 24 percent (Puri, 1999; Puri, 2000). Pesticide residue in crops is also a serious trade barrier.

Pest identification, understanding the changing dynamics of pest attacks and its severity over time and strategies to manage pests requires an integrated pest management approach. Integrated Pest Management (IPM) involves proper choice and blend of compatible tactics (cultural, mechanical, host plant resistance, biological and pesticide use) so that the components complement each other to keep pest population at manageable levels.

This article reviews key pests of rice and wheat in the rice-wheat cropping system, and discusses their past and present status, distribution, outbreaks, and management. The adoption and spread of IPM packages are uncommon in cereal crops. But IPM modules for rice and wheat crop have been developed over the years in India and its success stories, impact and the constraints to expanding the IPM packages are also discussed throughout this article.

PEST STATUS IN RICE-BASED CROPPING SYSTEM

Insect pests are more serious on rice than on wheat in the rice-wheat cropping sequence. More than 100 insect species have been reported to damage the rice crop. The changing dynamics of the rice and wheat pests are shown in Tables 1 and 8.

Brown plant hopper (BPH, *Nilaparvata lugens*), White backed plant hopper (WBPH, *Sogatella furcifera*), Green leafhopper (GLH, *Nephotettix virescens*), Leaf folder (LF, *Cnaphalocrocis medinalis*), Rice bug (RB, *Leptocorisa oratorius*), Rice hispa (RH, *Dicladispa armigera*), were minor insect pests in the pre-green revolution era. But today these insects have assumed importance as major pests in localized areas. Gall midge (GM, *Orseolia oryzae*),

TABLE 1. Past and present status of rice pests in the Indo-Gangetic Plains.

PEST	INTENSITY	CHANGE
Major Insect Pests		
Leaf Folder, *Cnaphalocrocis medinalis*	******	(×)
Brown Plant Hopper, *Nilaparvata lugens*	***	↑↑↑
Whitebacked Plant Hopper, *Sogatella furcifera*	******	↑↑↑
Green Leaf Hopper, *Nephotettix virescens*	******	↑↑↑
Gall Midge, *Orseolia oryzae*	******	↑↑↑
Rice Hispa, *Dicladispa armigera*	**	↓↓
Other Common Insect Pests		
Yellow Stemborer, *Scirpophaga incertulas*	**	↓↓
Whorl Maggot, *Hydrellia* sp.	**	(×)
Thrips, *Baliothrips biformis*	**	(×)
Caseworm, *Nymphula depunctalis*	**	(×)
Armyworm, *Mythimna separata*	**	(×)
Cutworm, *Spodoptera litura*	**	(×)
Gundhibug, *Leptocorisa varicornis*	**	(×)
Root Weevil, *Echinocnemus oryzae*	**	(×)
Diseases		
Bacterial Blight, *Xanthomonas oryzae*	******	↑↑↑
Blast, *Pyricularia oryzae*	******	↑↑↑
Sheath Blight, *Rhizoctonia solani*	****	↑↑
Sheat Rot *Acrocylindrium oryzae*	****	↑↑
Brown Spot, *Helminthosporium oryzae*	***	↑
False Smut, *Ustilaginoides virens*	***	↑
Rice Tungro Viruses (RTV)	**	↑
Nematodes		
White Tip Disease, *Aphelenchoides besseyi*	**	↑
Rice-Root Nematode, *Hirschmanniella* spp.	***	↑↑↑
Root Parasite, *Meloidogyne* spp.	**	(×)

↑↑↑: Major pest, increasing in economic importance; ↓↓: Major pest, declining in economic importance; ×: Minor pest, becoming a major problem. *Very low to ******Very high intensity.
Source: Chelliah and Bharathi, 1998, Sehgal and Gaur, 1999.

Ear-cutting caterpillar (*Mythimna separata*), Thrips (*Baliothrips biformis*) have also become important pests of high yielding rice varieties. Amongst diseases, the bacterial leaf blight (BLB, *Xanthomonas campestris*), sheath rot (*Acrocylindrium oryzae*), blast (*Pyricularia oryzae*), bacterial leaf streak and others have become major problems in the IGP and in certain years, can cause substantial yield losses in endemic areas (Table 1). Of several weed species, *Echinochloa colonum*, *Cyperus rotundus*, and *Ischaemum rugosum* compete vigorously with rice. Other common weeds are *Avena sativa*, *Convolvulus arvensis* and *Cirsium arvense*.

In wheat, insect pests like *Spodoptera* and termites, and diseases like

Helminthosporium sativum and *Alternaria* blight (Dubin and VanGinkel, 1991; Joshi, Gera and Saari, 1973; Joshi, Singh and Srivastava, 1988) cause noticeable crop losses (Table 8). Rust is effectively controlled by rust resistant varieties. Leaf blight is common in warmer areas. White-tip nematode (*Aphelenchoides besseyi*) and root-knot nematodes (*Meloidogyne triticoryzae* and *M. graminicola*) affect both the rice and the wheat crop, and *Phalaris minor* is now a major weed problem in the wheat crop in northwestern India (Chelliah and Bharathi, 1998; Gill, Walia and Brar, 1979; Gill, Sandhu and Singh, 1984; Sehgal and Gaur, 1999; Singh and Paroda, 1994; Taylor, 1969).

Yield Losses Due to Pests

Yield losses due to insect pests ranging from 20-50% were estimated because of moderate infestation of LF, GM, BPH, WBPH and other sporadic pests in extensive rice growing areas of the country. The period between 30 to 60 days after rice transplantation is the most vulnerable crop growth stage as maximum crop yield losses occur when infested by pests at this stage in majority of the rice growing areas. Comparatively, fewer losses occur during the first 30 days and beyond 60 days after rice transplantation (Abraham and Khosla, 1964; Chelliah and Bharathi, 1998).

It is estimated that about ten percent losses occur due to diseases in the IGP (Paroda, Woodhead and Singh, 1994). The losses due to individual pathogen may vary and may depend upon many ecological factors. Bacterial disease infecting the neck of the panicle at flowering stage may result in total crop failure. BLB may cause yield loss from 2-74% and RTV in epiphytotic proportions may cause losses up to 35%. Yield losses ranging from 10-50% due to nematodes were reported under different ecosystem of rice cultivation (Sasser and Freekman, 1987; Sharma, Price and Bridge, 1997).

INSECT PESTS OF RICE

Insect pest management in rice has largely been based on chemical control means in the IGP. The toxicity spectrum of spray formulations against insect pests is shown in Table 2. This section discusses the damage caused by these insects on the rice crop and options available for insect pest management in rice, in addition to chemical control means. Management options include rice varieties released for cultivation with resistance to specific insect pest (Table 3) or with multiple resistance against insect pests and diseases (Table 4).

TABLE 2. Spectrum of toxicity of spray formulations against insect pests of rice (S.B.: Stem Borer, L.F.: Leaf Folder, R.H.: Rice Hispa, B.P.H.: Brown Plant Hopper, W.B.P.H.: White Backed Plant Hopper, C.W.: Case Worm, G.L.H.: Green Leafhopper).

Insecticide	S.B.	L.F.	R.H.	B.P.H.	W.B.P.H.	C.W.	G.L.H.
Quinalphos	❖	❖	❖		❖		
Phosalone	❖	❖	❖	❖			❖
Monocrotophos	❖	❖	❖	❖	❖	❖	❖
Chlorpyriphos	❖	❖	❖		❖	❖	❖
Carbaryl		❖	❖	❖	❖		
Fenitrothion		❖					
Phosphamidon		❖			❖		
Fenthion			❖		❖		
Dichlorvos						❖	
Endosulfan	❖					❖	
❖❖ Ethofenprox				❖	❖		❖
❖❖❖ Cartap	❖						❖

Application rate is 350-500 g a.i./ha. ❖❖ = 50-100 g a.i./ha. ❖❖❖ = 300 g a.i./ha
Source: Reports of Directorate of Rice Research, Hyderabad, Several years.

TABLE 3. Sources of resistance and released varieties of rice against important insect pests.

INSECT PEST	DONORS	RELEASED VARIETIES
1. Brown Plant Hopper	ARC 5984, ARC 6650, Karivennel, Leb Mue Nhang, Manoharsali, Oorapandy, Ptb 10, Ptb 18, Ptb 21, Ptb 33, Triveni	Chaitanya, Krishnaveni, Vajram, Pratibha, Makom, Pavizham, Mansarovar, Co 42, Chandana, Nagarjuna, Sonsali, Rasmi, Jyothi, Bhadra, Neela, Annanga, Daya, Aruna, Kanakia, Remya, Bharathidasan, Karthika
2. White Backed Plant Hopper	Ptb 33	HKR 120
3. Green Leaf Hopper	Ptb 2, W 1263	Vikramarya, Lalt, Khaira, Nidhi
4. Stem borer	TKM6	Ratna, Sayasree, Vikas

Source: Reports of Directorate of Rice Research, Hyderabad, Several years.

Leaf Folder

The rice leaf folder complex that includes nine species of pyralids, viz., *Cnaphalocrocis medinalis, Marasmia exigua, M. bilinealis, M. patnalis, M. ruralis, M. suspicalis, M. trapezalis, M. venilialis* and *Bradina admixtalis,* and one species of gelechiid, *Brachmia arotraea* (Nadarajan and Rajappan, 1987) has gained major pest status in the last three decades. This has been due to the introduction of high yielding rice varieties and the changes in the

TABLE 4. Rice varieties with multiple resistance to insect pests or diseases.

Variety	Released in Indian States of	Resistant to
Sureksha	Andhra Pradesh, Orissa, West Bengal	BPH, WBPH, B1
Vikramarya	Andhra Pradesh	GLH, RTV
Shaktiman	Orissa, West Bengal	BPH, WBPH, B1
Rasmi	Kerala	BPH, B1
Daya	Orissa	BPH, GLH, B1
Samalei	Orissa, Madhya Pradesh	BPH, GLH, B1
Bhuban	Orissa	BLB
Kunti	West Bengal	B1
Lalat	Orissa	BPH, GLH, B1

BPH = Brown Plant hopper, WBPH = White Backed Plant hopper, GLH = Green Leafhopper, B1 = Blast, RTV = Rice Tungro Virus.
Source: Reports of Directorate of Rice Research, Hyderabad, Several years.

rice cultural practices, e.g., the use of more fertilizers and pesticides. There are a number of lepidopterous species which fold the rice leaves and cause similar type of damage to the crop. These species have been reported from different rice growing tracts of the world. But only *C. medinalis, M. patnalis, M. exigua* and *B. arotraea* (CRRI, 1982; Chatterjee, 1979; Gunathilagaraj and Gopalan, 1986; Jaganathan, Ramaraju and Venugopal, 1989) cause serious damage to the rice crop in the IGP. These pests were also recorded from Bangladesh, Nepal and Pakistan (Gill 1994), and from Punjab, Orissa, West Bengal, Madhya Pradesh, Bihar, Uttar Pradesh, Andhra Pradesh, Tamil Nadu, Kerala and Karnataka states of India.

Leaf folder also feeds on wheat, sorghum, maize and sugarcane besides several grasses that are found in and around paddy fields. More than 80 species of grasses and cultivated plants act as its host. Grasses that are its alternate host are *Cyperus difformis, C. rotundus, Fimbristylis miliacea, Scirpus, Cynodon dactylon, Dactyloctenium aegyptium, Echinochloa colonum, E. crus-galli, Eleusine indica, Leptochloa chinensis, Panicum repens* and *Paspalum distichum* (Gargav, Patel and Katiar, 1971; Lingappa, 1972; Yadava et al., 1972).

The larvae fold the leaves and remain inside, scraping the green tissues between the veins that make the leaves white and papery. Gradually the leaves dry up and turn brownish papery. Under severe infestation, the field presents a scorched appearance. The leaf folder attacks the crop in all stages of crop growth. Generally leaf damage is more in the vegetative stage of crop growth than in the reproductive stage. Damage to the flag leaf and the penultimate leaf causes severe yield losses. Economic threshold (ETL) of 0.75-1.40 and 0.50-0.70 3rd instar larvae/hill at tillering and early heading stage of crop growth, respectively, has been reported.

Intensive and extensive cultivation of rice for yield maximization has created favorable conditions for the spread of this pest by providing it year round food and shelter. Indiscriminate use of pesticides kills the natural enemies of the pest easily. As the rice leaf folder dwells in the leaf-fold, it is not eliminated completely by insecticides and some portion of the population that escapes the pesticide sprays, multiply rapidly in the absence of the natural enemies which are completely killed. Thus many insecticides have created the resurgence of rice leaffolder. High temperature (25-32°C) and high relative humidity (83-90%) favors development of *C. medinalis* whereas the other species require moderate temperature and low humidity for their development. These conditions prevail in the paddy environment. Depending upon these environmental conditions, the species complex is dominated by *C. medinalis* in the young crop while the other species tend to dominate the mature rice crop.

Management of Rice Leaf Folders

Source: Karuppuchammy and Goplan, 1986; Uthamasamy 1985; Mahabini et al., 1992; Panda and Shi, 1989; Raju, Gopalan and Balasubramanian, 1990; Raju, Saroja and Suriachandraselvan, 1988; Ramamoorthy and Jayaraj, 1977; Subramanian and Sathiyanadam, 1983.

- Proper sanitation of the field before transplanting of the rice crop.
- The sowing must be adjusted in accordance with the variety and abiotic factors.
- Use of resistance varieties, viz., TKM6, W1263, Kataribhog, Daruksail, Karpur kanti and Kamalabhog.
- One spraying of monocrotophos, phosphamidon, cartap hydrochloride, triazophos, quinalphos, BPMC, ethofenprox, cypermethrin or azinophos ethyl at boot leaf stage of the crop gives adequate protection against rice leaf folder.
- Egg parasitoids like *Trichogramma japonicum* are effective against rice leaf folder. Among the important larval parasitoids of these insects are *Cotesia, Cardiochiles philippinensis*. Among pupal parasitoids, only *Itoplectics narangae* is the most effective.
- *Beauveria bassiana, Nomuraea rileyi* and *Zoophthora radicans* are effective fungi reported to minimize the population of leaf folder complex.

Yellow Stem Borer

The rice stem borer *(Scirpophaga incertulas)* is also an important pest of rice throughout the IGP. Of several species, the yellow stem borer *Scirpophaga incertulas* is the key species in rice and one of the main causes for its low productivity. It has been recorded in Bangladesh, India, Nepal, and Pakistan. In India, this borer has been reported from Haryana, Punjab, Andhra Pradesh, Assam, Bihar, Gujarat, Himachal Pradesh, Jammu and Kashmir, Kerala, Madhya Pradesh, Maharashtra, Orissa, Sikkim, Tamil Nadu, Uttar Pradesh and West Bengal (Rao and Israel, 1964).

The yellow stem borer, *S. incertulas* is generally monophagous. However, *Cyperus rotundus* (Coco grass), *Coix lacryma-jobi* (Job's tears), *Cynodon dactylon* (Bermuda grass), *Ischaemum rugosum, Leptochloa panicoides, Saccharum officinarum* (Sugarcane), and wild rice species are its alternate hosts (Israel, Rao and Rao, 1964; Khan, 1964; Prakasa Rao, 1972).

The initial boring and feeding by neonate larvae in the leaf sheath causes broad longitudinal, whitish discolored areas at feeding sites, but only rarely does it result in wilting and drying of the leaf blades. About a week after hatching, larvae from the leaf sheaths bore into the stem and staying in the pith, feed on the inner surface of the stem walls (Israel, Vedamurthy and Rao, 1959). Such feeding frequently results in severing the apical parts of the plant from the base. When this kind of damage occurs during stem elongation, the central leaf whorl does not unfold, turns brownish and dries out, although the lower leaves remain green and healthy. This condition is known as "dead heart." Affected tillers dry out without bearing panicles. Occasionally, larval feeding above the primordia causes dead hearts.

During panicle emergence, borer infestation of plants from the base results in the drying of panicles. In severe cases, panicles may not emerge at all, and those that have already emerged do not bear any grains. The empty panicles become very conspicuous in the field because they remain straight and are whitish. They are called "White heads." Shriveled grains can be found when borer infestation occurs after grain formation is partially completed.

The damage potential is related to the inner diameter of the stem in relation to the diameter of the larvae. If the tillers are wider than the larva, the damage is less. There may be differences among species in this regard. Plant type and crop vigor determines eventual yield losses. Low tillering varieties have less opportunity to compensate for dead hearts than high tillering varieties. A high tillering variety can produce a replacement tiller for a dead heart. A vigorous, well-nourished crop can tolerate higher levels of dead hearts and white heads than a stressed crop.

The stem borer is polyvoltine, but the number of generations in a year depends on environmental factors. In most of the IGP States where rice is grown, the *S. incertulas* completes 3 to 4 generations. Temperature, day length and the growth stage of the host plants are the principal factors inducing diapause. *S. incertulas* larvae feed on mature plants and tend to enter diapause. The first generation of *S. incertulas* usually appears when the rice plants are still in the nursery or shortly after they are transplanted. The population increases in the subsequent broods and the second or later generations are the ones that build up to cause serious damage. In addition to seasonal fluctuations, distinct annual fluctuations in stem borer population also occur (Prakasa Rao, 1972).

Management of Yellow Stem Borer

Source: Israel et al., 1966; Khan, 1964; Chakrabarti, Kulshreshtha and Rao, 1971; Chatterjee and Prakasa Rao, 1975; Kalode and Israel, 1970; Raj and Morachan, 1973.

- Selection of crop varieties with even crop maturity. Susceptible, long duration, tall cultivars should be avoided.
- Planting time should be early to minimize stem borer abundance. Harvesting at ground level removes larvae in the stubble.
- Ploughing the field immediately after harvest kills larvae and pupae in the stubble.
- Using optimal rates of nitrogenous fertilizer in split applications.
- Applying slag increases the silica content of the crop making it more resistant.
- Use of resistant varieties (Table 3).
- Granular insecticides, viz., carbofuran, isazofos, diazinon, phorate, cartap hydrochloride, etc., @ 1.00 kg a.i./ha are effective against SBs. Amongst the sprayable insecticides–monocrotophos, chlorpyriphos and quinalphos @ 0.5 kg, phosphamidon @ 0.3 kg and triazophos @ 0.25 kg a.i./ha effective control of the SBs when applied at proper time.
- Augmentative release of *Trichogramma japonicum* @ 1,00,000 per ha after 30 days of transplanting and 3-6 subsequent releases are helpful for controlling *S. incertulas*.

White Backed Plant Hopper

The white backed plant hopper (WBPH), *Sogatella furcifera*, is an important pest of rice (Mochida, 1982). It is widely distributed in Bangladesh, India, Nepal and Pakistan. In India, WBPH attack on paddy was first reported from Surat, Pusa, Poona and Nagpur as early as 1903. Subsequently, it was observed in Bihar and Bengal, Jabalpur and other neighboring districts of Madhya Pradesh. WBPH appeared for the first time in 1966 in Punjab and in

Rajasthan in 1986. It is a major pest of rice in hilly tracts of Uttar Pradesh and in Haryana. It attains a higher level of incidence in Gujarat, Haryana, Rajasthan, Maharashtra, Madhya Pradesh and Punjab during mid-tillering phase in September/October (Dhakuwak and Singh, 1983). However, it was of no significance till early 1980s. In general, it is reported to be more severe in areas where resistant varieties of BPH have been grown.

S. furcifera has a wide host range. It has infested rice, finger millet, sorghum, wheat, wild rice and Leersia hexandra. The barnyard grass, Echinochloa crusgalli does not have antifeedant effect on WBPH as it does for BPH.

The WBPH is usually more abundant during the early stages of rice crop growth, especially in the nurseries. It attacks plants in fields with standing water and shows a marked increase with the age of the crop. The rice crop is more sensitive to attack at the tillering phase than at the boot and heading stages. Damages caused by the immigrants occur soon after their landing through feeding and egg laying. Under favorable conditions, WBPH produces several generations and can cause hopper-burn in the rice crop. Both nymphs and adults suck phloem sap causing reduced vigor, stunting, yellowing of leaves and delayed tillering and grain formation. It is seldom that rice plants are killed by WBPH except during the early stages of crop growth. However, yellowing of grown up rice plants have been reported. When the hoppers are present in large numbers late in the crop season, they are seen infesting the flag leaves and panicles. Gravid females cause additional damage by making ovipositional punctures in the leaf sheaths. Feeding punctures and lacerations caused by ovipositor predispose rice plants to pathogenic organisms and honey dew excretion encourages the growth of sooty mould such as Cladosporium spp., but not abundantly as in BPH.

Symptoms of rice plants attacked by WBPH are variable according to its population density, feeding duration, rice cultivars grown and crop growth stages. Damage in the form of "hopperburn" frequently appears uniformly over large areas of infected rice fields, compared to circular patches in the case of BPH. Fortunately, unlike BPH, S. furcifera is not a vector of any rice viral disease but was reported to be a vector of a viral disease of Pangola grass, Digitaria decumbens. Artificial infestation studies revealed that the grain loss varied from 11 to 39% when 15 insects/hill were released at varying stages of crop growth. Precise estimates on the damage caused by WBPH and the resultant losses are yet to be quantified in the field.

Outbreaks of WBPH have been recorded in recent times from several states in India. Over 1000 ha of rice were hopper-burned in Punjab during September 1983 and around 8000 ha in Cachar and Karimganj districts of Barak valley in Assam during May-June 1985. This outbreak was favored by high rainfall in early April followed by a prolonged dry spell with high temperature and humidity in the month of May. Damage to early rice during 1983 in Manipur was attributed to the unusually heavy rains and flooding in the Imphal valley at that time. In Karnataka, WBPH outbreak was reported during 1986 monsoon season in the Visveswaraya canal tract of Mandya. Higher N fertilization (130 kg/ha) with frequent heavy rains in September-October might have favored the WBPH outbreak. More than 80% damage to rice is common in West Bengal in areas of water stagnation. Weekly averages of 29°C temperature, 70% relative humidity, 8 hours of sunshine and 0-72 mm of rain were reported to favor WBPH outbreak in Delhi.

Management of White Backed Plant Hopper

Source: Chelliah and Bharathi, 1998; Israel, Kalode and Mishra, 1968; Rao and Kulshreshtha, 1985.

- Draining of water from rice fields.
- Dryland rice intercropped with cotton or pigeonpea had lower WBPH population.
- IR 48, IR 52, IR 60 and IR 62 rice varieties were found to be moderately resistant to WBPH.
- Use of parasites of WBPH, e.g., 11 hymenopterous insects as egg parasites, 5 hymenopterous, 2 strepsiptera, 1 dipterous insect, 4 nematodes and 1 mite as nymphal or nymphal-adult parasites.
- Use of predatory insects include egg predators: members of Lygaeoidea, Miridae (Hemiptera); and nymphal–adult predators: members of families Coccinellidae, Staphylinidae, Carabidae (Coleoptera), etc.
- Use of carbaryl WP 0.5-kg a.i./ha; endosulfan EC 0.35 kg a.i./ha; diazinon G 1.25 kg a.i./ha; phorate G 1.00 kg a.i./ha; carbofuran G 0.5 kg a.i./ha carbofuran and triazophos. Spraying of quinalphos, fenthion, chlorpyriphos, carbaryl or monocrotophos @ 0.5 kg a.i./ha or broadcasting of carbofuran 3G @ 0.75 kg a.i./ha or phorate 10 G 1.25 kg a.i./ha.

Rice Hispa

Rice hispa, *Dicladispa armigera* is a sporadic and occasional leaf-feeding pest and occurs in most of the rice growing areas. In certain areas, rice hispa is ranked as a major pest (Prakasa Rao, Israel and Rao, 1971). Sporadic outbreaks of rice hispa have been reported from many parts of Bangladesh, India, Nepal and Pakistan (Singh and Paroda, 1994).

Infestations of this pest has decreased in recent years due to the introduction of high yielding varieties and improved agronomic practices (Chelliah and Bharathi, 1998). A serious outbreak of hispa on rice hybrids and high yielding varieties of 30-35% infestation was reported in 1995-96 from western Orissa.

The larvae and the adult stages of the pest cause crop damage. In alternate years, rice hispa is found to cause serious damage in Punjab and Himachal Pradesh, coastal areas of West Bengal, Andhra Pradesh and Tamil Nadu. Under Orissa conditions, the severity of this pest has been marked in drought prone areas of Bolangir and Sundargarh besides its sporadic occurrence in inland districts.

Rice is the main host of this pest and other graminaceous weeds are its alternate hosts. During the off-season, rice hispa is found to breed on sugarcane, sorghum and wild grasses. Incidence of this pest usually occurs before flowering. Both beetle and grub feed on the green portion of the leaves causing characteristic linear patches along the veins. The yellowish grubs mine into the leaves presenting blister spots. It has also been reported that its grubs mine the rice leaves by feeding on the mesophyll between the veins and tunneling the tissue in the direction of the main axis of the leaf advancing towards the leaf-sheath (Acharya, 1967; Sen and Chakravorty, 1970). The adult feeds by scrapping the green matter and by removing the chlorophyll between the veins of the lamina, thus imparting white parallel streaks on the leaves. In certain cases adult beetles are also found to feed on the green portion of the leaves leaving only the epidermis. As a result, irregular/longitudinal white patches/blotches are produced. The pest prefers young plants mostly from the dorsal side of the leaves and so pest attack commences in the nursery itself. A single adult beetle may consume about 25 mm of leaf area/day. In severe cases of pest attack leaves look brown and field presents dried up appearance. Even rice re-transplanting may not be of much use as the pest present in the field may infest the newly planted seedlings. Average crop yield losses due to rice hispa varied from 6-65%. Economic threshold of rice hispa at early transplanting is 1 adult or 1 damaged leaf/hill and at mid-tillering is 1 adult or 1-2 damaged leaf/hill.

High humidity after rains and intermittent sunshine seem to favor hispa density. Rainfall was reported to show negative effect on the activity of hispa. Heavy rainfall in July followed by usually low rains in August and September, and uniform day and night temperatures was characteristic of epidemic years (Khan and Murthy, 1965).

Management of Rice Hispa

Source: Rao and Kulshrestha, 1985, Aggarwala, 1955.

- Removing grassy weeds in and around rice fields removes alternate hosts.
- Planting rice early at the start of the monsoon rains.

- Hand picking of damaged leaves removes grub from the field.
- Damaged leaves can be removed until booting.
- A piece of rope soaked in a mixture of 1 part kerosene and 1 part water can be pulled through the leaf canopy.
- Application of phorate to the rice nursery minimizes the infestation.
- In the main fields, application of monocrotophos or quinalphos or chlorpyriphos @ 0.5 kg a.i./ha is effective.

Rice Root Weevil

The rice root weevil (*Echinocnemus oryzae*) is a primary pest of the grasses and a secondary pest of rice particularly in the delta regions of the country during the monsoon seasons (Chakrabarti, Kulshreshtha and Rao, 1971). Earlier this was a minor pest of paddy but now has become an important pest in some river basins of India. The adults feed on newly transplanted rice, but seldom cause economic damage, while the grubs feed on fibrous roots during the monsoon season and infested plants become stunted and tillering is reduced. Plants attacked during tillering stage show more damage symptoms than plants attacked after tillering. Further, damage can be severe for a newly transplanted crop during July-August. In severe cases of infestation, plants wither and the fields show many patches that have to be filled by transplanting of new seedlings. Once the crop is established, the chances of withering are less. Yield losses may be up to 10%. Clay and heavy loam soils favor its incidence but the weevil does not occur in sandy soil. Chemical control using pre-sowing soil treatment or seed dressing is suggested for its management.

Brown Plant Hopper

The brown plant hopper (*Nilaparvata lugens*) is a common pest of rice (Das, Memmon and Christudas, 1972; Dyck and Thomas, 1979), however, on basmati rice, it has generally been recorded as a minor pest only. These insects congregate in large numbers, causing hopper burn. Infested plants become yellow and die. Hoppers transmit grassy stunt, ragged stunt, and wilted stunt virus diseases. Long-winged adults invade fields and lay eggs in leaf sheaths or midribs. Eggs have broad egg caps. Nymphs hatch in 7-9 days. There are 5 nymph instar. Nymph period lasts 13-15 days. First instar nymphs are white, later stages are brown. Short-winged and long-winged adults are produced. Short-winged adults dominate before flowering, and the females are found among tillers at the extreme base of hills. As the crop ages, long-winged forms capable of migrating are produced. Lot of work on their management has been reported which includes use of resistant varieties, biocontrol agents and pesticide control (Tables 2, 3 and 4).

Green Leafhopper

The green leafhopper (*Nephotettix virescens*) is widely distributed. They are important vectors of viruses that cause rice dwarf, transitory yellowing, tungro and yellow dwarf diseases. Adults are 3-5 mm long, bright green, with variable black markings. Eggs are deposited in the midrib of a leaf blade or sheath. There are 5 nymphal stages, also with variable markings. It can be managed by chemical and biological means (Tables 2, 3 and 4; Chelliah and Bharathi, 1998).

Gundhi Bug

The Gundhi bug (*Leptocorisa varicornis*) feed on milky rice grains. Both adults and nymphs pierce rice grains between the lemma and palea. Feeding during the milky stage results in empty grains. Feeding during the soft dough stage results in lower grain quality and broken grains. Adults are brown and slender with long legs and antennae. Eggs are deposited in rows on leaves and panicles. Both green nymphs and adults have a characteristic foul odor. Management is possible by chemical means, viz. dust with carbaryl, malathion or endosulfur.

DISEASES OF RICE

Rice diseases have also been primarily controlled either by the use of commercial fungicides (Table 5) or by the cultivation of disease resistant rice varieties in the IGP (Tables 4 and 6). In addition to these two approaches, management of water, fertilizers, weeds and soil amendments can greatly enhance disease management options.

Bacterial Blight of Rice

The bacterial blight of rice (Ishiyama 1922), *Xanthomonas oryzae*, has assumed a major pest status in recent years. In India, the introduction of modern rice cultivars and their response to nitrogenous fertilizer has favored the spread of bacterial blight in rice (Premalatha Dath, 1974). The disease was considered to be localized in Maharashtra and was not taken seriously until the epidemic of 1963 in Shahabad district of Bihar and other parts of IGP (Srivastava and Rao, 1963). This sudden flare up was due to the introduction of highly susceptible varieties. With the cultivation of large areas with high yielding but susceptible rice cultivars, the disease is now of common occurrence all over the IGP and is considered to be major hurdle in increasing

TABLE 5. Bio-efficacy of commercial fungicides available in India against major rice diseases.

Fungicide	Blast	Sheath Blight	Brown Spot	Sheath Rot
Organophosphorous group				
Hinosan 50 EC	***			
Kitazin 48 EC	**			
Kitazin 17 G	***			
Carbendazim group				
Bavistin 50 WP	***	***	**	**
Denosal 50 WP and others	***			
Others				
Dithane M 45 (75 WP)	*	*	***	**
Pyroquilon (Fongorene 50 WP)	***			
Tricyclazole (Beam 75 WP)		*		
Contaf 5 EC		***		
Tilt 25 EC		***		

Source: Reports of Directorate of Rice Research, Hyderabad, Several years.

rice yields. In 1979 and 1980, major epidemics occurred in Punjab, Haryana and Western Uttar Pradesh where severe Kresek was observed and total crop failure was reported.

Bacterial blight is a typical vascular wilt disease, while the leaf blight is a mild phase resulting from secondary infection. The partial or total blighting of leaves or complete wilting of affected tillers leads to unfilled grains. The wilt phase resulting from early nursery infection may result in complete crop death before maturity. It is not easy to have precise estimates of losses in grain yield due to lack of standard rating of disease severity, however, crop losses can be as high as 66% (Srivastava, Rao and Durgapal, 1966).

Bacterial blight of rice is of common occurrence both in temperate and tropical regions of the country. In the temperate region, the leaf blight phases, and in tropical region two additional types of symptoms is also common, i.e., Kresek (wilting phase) and pale yellow phase. The leaf blight phase is the predominant phase of the disease found to occur between tillering and head-ing stage of the crop and is mainly noticeable at the maximum tillering stage and is conspicuous at the heading stage. The appearance of dull greenish and water soaked or yellowish spots at the tip of the leaf margin are the early symptoms of the blight phase. With the progress of the disease, the lesions enlarge both in length and width, several lesions may coalesce to form straw or brown colored large lesions or blighted portions. The infected leaves turns yellow, dry rapidly and wither as the spot enlarge. Turbid ooze of the bacteri-um, streaming from the vascular bundles can be observed on immersing the cut affected leaves in clear water. In the Kresek (seedling blight or wilt

TABLE 6. Sources (donors) of resistance to blast, sheath blight, brown spot, BLB and RTV diseases, and released varieties of rice.

Disease	Donors	Released Varieties
Blast	Tetep, Tadukan, Zenith, Co 4, Dawn, Moroberekan, Correon, Dissi Hatif, Taride 1, Iac 25, Irat 3	Rasi, Akashi, Sasyasree, Vani, Improved Sona, North 18, Himdhan, Himalaya 1, Himalaya 2, K 332, K333, HPU 741, VL8, VLK 39, NLR 9672, NLR 9674, IR 8, IR 20, IR 36, IR 64, VL Dhan 221, Pant Dhan 10
Sheath Blight	T 141, OS 4, BCP 3, Saibham, Bhuhan, Saduwee, Laka, Ramedja, Tapoo-Cho-Z, Athebu Phourel, ARC 15368	Pankaj Swarnadhan, Manasarovar, Vikramarya
Brown Spot	T 141, BAM 10, Ch 13, Ch 45	Rasi, Jagannath, IR 36, IR 42, CNM 529, CNM 540
Bacterial Blight (BLB)	BJ 1, TKM 6, Lacrosee Zenith Nira, Java 14, Wase aikoku and several ARC	Mahsuri, Prasad, Ramakrishna, Saket 4, Sasyasree, IET 4141, CNM 540, IR 20, IR 54, IR 64, Ajay, Asha, Daya
Rice Tungro Virus (RTV)	PTB 18, ADT 21, ARC 10599, ARC 14320, ARC 14766 and others	Vikramarya, Radha, Nidhi

Source: Reports of Directorate of Rice Research, Hyderabad, Several years.

phase), the disease causes serious damage and is mainly of a sporadic occurrence. Infection starts one or two weeks after transplanting of rice. Pale yellowing and blighting of the leaves are the symptoms produced sequentially. The rolling and withering of the entire leaf, including leaf sheath follows. The bacterium reaches the growing point of young leaves, thereby killing the young plant. Kresek being the most serious form of the disease may result in the death of the whole plant or wilting of only a few leaves. The disease may also impart pale yellow color in leaves and may be found on mature plants, resembling iron deficiency (Goto, 1964).

The young leaves remain pale yellow or have a yellow or greenish yellow broad stripe, while the older leaves are green. Bacteria are not observed in pale yellow leaves but the bacterium can be deducted from the crown of the stem and from the internodes immediately below the infected leaves. The affected leaves remain intact but the symptoms resembling chlorosis are observed on leaf blade, indicating a chlorophyll deficiency (Srivastava and Rao, 1966).

Excess utilization of nitrogenous fertilizers, deficiency of phosphate and potash and excess silicate and magnesium favors the disease development

(Howlidar et al., 1984; Inoue and Tsuda, 1959; Kojima, Iwase and Inoki, 1955). The disease incidence is also influenced directly by total precipitation, the occurrence of heavy rain, and the presence of deep irrigation water and strong winds. The relatively high temperatures during growth of the rice crop increases incidence of this disease, but severe summer and drought suppresses the disease. Soil moisture at and above saturation favors development of Kresek phase of the disease. It was observed that pruning triggers blight infection and darkness favors the spread of the fungus inside the host.

Management of Bacterial Blight

Source: Devadath and Padmanabhan, 1970; Mehra, Thind and Mehra, 1994; Mondal et al., 1992; Padmanabhan, 1983; Srivastava, 1969; Thind and Mehra, 1992.

- Follow clean cultivation, destroy weeds around the rice fields.
- Avoid flooding or use of deep waters in the nursery.
- Avoid excess use of nitrogenous fertilizers, use balanced fertilization, potash application reduces the disease incidence.
- Use of resistant varieties (Table 6).
- Use seeds from disease free crop or disinfect seeds. Soaking seed for 8 hours in Agrimycin (0.25%) and wettable Ceresan (0.5%) followed by hot water treatment for 30 minutes at 52-54°C eradicates the bacterium in the seed.
- Spraying of copper fungicides alternatively with streptocycline (250 ppm).

Rice Blast

Blast of rice, *Pyricularia oryzae* is one of the earliest known important and widely distributed rice plant disease responsible for significant yield reduction. The first recorded report of the disease was from China in 1637. The disease was first recorded in India in the year 1913. A devastating epidemic was reported in the year 1914 from Tanjore delta area of Tamil Nadu, later from Maharastra, and subsequently from several other parts of India. It is now of common occurrence in almost all rice growing states of the country. The outbreak of the disease in epiphytotic proportions was reported in 1986 and 1989 from Haryana.

The blast disease reduces the number of mature panicles, grain and straw weight. Plants attacked by the disease between seedling and maximum tillering stages often die. In leaf blast phase loss in yield is due to reduction in spikelet number, grain weight and partially filled grains. When infected at early neck stage, grain filling does not occur and the panicle remains erect like a dead heart. Infection at a later stage results in partial grain filling. Loss in yield may be up to 65% due to blast disease. Its incidence on a large scale has been reported from the plains and peninsular India. A 1% panicle blast may result in a yield loss of 0.3%.

The fungus attacks all aerial parts of the plant and at all stages of growth but very rarely the plant roots. Leaves and neck of the panicle are more commonly affected. The lesions on leaves begin as small water soaked yellowish dots, which subsequently enlarge into spindle shaped spots varying in length from 0.5 to several cms. These spots enlarge quickly under moist conditions. Lesions may coalesce and kill the entire leaf in susceptible varieties. However, it may remain as minute, pin sized, brown specks on the resistant variety. Spots on susceptible cultivars show very little brown margin when grown under moist condition and on the contrary sometimes develop a yellow halo around the spots. The development of brown color is usually the resultant of a resistant reaction of the variety or the existence of some conditions unfavorable for lesion development. Seedlings or plants at the tillering stages are often completely killed in the field. An early seedling infection gives the crop a burnt appearance. Spots on leaf sheath resemble those on the leaf.

In case of neck blast, the panicle (neck) is infected during ear emergence, the neck region is blackened and shriveled, the culm below usually breaks and the upper part of the plant turn grayish brown, looking as if it has been burnt. Nodal blast is usually noticed after heading in which the sheath pulvinus rots and turns blackish and the culm may break at the infected node. In severe infections, the lesions may be noticed on internodes as well.

The development and suppression of the disease to a greater extent is directly influenced by the environmental factors, e.g., temperature, humidity and moisture. The conidia exhibit a nocturnal pattern of diurnal periodicity, with peak concentration of spore dispersal occurring around 4 a.m. favored by a night temperature of 25-27°C and RH of 86 to 98% (Sadasivan, Suryanaryanan, Ramakrishnan, 1965; Suryanaranyanan, 1966; Suzuki, 1975). In general sunshine inhibits development of blast fungus and spores fail to germinate in direct sunlight. Infection is favored more in darkness and is suppressed under diffused light. Cloudy overcast weather is reported to encourage blast spread. Leaf wetness has a direct effect on its development, longer the wet period greater is the infection. There is an inverse relationship of blast susceptibility to soil moisture conditions. Rice plants become susceptible when grown in dry soil, moderately resistant in moist soil and resistant under flooded conditions.

Heavy dose of nitrogenous fertilizers and soils deficient in silica enhance the incidence of blast. In Zn deficient soils, the disease incidence is reported to be high. Seedling and tillering stages are more prone to foliar blast and the flowering stage to neck/panicle infection.

Management of Rice Blast

Source: Chakrabarti, 1976; Govindhu, 1977, Tables 4, 5, and 6.

- Use seed from disease free crop.
- Treat seed with Cerasan or Agrosan GN before sowing @ 2-2.5 g/kg.
- Use of resistant or tolerant cultivars, e.g., Jaya, Vani, Akashi, Rasi, IR 8, IR 36, etc.
- Proper date of planting.
- Nitrogen and phosphate should be used judiciously. Application of lesser amount of nitrogenous fertilizer and in split doses reduces severity.
- Neem coated and neem blended urea release N slowly and reduces blast disease.
- Coating rice seed with *Cheatomium globosum* prevented the early infection.
- Use of systemic fungicides like Carbendazin, Benlate, Bavistin and MBC are effective in controlling blast.
- Rabcide 20% solution at 1.5 kg/ha spray controls both leaf and panicle blast. This fungicide offers inhibition to penetration. It has a long residual effect.
- The use of newer fungicides like Coratop 5G (Pyroquilon) and Kitazin 17G (IBP) may also prove promising in coming years, as they are easier to apply compared to conventional spray formulation.

Sheath Blight of Rice

The sheath blight of rice is caused by the fungal pathogen *Rhizoctonia solani*. The sheath blight was first reported in 1910 from Japan, between 1930-40 from Sri Lanka, China and India. The susceptibility of high yielding semi dwarf rice varieties to the disease is mainly responsible for it's becoming one of the important diseases of rice. With the increased use of fertilizers and fertilizer responsive high yielding cultivars, the sheath blight of rice which was earlier a lesser known disease has assumed significant importance in recent times in most of the rice growing regions (Kulshrestha et al. 1970; Kohli, 1966; Paracer and Chahal, 1963). Its distribution in Asia is the widest of all the rice diseases. Internationally, it is regarded as one of the most important disease of the rice crop. An epidemic of this disease in the year 1978 occurred in the Punjab State of India.

Seedling may be infected in the nursery if planted in infested soils. In the nursery, infection starts at the base of the plant and death of the seedling is observed in patches. The formation of high molecular weight compounds because of host pathogen interaction results in the blocking of cell membrane pores. In the later stages of infection, loss of chlorophyll is a common phenomenon and liberation of phenylacetic acid, meta- and para-hydroxyphenol acetic acids by the mycelia might result in the inhibition of secondary root formation. The rice seed may be infected and serve as a source of inoculum and may result in 4-6.6% seedling infection (Kannaiyan and Prasad, 1979). In India, losses of 5.2-50% have been reported but are dependent on the cultivar grown and the cultural conditions. In cases of high incidence of seed borne inoculum, the seedlings may rot at the base after 15 days of its sowing.

In transplanted rice fields, spots are first noticed near the waterline. The disease in the beginning is observed in the form of spots or lesions mostly on the leaf sheath. During favorable conditions, these spots extend further to the leaf blade. The spots are initially greenish gray, ellipsoidal or ovoid, irregularly elongated, and later become grayish white with a brown margin, varying in size from 1-3 cm long. In advanced stages the sclerotia are formed on or near these spots and matured sclerotia are detachable. Unfavorable conditions restrict the lesions to the lower leaf sheath, while in favorable conditions the disease progresses fast. Humid conditions trigger the mycelial growth of the fungus and also favor the coalescing of spots. The formation of large spots (as the spots coalesce) on a leaf sheath may result in the death of the whole leaf and in severe conditions all the affected leaves appear blighted.

The factors favoring vegetative growth of the plant, the dwarf, broadleaf and closed canopy characteristics of modern varieties, and congenial warm humid microclimate facilitates the physical contact of inoculum. The number of sclerotia that come in contact with the plant can directly be correlated with the severity of primary infection. The disease is severe under high humidity and warm temperatures. A temperature range of 28-32°C and RH of 81-97% is conducive for disease development. As per the work at CRRI, high N and/or P use increases the disease incidence and high K lowers it. Late maturing varieties are severely affected.

Management of Sheath Blight

Source: Das and Mishra, 1990; Gangopadhyay and Chakrabarti, 1982; Kasehm et al., 1994; Rao 1996; Savary et al., 1995.

- Draining the field will control the disease to a certain extent.
- Destruction of collateral hosts during off-season.
- Burning or destruction of crop residue also helps in reducing sclerotia from the soil.

- Balanced NPK, fertilization, split dose of N or use of its slow release form can help reduce the disease incidence.
- Green manuring, e.g., *Sesbania aculeata*, have been found to reduce the survival period of the sclerotia.
- Two foliar sprays of micronutrients, e.g., borax, zinc sulfate, copper sulfate and ferrous sulfate at 0.05% has been reported to reduce the disease incidence and to increase the grain yield.
- Soil amendments like oil cake, sawdust and rice husks have been found to be very effective in suppressing sheath blight.
- Use of resistant varieties (Table 6).
- Seed treatment with Thiram 75% a.i. @ 100 g/kg of rice seed controls the seed borne infection by controlling the disease and by increasing the germination percentage.
- In heavily infected patches, soil drenching with 0.1% wet Cerasan will be useful.
- Soil application @ 2.5 kg a.i./ha of Thiram or PCNB before transplanting.
- Various fungi such as *Gliocladium virens*, some species of *Trichoderma*, *Aspergillus* and others, and florescent bacteria can inhibit the growth and restrict the colonization of sheath blight pathogen in soil.
- Foliar sprays of IBP (Kitazin), edifenophos (Hinosan), benomyl (Benlate) carbendazim (Bavistin) and carboxin (Vitavax) has been found to reduce the rhizosphere population of *R. solani*.
- Pentachlorophenol used as weedicide in rice fields can also control sheath blight.

Brown Spot of Rice

Brown spot of rice is caused by *Helminthosporium oryzae* (= *Cochliobolus miyabenus*) and was found associated with the Bengal famine of 1942-43. The disease may result in heavy losses particularly in leaf spotting phase, which is of wide occurrence. Except during epiphytotic years, it is not likely to cause appreciable losses in crop yield. Cloudiness, humid weather, poor growth of plants and temperature range of 25-30°C favors the disease outbreak. Boot or flowering stage is more susceptible than other stages of the plant growth (Mishra, 1959; Mishra and Mukherjee, 1962; Sundararaman 1922; Suryanaranyanan, 1958).

Innumerable dark brown elliptical spots are of common occurrence on leaves, stem and glumes. At maturity these lesions/spots may exhibit a dark or reddish brown margin with light brown or gray center. In severe infections, the whole grain surface becomes blackened and the seeds are shriveled and discolored, inflorescence in a mature plant becomes distorted and grains fail to form. This disease is also known to cause blight of seedlings. Seedlings are often heavily attacked with numerous lesions and as a result the leaves may dry out.

Management of Brown Spot

Source: Chattopadhay, 1952; Mukherjee and Bagchi, 1965.

- Use of healthy or treated seeds.
- Complete destruction of collateral hosts.
- Removal of stubbles.
- Practicing soil amendments and proper fertilization.
- Use of resistant cultivar as per the recommendations for the local region (Table 6).
- Seed treatment with *Bacillus subtilis* enriched soil followed by two foliar sprays of bacterial suspension at 30 and 60 days of plant age.
- Use of boron, copper, manganese and zinc mixture as foliar spray is advocated for the management of air borne infection.

False Smut of Rice (Green Smut)

The causal agent of false smut of rice is *Ustilaginoidea virens*. This disease seldom reaches serious proportions and is regarded to be a disease of minor importance in rice growing areas. Plants are most vulnerable to infection at booting stage. Dampness promotes severe infection. Rainfall accompanied by cloudy days between flowering and maturity boosts the disease incidence. As a matter of belief, this disease is known to be more prevalent in seasons favorable for good growth and high yields, and the farmers in some of the areas consider its incidence as an indication of good harvest. The affected grains are transformed into greenish black masses and in general only a few grains in a panicle are infected (Govinda Rao and Gopala Raju, 1955; Manibhushan Rao, 1964).

Management of False Smut of Rice

Source: Manibhushan Rao, 1964; Mehrotra, 1980.

- Seed disinfection.
- The complete destruction of collateral hosts and spraying or dusting of copper fungicide on rice crop a few days before heading are recommended management measures.

Rice Tungro Virus

Stunting of the plant and discoloration of leaves characterize rice tungro virus. Its infection reduces tillering, number and length of panicles and number of spikelets and also delays maturation. The infected leaves accumulates high amounts of starch, this property of starch accumulation is useful in routine diagnostic tests. Tungro virus infection is generally known to delay flowering. The viral infection also results in incomplete grain filling and reduces grain yield. The two types of virus particles–spherical and bacilliform are found to be associated with the disease.

The incidence of rice tungro virus is dependent on the population dynamics of its vector, the rice green leafhopper (GLH), *Nephotettix virescens*. In Indian conditions, early summer showers favored the build up of the vector population. The vector population is reported to be abundant in post monsoon warm and humid climate (Raychaudhari, Mishra and Ghosh, 1927a and 1927b; John, 1968).

Management of Rice Tungro Virus

Source: Anjaneyulu, 1975; John, 1970; Mukhopadhyay and Chawdhury, 1973.

- Complete elimination of natural virus sources (e.g., rice stubbles–the main source of the virus, and should be ploughed under).
- Agronomic adjustments (susceptibility of the plant declines with age) and vector control (which can be viewed as one of the main option of the control strategy).

RICE NEMATODES

Nematodes cause unnoticeable losses because of the concomitant prevalence of pests, pathogens, weeds and other abiotic and biotic stresses in rice crop in the system. A number of ecto- and endoparasites of root, stem and foliar parts, e.g., *Aphlenchoides besseyi, Ditylenchus angustus, Meloidogyne* spp. *Hirschmanniella* spp. have been reported in the rice crop from the rice-wheat cropping system of the IGP (Bridge, Luc and Plowright, 1990). However, more than thirty nematode species have been recorded from rice fields (Dhawan and Pankaj, 1998; Panwar and Rao, 1998) and may cause losses of US$ 5.2 billion (Prot, 1993) in South and Southeast Asia. In India losses of 8% have been estimated. Several rice varieties have been recorded to be either tolerant to, or resistant to parasitic nematodes in India (Table 7). The next chapter of this publication (by S. B. Sharma) discusses the details of rice nematode spectrum in the IGP.

White Tip Nematode

White tip nematode, *Aphelenchoides besseyi* feeding ectoparasitically on leaf-sheath and flowers, is an important disease of rice in the rice-wheat

TABLE 7. Rice varieties recorded as resistant or tolerant to parasitic nematodes.

Nematode	Rice Varieties
Rice-Root Nematode	RP 1155-128-1, C1R 142-2-2, CR 141-6058-1-35 (Rao & Rao 1971), Ptb.27, Annapurna, CR 130-203, CR 44-140-2-1051, CR 320, CR 294-548, Mtu. 28, CR 52, N 136, 2 113, Tin Pakia, CR 44-32.
Root Knot Nematode	TKM-6, Patnai-6, N 126, AC-13, Ptb. 10, CR 3047, Cult. 377, CT 426, CR 289-1008, CRM 13-3511-1, CR 194-523, CR 407-1, CT 401-8, CR 294-28, Mtu mut 288, CR 329-36, CR 256-10-4-34, CR 266-15-440-534-196-57, CR 256-30-393-19, CRKR-2, CR 256-30-978, CR 1014 mut, CR 208-5340, CR 147-2-2-2 (Rao et al. 1984), Hamsa, Akashi Lalnakanda -41, TNAU 13615, CR 288-5387, CR 146-4005, CR 146-7001, CR 146-4005, CR 401-7, CR 407-19, CR 407-6-2, CR 407-4, CR 260-312, CR 254-43-1-6 MW-10, Udaya, IR 36, Daya.
White Tip	Gurmatia, Chinoor (dastur 1936), Co 13, Rohini. T. 292, Rp. 113-31 (1%), Ratna, Triveni, IET 338, TKM 7, T.482, Asd 14, Yamuna, Padma IR 5 (2%) SR 26B, Cult. 7711, Vijjaya, Sona, Soorya, IR 30 (3%), Krishna, Bala (4%), Cult 377, CR 292-7050, CR 309x, 1.W., CRM 10-4078-68-73-36, CR 260-40, CR 295-81-171, (Anon 1971-76), Annapurna, Mdu 1, Asd 15, Bhagyavati, Kaveri, BK 190, Chambal, IR 579, Himalaya 1, PNB, Basumati, SSM 591, PNL 530, KJT. 14, KJT. 24, IGPI-37, RTN 68, R.24, GR.11, Indira (Anon 1984), TPS.1, TP. 1974, TKM.9, Patyur-1, HKR-1.
Stem Nematode	Rayada 16-06 and 16-08.

Source: Reports of Directorate of Rice Research, Hyderabad, Several years; Prasad, Panwar and Rao, 1992.

cropping system (Bridge, Luc and Plowright, 1990; Srivastava, 1966). This nematode has been reported from Orissa, Tamilnadu, Uttar Pradesh, and West Bengal in India (Rao, 1985; Sehgal, 1999; Rao and Yadava, 1971). The nematode was also recorded from deep-water rice in Bangladesh (Rahman and McGeachie, 1982), Nepal (Karki, 1997), and Pakistan. The infected plants have withered leaves from the tip downwards, giving a whip-like appearance to the leaf tops during the wet season, whereas in the dry season, the leaves turn bronze in color. The plants appear stunted and tillering is reduced. At the heading stage, the margins of the leaf and the sheath, especially those of the leaf flag are curled, and the emergence of the earheads is incomplete. Flowers do not open and the ovary remains shriveled (Subramaniam et al., 1973). The length and the number of spikelets are reduced. Temperatures of 26 to 32°C (maximum) and 15 to 22°C (minimum), with no rainfall, and dew deposits of 21 mm on the foliage during the heading stage of the crop favored the migration of the nematodes to the glumes (Rao and Nandakumar, 1973). The earheads emerging from the stubble of the ratoon crop and the weed *Echinochloa colona* carry these nematodes.

Management of White Tip Nematode

S Avoid use of nematode infested seeds.
S Immersing seeds in hot water at 45_C for 15 minutes. The seeds should be dried properly after the treatment so that the germination is not affected.
S Seed treatment with acetic thiocyanatoester REE 200, benlate 50, diazinon, malathion, parathion, phosphomidon or thiabendazole.
S Use of available resistant germplasm may be effective (Table 7).

Rice Root Nematode

Rice root nematode, *Hirschmanniella* spp. is an important pest of the rice crop, widely distributed in rice growing tracts (Figure 1), especially the low land and irrigated paddy fields (Haii 1969; Mathur and Prasad, 1971; Prasad, Panwar and Rao, 1992; Sehgal, Kalra and Gaur, 2000). Three species namely, *H. oryzae, H. mucronata* and *H. gracilis* are more commonly encountered (Panwar and Rao, 1998; Sehgal and Gaur, 1999). There are no specific symptoms caused by this nematode, however, the infested roots show small brownish spindle shaped water soaked lesions at the point of penetration of juveniles and the adult nematodes. The nematode damages the cortical cells and migrates to new cells creating numerous cavities in the cortex (Jairajpuri and Baqri, 1991). Infected plants gradually turn yellow and wither. The roots harbor large number of nematodes. Other pathogen soon invades the injured roots. The rice plants remain stunted with few tillers and poor grain filling reuslting in 10-36% yield loss (Ahmad, Das and Baqri, 1984; Panda and Rao, 1969; Rao and Panda, 1970). These nematodes have also been isolated from rice leaves affected by bacterial blight. The prevalence and incidence of this nematode is influenced by the soil, crop variety, and management strategies followed (Birat, 1965; Rao et al., 1971).

Management of Rice Root Nematode

Source: Rao et al., 1984; Prasad and Rao, 1984; Sehgal and Gaur, 1999.

S Deep summer ploughing, rotation with groundnut, wheat, jute, potato or blackgram.
S Pre-sowing or pre-planting soil treatment with carbofuran or phorate 2 kg/a.i. either once or split at planting.

Stem Nematode

Stem nematode, *Ditylenchus angustus* is an obligate parasite and a serious pest of rice in the rice-wheat cropping system (Bridge, Luc and Plowright,

FIGURE 1. Schematic maps (not to scale) showing distribution of *Hirschmanniella* (Figure 1A) and *Meloidogyne* (Figure 1B) nematode species in India (ETL refers to Economic Threshold Levels).

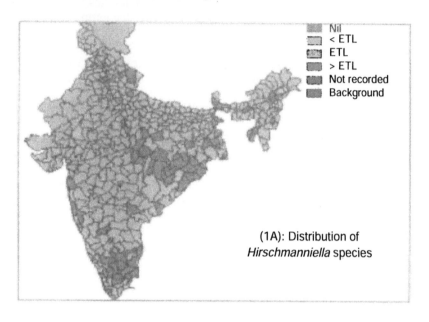

Nil
< ETL
ETL
> ETL
Not recorded
Background

(1A): Distribution of
Hirschmanniella species

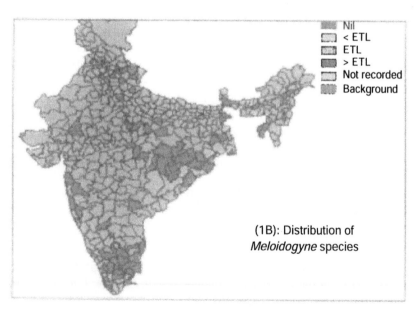

Nil
< ETL
ETL
> ETL
Not recorded
Background

(1B): Distribution of
Meloidogyne species

1990; Sharma, Price and Bridge, 1997). In 1919, Butler reported the disease as "ufra" disease from the IGP in northern India. Yield losses of 50% due to its infestation have been recorded. It is widespread in northeastern and eastern India. The nematode feeds ectoparasitically on young stem and leaves, leaf sheath, panicles and spikelets. Symptoms are difficult to detect in the first eight weeks of crop growth. Chlorosis may be observed in the infested young seedlings within a week after infection. The chlorotic portion of leaf becomes brown to dark brown. The twisting of leaf and leaf sheath is common. Sometimes the leaf margins become corrugated and many branches may grow from the infested node, which gives a bushy appearance to the plant. The panicles may remain enclosed within the flag leaf sheath or may emerge partially or fully from it. Fertility and grain filling are adversely affected and grains may also be deformed with a brownish color.

Management of Stem Nematode

Source: Prasad, Panwar and Rao, 1992; Sehgal and Gaur, 1999.

- Burning of all stubbles, ratoons, wild rice and weeds after harvesting the crop.
- Crop rotation with mustard reduces its population (Rao, Prasad and Panwar 1986).
- Growing jute in the ufra infested field also reduces the nematode population.
- Root-dip treatment of seedlings in Miral 3% and Tecto 40FL @ 7.5-10% may reduce the nematode population.
- Varieties of deep-water rice and wild rice have also been reported to be resistant to ufra disease (Table 7).

Root-Knot Nematode

The root knot nematode, *Meloidogyne* spp. has been reported (Figure 1) from various parts of India. *M. graminicola* and *M. triticoryzae* are mainly distributed in the rice growing states. *M. triticoryzae* have been reported from Delhi, Haryana and some parts of Uttar Pradesh (Gaur, Khan and Sehgal, 1993; Gaur, Shah and Khan, 1993) as important pests of the rice-wheat cropping system. Occurrence of other root nematode spp. *M. triticoryzae, M. incognita, M. javanica* and *M. arenaria*, etc., have been also reported from rice growing areas of India and neighboring countries.

The leaves of the infected plants become discolored. Leaves or leaf tips dry or change to bronze color from the margins towards the midrib (Israel, Rao and Rao, 1963). The infected plants are stunted and there is a delay or reduction in flowering. The root-knot nematode spp. causes swelling and gall formation in the root of the crop. The nematode spp. can survive up to 900 days even in the absence of its hosts (Sehgal, 1991). The nematode activity was more pronounced on soil with acidic pH. An increase in the production of root knots and egg masses occurred with the application of 40 kg of N/ha but at higher doses the nematode frequency and density was reduced.

Management of Root-Knot Nematode

Source: Rao and Panda, 1970; Sehgal and Gaur, 1999; Gaur et al., 1998.

- Soil solarization of the nursery.
- Flooding of the field.
- Crop rotation.
- Use of soil amendments.
- Use of Carbofuran and Dasanit has been effective for its control.

INSECT PESTS OF WHEAT

The status of the common insects, diseases and nematode pests of wheat is shown in Table 8. The next chapter of this publication (by S. B. Sharma) also discusses details of nematode damage in wheat. In general, pest attack is higher during the hot and humid rice season, compared to the cooler wheat season (Dubin and VanGinkel, 1991).

Twenty-three insect pests have been recorded in the wheat crop, of which termites and to some extent Gujhia weevil were recorded as economically important pests. But in the last few years there have been several outbreaks of otherwise sporadic pests like cutworm, armyworm, flea beetles, and sugarcane leafhopper. Pests like wheat mite, aphids and shootfly have become regular in occurrence. In the rice-wheat rotation, termites continue to be serious pests in the sub-tropical areas of the IGP while the Gujhia weevil has assumed a non-pest status (Bhatia 1978; Singh 1999).

Termites

The termites, *Microtermes obesi, Odontotermes obesus* and 14 other species are serious traditional polyphagous pests of the wheat crop causing damage at all crop growth stages from planting till harvest, especially in unirrigated areas with light soils. The feeding workers feed on roots and underground parts of the stem and result in the drying up of the plant and tillers.

TABLE 8. Common insects, diseases and nematodes of wheat.

Pest	Status
Insect	
Termite, *Odontotermes obesus*	***
Aphid, *Macrsiphum avenae*	**
Pink stem borer, *Sesamia inferens*	**
Army worm, *Mythimna separata*	**
Disease	
Loose smut, *Ustilago tritici*	***
Karnal bunt, *Neovossia indica*	***
Leaf blight, *Alternaria triticana, Helminthosporium* sp.	***
Leaf rust, *Puccinia recondita*	*
Stripe rust, *P. striiformis*	**
Stem rust, *P. graminis*	**
Powdery mildew, *Erysiphe graminis tritici*	**
Head blight, *Fusarium* spp.	**
Nematode	
Cereal cyst nematode, *Heterodera avenae*	***
Seed gall nematode, *Anguina tritici*	**
Root-knot nematode, *Meloidogyne triticoryzae*; *M. incognita*	**

*Low; **Low to Moderate; ***Moderate to Medium.
Source: Garg and Jeswani,1999; Sehgal and Gaur, 1999.

Management of Termites

Source: Bhatia, 1978; Sahni and Bhutani, 1966; Singh, 1999.

- Destruction of above ground termitaria.
- Pre-sowing soil treatment with chlorphyriphos @ 800 g a.i. followed by light irrigation.
- Seed treatment with endosulfan or chlorphyriphos @ 240 and 90 g a.i./ quintal of seed.
- Use of rottened farmyard manure.

Gujhia Weevil

Gujhia weevil, *Tanymecus indicus*, causes injury to the plant that resembles termite injury. This pest is a sporadically serious pest of wheat seedlings, roots or stems below the soil surface, the injury being more severe at the seedling sage. Adults occasionally feed on the leaves. The pest threatened the wheat growing areas of Punjab, Haryana, Madhya Pradesh and Uttar Pradesh. But the introduction of dwarf varieties and irrigation resulted in its decrease over time (Bindra and Singh, 1970).

Management of Gujhia Weevil

Source: Singh, 1999.

- Irrigation at the crown root stage of crop growth.
- Early sowing of the crop.

Cutworm

Three species of cutworms, i.e., *Cirphis loreyi, Agrotis ipsilon* and *Euxoa spirifera*, are polyphagous pests. Their serious outbreaks were observed only in the post-green revolution period in all the wheat-growing areas of the country. The caterpillars that remain hidden under the crop debris and soil clods during the day, cuts and feeds on the leaves and shoot during the night time. Cultivation of dwarf wheat varieties and high soil moisture status is conducive to the multiplication of the pest.

Management of Cutworm

Source: Singh, 1999.

- Mid-October to mid-November sowing of the wheat crop.
- Balanced use of fertilizers.
- Foliar sprays with fenitrothion or dichlorvos @ 50ml a.i./ha or carbaryl @ 1 kg a.i./ha.

Shoot Fly

Shootfly species include *Atherigona naqvii, A. oryzae, and A. bituberaculata* dipterous insect species. The predominant species, *A. naqvii* has been recorded as an important pest of wheat crop during the post-green revolution era. The insect flies are active during the daytime and lay their eggs on the stem base. Maggots that grow with the leaf sheath, damage the crop by cutting the meristematic portion of the stem (Khan and Sharma, 1974).

The crop variety, irrigation schedule, nitrogen application rates and sowing date influence the level of its infestation. Dwarf varieties are more susceptible than taller varieties.

Management of Shoot Fly

- Avoid late sowing of the crop.
- Application of carbofuran granules or the foliar spray of cypermethrin (50 g a.i./ha).

- Use of resistant varieties: UPD8, UP 85042, PDW 215, PDW 206, Raj 3214 and HUW 206.

Army Worm

Armyworm, *Mythimna separata* has become an economically important pest in all the wheat growing areas of the IGP with the introduction of dwarf wheat varieties. In the last three decades, several outbreaks of the armyworm were recorded in Punjab (1970, 1974 and 1980), Rajasthan (1971), Haryana (1970) and Delhi (1968, 1969). Its appearance in large numbers have been reported from other states and also crops like sugarcane, rice, maize, sorghum, barley and oats. The larvae that thrive and feed on the leaves may reduce the leaf number and sometime completely defoliate the leaves during night. It can also damage the earheads.

Sowing date, soil moisture, heavy rains followed by drought, nitrogen application, and flooding has been found to be conducive for the growth, development and multiplication of the pest. Army worm can be controlled by foliar sprays with fenitrothion or dichlorovous @ 500 ml a.i./ha.

Mites

The brown mite, *Petrobia latens*, which is found to be widely distributed in the wheat growing areas, emerged as a pest in the IGP only after the introduction of dwarf, high yielding wheat varieties. It may cause severe crop loss in the non-irrigated crop area and is a phytophagous pest but remain in the soil and feed only during the daytime.

The pest feeds on the plant leaves, stem and earheads, the infestation starts from the lower leaves and spreads upwards. The stunted growth and grain shriveling can be mistaken for water stressed conditions. The mite incidence usually starts December-end and first week of January and continues till crop maturity. The level of its incidence depends upon the hatching of eggs which coincides with the fall in temperature below 15°C and R.H. of 75% and above. The mite population builds up during the dry, cold and sunny winter weather.

Management of Mites

Source: Doel and Sandhu, 1974; Singh, 1999.

- Use foliar sprays, e.g., Formothion (@160 ml a.i./ha), Phosphamidon (250 ml a.i./ha), Oxydementon methyl (160 ml a.i./ha) or Monocrotophos (200 ml a.i./ha), two-three times at an interval of 15 days, depending upon the ETL.
- Use resistant lines C306, AP-30-1, WH 629, CPAN 2092 and ISW 38, etc.

Aphids

A number of sucking pests, e.g., *Macrosiphum avenae, Rhopalosiphum maidis*, etc., have emerged as sporadic pests on the high yielding dwarf varieties which are grown with high fertilizer doses and irrigation (Chaudhary, Ramzan and Atwal, 1968). Of these pest species, the foliar feeding species *Macrosiphum* and *Rhopalosiphum* are the predominant pests. In this present rice-wheat system, they are recorded as regular pest from the Punjab, Haryana, Uttar Pradesh, Delhi, Rajasthan, Bihar, Madhya Pradesh and Maharashtra states of India. Delayed sowing and cloudy conditions favor the multiplication of the aphids, but its population decreases with rainfall (Dhaliwal and Singh, 1975). If the infestation continues till crop maturity and the damage to the developing grain is serious, wheat yields can be reduced by 26%, depending upon the aphid population (Grewal and Bains, 1975). Adjusting sowing date of crop, and use of phosphamidon, oxydementon methyl, formothion and dimethoate effectively controls aphids.

Pink Stem Borer

The pink stem borer, *Sesamia inferens*, is a lepidopterous pest which feeds on the major cereal crop, e.g., wheat, rice, maize, millets and sugarcane. This pest has been recorded from all the wheat growing areas and is of regular occurrence in Punjab, Haryana, Uttar Pradesh, Bihar, Madhya Pradesh, Rajasthan, Gujarat, Maharashtra and Karnataka states of India. The level of its incidence is increasing in the rice-wheat zone of the IGP. The larva feeds within the stem resulting in empty white heads at crop maturity. The level of its incidence varies with the crop variety and the date of sowing, and the cropping pattern. Staggered sowing and uneven fallowing may cause more incidences in the late sown crops. Pink stem borer infestation can be controlled by hand picking of infested tillers and burning of stubbles after harvest.

Gram Pod Borer

The gram pod borer, *Helicoverpa armigera*, a polyphagous pest of cotton, chickpea and pigeonpea has now become a regular pest of the wheat crop in Punjab, Haryana, Delhi, Uttar Pradesh, Madhya Pradesh, Rajasthan, Gujarat and Jammu and Kashmir (Doval, Bhora and Sharma, 1972). However, in 1968-69 and 1973, it was a serious problem in wheat in Rajasthan and Punjab, respectively. The larva feeds on the earhead and reduces the quality of grains. Staggered sowing and frequent irrigation favors its incidence.

Semiloopers

The infestation of the semilooper species, e.g., *Plusia orichalcea* and *P. nigrisignis*, which are distributed throughout the country, is sporadic in occurrence. They remain hidden in the soil and crop debris and feed on the plant at nighttime and may cause 1-2% economic loss of the crop. The management practice recommended against the armyworm is also effective for controlling the semiloopers.

Sugarcane Leaf Hopper

Pyrilla perpusilla, i.e., the sugarcane leaf hopper which is a serious pest of the sugarcane crop, have been reported from the wheat fields in 1973 and 1985 and has seriously affected wheat cultivation in the Uttar Pradesh state of India. The pest has a tendency to shift from the sugarcane fields to the nearby wheat fields, after the sugarcane crop is harvested. The nymphs and adults feed on the leaves and earheads and can cause severe damage to the crop in the initial growing period. Foliar sprays of fenitrothion or dichlorovous @ 500 ml a.i./ha is an effective control measure against the sugarcane leaf hopper.

DISEASES OF WHEAT

Karnal Bunt

Karnal bunt of wheat which is caused by *Neovossia indica*, occurs sporadically, but may cause huge losses in epidemic years reducing wheat yields by 10 to 30 percent (Joshi et al., 1983). The disease was first detected in experimental seed material at Karnal, Haryana, India by Mitra, 1937, and was named as Karnal bunt. The disease was localized till 1970 to Western Uttar Pradesh and Punjab but spread with the adoption of high yielding, dwarf mexican wheat varieties in India and Pakistan and was also reported from Nepal (Hasan, 1973; Singh, Dhaliwal and Metzgar, 1988, Swaminathan et al., 1971; Singh et al., 1989). The causal organism also affected the seed germination, plant vigor, seed viability, wheat flour quality (Bansal, Singh and Joshi, 1984; Bhat et al., 1981; Warham, 1984, 1990 a&b; Warkham, Prescott and Griffith, 1989; Sekhon et al., 1980).

The disease is first visible only when the grains are formed. The grains are partially or wholly converted into a black powdery mass enclosed by the pericarp. All the grains in an ear may not be uniformly infected. Hence it was also known as "Partial bunt" (Singh, 1986).

This disease was reported to be soil borne but later was also shown to be air borne (Bedi, Sikka and Mundkur, 1949; Chona et al., 1963; Mundkar, 1949; Mitra, 1931a and 1931b; Mitra, 1937). The spore remains in the soil (Singh, 1986; Singh, Joshi and Srivastava, 1983) and germinates in the month of February or March under favorable moisture and temperature conditions. High dose of nitrogenous fertilizers predisposes the plants to infection (Munjal, 1971; Singh, Srivastava and Joshi, 1980). The sporidia which develops under these conditions are spread by wind and infects the wheat grains through the ovary wall (Dhaliwal and Singh, 1989; Aujla, Kaur and Gill, 1990; Aujla et al., 1990; Munjal and Chatrath, 1976). The boot leaf serves as the catchment area for the sporidia and serves as a reservoir (Nagarajan, 1991). Irrigation or rains may enhance the probability of infection (Bedi, Sikka and Mundkar, 1949). The causal organism, *N. indica* thrive on *Triticum aestivum* but can also survive on other *Triticale* spp., e.g., Durum wheat, wild species of wheat (Aujla et al. 1985 and 1986; Aujla, Sharma and Grewal, 1979; Royer and Rytter, 1988).

Management of Karnal Bunt

Source: Aggarwal et al., 1996; Amer, 1995; Mitra, 1937; Bedi, Sikka and Mundkar, 1949; Singh and Rai, 1986; Singh and Singh, 1985; Singh and Prasad, 1978; Singh, Srivastava and Joshi, 1979; Singh et al., 1999; Singh, Dhaliwal and Gill, 1989; Singh, Tewari and Rai, 1985.

- Non-cultivation of wheat for two consecutive years.
- Proper adjustment of water and balanced use of nitrogenous fertilizers.
- Avoid irrigation at the time of flowering.
- Adjustment of sowing date as per agro-climatic conditions.
- Cultivation of resistant varieties.

Loose Smut of Wheat

Loose smut of wheat, caused by *Ustilago tritici*, is widely distributed in the IGP and is a severe problem in the humid and semi-humid areas. The smuts are considered next to rusts in causing economic damage to crops. It may cause 3-100% losses in the newer high yielding varieties and its incidence increases with the cultivation of susceptible varieties (Mishra and Singh, 1969; Patel, Dhande and Kulkarni, 1950; Srivastava et al., 1992; Tandon and Sethi, 1991; Joshi, Singh and Srivastava, 1985; Person, 1954).

Prior to booting and ear emergence, the infected plants appear similar to the healthy plants. The symptoms of the disease only appear at the time of ear emergence. The diseased ears emerge from the boot leaf slightly earlier than the healthy ears and form a black powdery mass. A smooth silvery membrane

covers this powdery mass of spores. The spores are released with the ruptur-
ing of the membrane. The infected plants have reduced numbers and heights
of tillers (Goel et al., 1977). In high yielding varieties, prominent yellow and
chlorotic streaks are visible on the flag leaf (Aggarwal et al., 1982; Chatrath,
Thomas and Mohan, 1976; Singh, Yene and Chaube, 1974). The transmission
of loose smut was reported to be maximum at 23°C (Dean, 1969) and higher
temperatures adversely affected the disease expression.

Management of Loose Smut

Source: Aggarwal, Chahal and Mathur, 1993; Khanzada and Mathur,
1984; Srivastava, Singh and Aggarwal, 1998.

• Use of resistant wheat varieties.
• Eradication of loose smut infected seeds.
• Hot water treatment of seeds at 49°C for 2 minutes.
• Chemotherapy with Benomyl, Carboxin, Carbendazion, Fenfuram, Tri-
  diminol, Triademfean, Baytan, Dichlopentazole or Raxil @ 2 g/kg of seed.
• Use of *Trichoderma viridae*, *Gliocladium deliquescense* for seed dress-
  ing.

Rusts

Wheat rusts are destructive pathogens and may cause 6% crop loss at
harvest (Mehta, 1940) and under favorable conditions, the crop losses are
augmented (Joshi 1982; Mehta, 1942; Nagarajan and Joshi, 1975; Saari and
Prescott, 1985; Stakman and Harrar, 1957). In India, rust epidemic occurred
as early as 1827, in Punjab (1907), and again in Bihar (1957), and northwestern part of India (1971), causing huge yield losses (Joshi et al., 1980). In
1947, the rust epidemic in Madhya Pradesh almost resulted in a famine. All
the three types of rusts, i.e., stem, leaf and stripe rusts are known to attack the
wheat crop in the IGP.

Stem Rust

Stem rust or black rust caused by *Puccinia graminis* occurs in almost all
the wheat growing areas of the IGP. The stem rust is macrocyclic, heteroecious disease and completes its life cycle on wheat and its alternate host
Berberis vulgaris. In northern India, the rust may appear in the beginning of
March at dough stage of crop growth and therefore may not be very severe.
But in the peninsular and southern parts of the country the disease may
appear at the end of November or first week of December and may cause

huge crop losses. In this type of rust infection, the stems are most severely attacked followed by the leaf sheaths, leaves, ears, glumes and awns. There are two types of pustules, i.e., the uredio and the telio pustules. The uredio pustules first appear on the culm, leaf sheath, are elongated, brick red to brown in color. These pustules coalesce and dehisce early by breaking the epidermis and spreading by wind. Later in the season the telia pustules are produced which are black in color and have a long resting phase before they germinate to form promycelium bearing single celled basidospores of two different strains namely + and −, each being self-incompatible. This mycelium proliferates and forms cup shaped aecia by a peridium. Under high humidity conditions, the aeciospores are ejected from the lower portion of the leaves and are incapable of infecting *Berberis*, but capable of infecting a number of grasses, e.g., *Aegilops squarrosa, A. trinecilis, A. ventricosa, Agropyron semicostatum, Bromus carinatus, B. coloratus, B. mollis, Hilaria jamesii, Hordeum destichum, H. murinum, H. stenostachys* and *Lolium perenne*, etc. (Prasada, 1951). Under Indian conditions, the organism can survive, multiple and develop under repeated urediospores (Mehta, 1940).

Leaf Rust

Leaf or brown rust caused by *P. recondita* is also heteroecious, occurring in almost all the wheat growing areas of India. The severity of the disease varies with the climatic conditions. Epidemic of this rust occurred in the northwestern parts of the country in 1970-71 and 1972-73. Leaf rust attacks leaves almost exclusively, but rarely the leaf sheath, stem, awns, glumes, peduncles and internodes. The uredopustule develop on the leaf sheath and the stalks burst on the upper surface. The process of penetrations is the same as in black rust.

The organism is normally observed in January. The rust appears slightly earlier and establishes itself in the foothills of the Himalayas in northern parts of the country and in the plains of Tamilnadu and Karnataka in the southern part. The organism, which develops and multiplies in the southern parts of the country moves north into the states of Madhya Pradesh and Maharastra. Simultaneously, the organism from the northern part of the country spreads to the northern plains and further south. During favorable climatic conditions, the populations from the north and southern parts merge and can cause a rust epidemic. The spread of the rust can also be predicted on the basis of Indian stem rust rules (Nagarajan and Joshi, 1985; Nagarajan and Singh, 1975). The leaf rust pustules are smaller than the stem rust pustules, circular to rectangular in shape. The pustules are produced from both the leaf surfaces, but many from the upper leaf surface. These uredospores are wind borne and they burst early.

Stripe Rust

Stripe or yellow rust, one the most destructive disease is caused by *P. striiformis*, is heteroecious, yet no alternative host is known. However, many new pathotypes have been identified (Nayar, Nagarajan and Bahadur, 1988; Sharma and Prasada, 1967). The disease is mainly restricted to the north and northwestern part of the IGP and is prevalent in the low temperature region. The disease is susceptible to hot climatic conditions, is partially systemic, and can spread from one leaf to another. It appears in the foothills of India in the last week of December or the first week of January every year, and spreads along the wheat belt (Nagarajan et al., 1980).

The pustules appear as stripes on leaves, stalk and glumes. The color of the leaves fades, on which small uredosori appear. The uredosori are oval, lemon or yellow color. With time and under favorable climatic conditions, these pustules rupture and are blown by wind. Black patches in rows of teliospores are produced. When the temperature is low, the infection remains dormant but becomes apparent under favorable climatic conditions (Nayar and Bhardwaj, 1998). During January-February, the spread of this rust is much faster compared to brown rust. However, with the increase in temperature, the spread of the disease is checked. More than 300 grass species are infected by stripe rust but the exact role of these collateral hosts in the perpetuation and spread of the disease is not clearly understood in India (Stubbs, 1985).

Management of Rusts

Source: Joshi, 1982; Nagarajan, Nayar and Bhardwaj, 1992; Nayar et al., 1990; Rowell, 1985.

- Use of recommended resistant cultivars as per agroclimatic conditions.
- Eradication of alternate and over-summering hosts.
- Mixed cropping of wheat with pea, lentil, gram, linseed, mustard.
- Balanced use of fertilizers. Reduced N in NPK ratio and use of phosphate fertilizers.
- Use of biological control agents, e.g., *Sphaerellopsis filum*, *Verticillium lecanii*.

Leaf Blight of Wheat

Leaf blight of wheat, caused by *Alternaria triticana*, is a common and economically important disease in the entire rice-wheat growing IGP region. The disease was first reported as early as in 1924. The disease starts as a small, discolored, irregular lesion brown to grey in color, and varies in size.

The lesions can attain a size of 1 cm or more, and be surrounded by a light yellow zone, which later coalesce to cover lateral areas of the leaf. When severely infected, the whole leaf may die. Black colored masses of conidia are produced on the surface of the leaves. The estimated crop loss due to *A. triticana* may vary from 20 to 30% (Parashar et al., 1995) and under artificially created severe infection, total crop failure for susceptible varieties may occur (Chenulu and Singh, 1964).

Management of Leaf Blight

Source: Goel et al., 1999; Mahto 1999.

* Cultivation of resistant varieties.
* Three sprays of Tilt (Propiconazole) 250 EC @125 ml a.i./ha at two week interval.

Spot Blotch of Wheat

The spot blotch or foliar blight caused by *Helminthosporium sativum* is a highly virulent pathogen of the IGP region where rice-wheat crop rotation is practiced. The disease has been reported from Bangladesh, India and Nepal (Alam et al., 1993; Bimb and Mahto, 1992; Goel et al., 1999). In Bangladesh, yield losses due this disease are about 20% (Razzaque and Hossain, 1991) and in Nepal as high as 27% (Devkota, 1993).

The disease appears as small light brown spots distributed throughout the leaf blade, leaf sheath, nodes, internodes and glumes, slowly increasing in size and into oblong or elliptical shapes. The infections then changes into light brown color with dark brown margins. On maturity these results in dark brown margins with light brown center. The size of the infections changes with the crop variety, climate and the age of the plant. In susceptible varieties and under favorable conditions, the infected leaves dry prematurely and may shed off.

Head Blight

Head blight or Fusarium scab caused by *Fusarium,* can result in wheat crop loss of 20-40% in Punjab. The level of incidence increases in the rice-wheat or maize-wheat rotation. The outbreak of the scab was first reported from Arunachal Pradesh and Assam in 1973.

Flag Smut of Wheat

Flag smut of wheat is caused by *Urocystis agropyri,* which attacks the stem, culm and the leaf. This disease is prevalent in almost all the important wheat growing areas of the world. It was first reported from Australia in 1868 and later on from India, Pakistan, Japan, USA and Egypt. The spores appear on the leaf and the leaf sheath as tiny balls, bright brown in color. The disease manifests itself from the late seedling stage till the crop maturity. The infected leaves may get twisted and begin to droop and finally wither. The plant ultimately dies and very often the culm becomes sterile, or even the grain formed may not germinate.

Powdery Mildew of Wheat

Powdery mildew of wheat caused by *Erysiphe graminis* causes considerable losses in the area where temperature is low and cloudy. It is distributed in some parts of Haryana, Punjab, and Rajasthan. The occurrence of the disease depends upon the climate and cropping pattern. A greyish white powdery growth appears on the leaf, leaf sheath and floral parts, as superficial colonies of the pathogens develop on them as a result of which there is reduction in the number of leaves.

NEMATODES IN WHEAT

Cereal Cyst Nematode

The cereal cyst nematode, *Heterodera avenae*, is the causal organism of a serious disease commonly called 'Molya' of wheat and barley in the IGP (Bhatti and Dahiya, 1992). In India it was first observed in Rajasthan. It is now known to occur in the states of Haryana, Punjab, Jammu and Kashmir, Himachal Pradesh, Delhi, Madhya Pradesh and Uttar Pradesh of India. Cyst nematodes exhibit sexual dimorphism. Males are vermiform and live in the soil. The second stage juveniles are endoparasitic and after the development of lemon shaped females, they are sedentary and protrude out on the root surface. After its death, the female body encloses the eggs to form brown lemon shaped cysts. Usually, one generation is completed in each season. During the off-season, the nematode survives in the cyst stage. The threshold level for economic damage to wheat was reported to be 20 eggs and juveniles/g of soil. The disease is characterized by patchy growth, stunted and yellowish plants. Infected plants exhibit fewer leaves, reduced tillering, delayed emergence of ears, and reduced number of spikelets and grains. Presence of cysts on the roots or in the soil is the only confirmatory indication of nematode infestation.

The development of the disease depends upon environmental factors, which influence the survival and emergence of larvae from cysts. Open textured and well-drained soil enhances the survival and infestation of the nematode population. Larval emergence takes place in a wide range of soil pH (3.5 to 11.5).

Management of Cereal Cyst Nematode

Source: Dhawan and Pankaj, 1998; Bhatti and Dhaiya, 1992; Sehgal and Gaur, 1999; Sharma, Johansen and Midha, 1998.

- Crop rotation by replacing the wheat crops with mustard, pulses or fenugreek, etc.
- Summer deep ploughing.
- Use of trap crops.
- Additional irrigation.
- Late autumn sowing should be preferred.
- Use of soil amendments, e.g., mustard cake, etc.

Wheat Gall Nematode

Anguina tritici, also known as the wheat gall nematode is one of the important pests of wheat crop in the IGP (Paruthi, 1992). The nematode galls have been disseminated through contaminated seeds to practically all wheat producing regions of India. Whenever infested wheat seed is sown, and on contact with soil moisture, the nematode larvae emerge from the galls through its softened walls. The second stage larvae climb to the growing points of the seedlings and reaches the inflorescence. The larva moults within 3 to 5 days and develops into males and females in the ratio of 1:2 (Paruthi, 1992). Enlargement of basal stem parts, near the soil base is the first visible symptoms that can be seen in 20 to 25 day old seedlings. The infected plants have more number of tillers with a comparatively faster growth rate than the healthy plants. The affected ears are short and broad, with very short or no arms on the glumes. In such ears, the nematode gall replaces either all or some of the grains. Each spikelet may contain 1 to 5 galls.

The bacterial disease of wheat commonly associated with it is popularly known as 'Yellow slimy disease' or 'tundu' or 'tannan.' The second stage larvae present on the growing point of the seedlings may enter the leaf tissues in its early stages or enter the glumes or awns tissues at the time of flowering. Small galls may also be formed in these tissues but the number of larvae present in such galls is few (Paruthi, 1992).

For the yellow slime disease, the characteristic symptoms is the production of a bright yellow slime or gum on the leaf surface of the young plants and on the ears and leaves in contact with them while they are still in the boot leaf stage. The yellowish slime, which can be seen trickling down the tips under humid condition, become hard, brittle and brown in color on drying.

Management of Wheat Gall Nematode

- Use certified seeds.
- Dipping of contaminated seed lots in 10% brine solution in which nematode galls float and can be skimmed off, followed by fresh water washing.

IPM MODULES TESTED ON WHEAT AND RICE IN INDIA

IPM Module for Wheat in the Rice-Wheat Cropping System

Pre-Sowing

- Identify the termitaria in the wheat field and treat it with chlorpyriphos spray.
- Use only well rotted manure in the field.
- Seed treatment with endosulfan or chlorpyriphos two days before sowing.
- Advocating normal sowing of wheat on a community basis.

Post-Sowing

- Spray 0.8 liter of chlorpyriphos/5 liter of water/ha, followed by irrigation.
- For the late sown crop, seed rate could be increased by 5-7%.
- Periodical removal of shootfly dead hearts.
- For above ETL level of shootfly (more than 10% incidence), spray cypermethrin @ 50 g a.i./ha two times at 15 days interval.
- For sucking pests, spray formothian, oxydementon methyl or phosphamidon @ 160, 160, 250 ml a.i./ha that can be repeated a fortnight later.
- Use of zinc phosphide.
- Cutworm, semilooper, pyrilla, population was reduced with the use of DDVP or fenitrothion @500 g a.i./ha.

Development and Implementation of IPM in Rice: Case Studies from India

A project entitled "The Operational Research Project on IPM" was conducted at different locations in six states and in different agro-climatic zones of India, e.g., Orissa (Cuttack); West Bengal (Hoogly); Madhya Pradesh (Raipur); Andhra Pradesh (Warangal and Bapatla); Kerala; Maharashtra (Thane). Under this project the following objectives were addressed:

- Comparative assessment in the farmers' fields on the efficiency of traditional (chemical) methods of insect control versus the use of IPM by involving cultural methods, varietal resistance, pest avoidance, biological and need-based insecticidal application.
- Spread in the adoption of IPM for high yielding rice varieties.
- Demonstration of the economic threshold levels (ETL) of rice pests.
- Minimizing pest populations by avoidance of excess use of chemicals.
- Determination of cost benefit ratios for the IPM versus chemical control methods.
- Comparison of grain yields.
- Effect on socio-economic conditions of rice farmers.

Brief note of experiments undertaken and their results are as follows:

I. State: Orissa; Locations: CRRI, Cuttack; Area: 2500 acre; Year: 1975-76.

Target Pests: Stem borers, Gall midge, BPH, GLH and Rice leaf folder.

Component:

- Use of a pest tolerant variety.
- One application of diazinon granules at 1.5 kg a.i./ha, ten days after planting.
- Application of *Bacillus thuringiensis* spore dust at weekly intervals.
- Spraying with larval suspension of insect parasitic nematode DD-136.

Result: Number of insecticides spray decreases from 3 or 4 to 1 only.

II. State: West Bengal; Locations: Pandua block, Hoogly; Area: 2100 acre.

Target Pest:

Insects: Stem borers, Gall midge, BPH, GLH and Rice leaf folder, Gundhi bug, Whorl maggot, Case worm and Cutting caterpillar.

Diseases: Bacterial leaf blight and Sheath blight.

Component:

- Summer ploughing, crop rotation, tolerant rice varieties, manual weeding.

- Nursery treatment with pesticides, need-based application of foliar sprays, to make fire for nocturnal insects, insecticidal dipping of rice seedlings.
- Cultivation of early maturing varieties.
- Balanced fertilizer application, timely transplanting, water regulation for case worm and BPH.

Result: Application of the insecticide only on the basis of pest damage.

III. State: Andhra Pradesh; Location: Buddam, Bapatla.

Target Pests: Yellow stem borer, Brown planthoppers, along with a few other rice pest complex were endemic in this area.

Component: Monitoring of pests through light traps, field sampling, nursery protection and single treatment with insecticides was advocated to farmers.

Result: Reduction in the number of pesticides and increase in rice yield.

IV. State: Andhra Pradesh; Location: Kolakaluru, Halfpet and Kopalle Gudivada, Tenali in Guntur District; Area: 1920 ha.

Target Pests:

Insects: BPH, Leaf folder, Ear cutting caterpillar, Yellow stemborer.

Disease: Sheath blight and blast.

Component:

- Resistant varieties.
- Cultural methods.
- Conservation of natural enemies.
- Need-based applications of insecticides were used against the Brown plant hopper at Kalakaluru village.

Result: Increase in average yield by 36% and reduction in the use of insecticide.

V. State: Kerala; Location: Kuttanad; Area: 40 ha; Year: 1980-85.

Target Pests: Stem borer, Cutworm, Leaf folder, Rice bug, BPH.

Component:

- Advancing the cropping season between October-February instead of November-March to avoid peak population of BPH.
- Growing BPH resistant varieties like Bhadra, Asha and Pavizhan.
- Balanced fertilizer use of N:P:K @ 90, 40 and 45 kg/ha.
- Clean cultivation involving weed control and stubble destruction.
- Draining water to field capacity to avoid build-up of BPH and case worm.
- Need-based application of insecticides at recommended doses.

Result: In the 1980-81 season, the average yield of the IPM block was about 40% higher than that of farmer's practice block. However, during the other years of conducting these demonstrations, there was no significant differences in the grain yield of the two field blocks, but there was a substantial reduction in the cost of plant protection in the IPM block (Table 9). The combined effect of the savings from plant protection and increased yield became more pronounced when expressed as cost of production per quintal of rice.

IPM of Rice in the State of Jammu and Kashmir

Integrated management of insect pests of rice was evaluated in a demonstration and training programme in 2 villages in Jammu and Kashmir state in 1992. The various control measures included releasing the egg parasitoids, *Trichogramma japonicum* and *T. chilonis*. The cost benefit ratio averaged 1:1.97 for IPM as compared to 1:1.55 for non-IPM treatments.

IPM of Rice in Haryana

In the field trials conducted in the Haryana villages of Kurali Rembha, Darar, Araipura and Salaru during 1983-90, the effect of IPM on pesticide

TABLE 9. Results of demonstration of IPM in Kuttanad, Kerala, India.

	1980-81		1981-82		1982-83		1983-84		1984-85	
	a	b	a	b	a	b	a	b	a	b
Cost of plant protection (Rs. ha^{-1})	625	900	393	613	447	610	418	692	450	768
Savings in using IPM method (Rs. ha^{-1})	275	–	230	–	161	–	274	–	318	–
Yield of grain	3750	2620	3900	3800	4200	3900	3745	3500	3715	3450
Plant protection cost Rs. qtl^{-1} of grain	16.67	34.35	7.82	16.13	10.67	15.64	11.16	17.77	12.11	22.26

a = IPM demonstration; b = Farmers practices control plot.

expenditure and yields of rice with varieties PR-106, PR-107, PR-109, HKR-120, Jaya and Basmati-370 were evaluated.

The cost of plant protection was 58.2% higher in non-IPM fields than in IPM fields. Moreover, rice yields increased by 3.3% in IPM over the non-IPM fields.

Success Story of IPM Use in Basmati Rice

Many insect pests attack the basmati rice in the predominantly basmati rice growing areas of Haryana, western Uttar Pradesh and Punjab. Leaf folders, *Cnaphalocrocis medinalis* and yellow stem borer, *Scirpophaga incertulas* are the major insect pests in Haryana besides many minor pests. Among diseases, blast is the most important followed by bacterial leaf blight and other minor diseases. For controlling these pests, farmers follow chemical control methods, which is quite expensive and many times leads to pesticide residue problems. Being an export-oriented commodity, residue problem hampers its export potential. To overcome these problems, an IPM module was developed. The module was field tested between 1994 to 1996 by the National Center for Integrated Pest Management, New Delhi, in collaboration with CCS Rice Research Station, Kaul, in the Haryana State of India. Taraori, a popular basmati rice variety in the area, was chosen for experimentation and raised according to normal agronomic practices followed in the region. The IPM strategy consisted of the release of *Trichogramma japonicum*, spraying of neem-based pesticide, and use of insecticidal spray only as the last resort. For blast, application of burnt rice husk, which induces resistance to the disease, and need-based application of fungicide was the main component. The IPM treatment was compared with sole pesticide treatment and untreated control. The results showed that the IPM approach reduced the infestation of leaf folder and stem borer effectively and it was almost at par with the insecticidal control during all the 3 years. The chemical control gave the highest yields of 31.51, 34.33 and 28.25 q/ha, compared to the IPM treated fields of 29.89, 31.96 and 28.55 q/ha during 1994, 1995 and 1996, respectively. However, economic analysis indicated that the IPM method was superior to the chemical control method, as the mean cost benefit ratio of IPM over untreated control was 1:5.7 as compared to 1:5.03 of the chemical control method.

On farm trial of this IPM technology was carried out at Baraut during the monsoon season of 1997 in western Uttar Pradesh State of India, which is emerging as a potential basmati producing area. The continuous monitoring of pests showed moderate to high incidence of leaf folder and low incidence of stem borer in this area. The incidence of sheath blight was also noticed but did not warrant fungicidal application. However, timely field release of *Trichogramma japonicum* in IPM fields suppressed the incidence of leaf folder

TABLE 10. Economics of IPM for basmati rice, monsoon 1997 and 1998.

Treatments	Cost of PP (Rs)	Yield (Q/ha)	Monetary gain over FP	Cost benefit ratio (CBR)
		IPM		
1997	1020	58.38	14332.00	1:14.05
1998	1445	47.71	10716.00	1:7:41
		Chemical control		
1997	1260	47.75	3968.00	1:3.14
1998	1705	35.96	− 1588.00	1: − 0.93
		Farmer's practice (FP)		
1997	420	43.68	–	–
1998	445	36.28	–	–

Note: Price of paddy was @ Rs. 975/Q in 1997 and Rs. 1025/Q in 1998.
PP: Plant protection measures; FP: Famers practice.

and stem borer to a bare minimum. Overall results showed the superiority of IPM over chemical method or farmers' own practices as indicated by the yield data and economic analysis (Garg and Jeswani, 1999; Table 10).

Implementation of IPM in Rice

Although IPM has been accepted in principle as the most attractive option for the protection of agricultural crops from the ravages of pests, yet implementation at the farmers' level has been rather limited. Pesticides remain as the means of intervention and as an essential component of IPM strategies.

Even though it may be impossible to avoid chemical pesticides altogether, integrating non-chemical methods in pest management can reduce dependence on chemical control. This would reduce the costs considerably apart from offering protection in an ecologically sound manner. In rice, the number of pesticide applications has been reduced to 0.8 per season in case of IPM trained farmers as compared to 2.4 per season for untrained farmers, and the cost of average pesticide application was Rs. 163.50 and Rs. 447.90 per hectare, respectively.

MAJOR CONSTRAINTS FOR ADOPTION OF IPM AT FIELD LEVEL

The major constraints in the implementation and adoption of Integrated Pest Management technologies at the field level in the Indo-Gangetic Plains are as follows:

- The natural enemies have a short life span, are temporal, and of specific efficacy.
- Lack of knowledge for the mass production of the natural enemies of pests.
- Delays in transshipments of biological control agents.
- Lack of adequate training in application methods.
- A slow result-oriented process.
- The use of bio-pesticides is also restricted due to moderate toxicity, slow action, specificity, cost and photo-instability.
- Farmers are not aware of the ecological benefits, its proper usage, the source of availability, and suppliers of bio-agents and bio-pesticides.
- The necessity for repeated applications, low toxicity and persistence, cumbersome procedures of collection and extraction coupled with low yields have discouraged use of botanicals by the farmers.

Thus, IPM technology adoption is influenced by cost versus efficacy of products, need for sophisticated information for the decision-making process, ability to integrate new products and techniques into existing farm management practices and managerial skills. Apart from these the other constraints for wider adoption of IPM include the institutional, informational, sociological and ecological constraints.

CONCLUSION

The systems and technologies that are being used now may have to be modified and new technologies developed to achieve the goals of IPM. Though pesticides continue to be the key component of pest management for the foreseeable future as there are no practical alternatives, their usage can be reduced with the use of botanicals and bio-pesticides on a large scale. This also promotes management in an eco-friendly manner.

For the adoption of IPM practices, it is imperative to understand the causes of pest outbreaks and to modify the design and management of systems to control pest damage. The coordinated efforts of all the research institutes and extension personnel to motivate the farming community to practice IPM will have to continue.

REFERENCES

Abraham, T.P. and R.K. Khosla. (1964). *Estimation of loss due to pests and diseases on the rice crop, results of a pilot survey.* All India Ent. Conf. New Delhi (Mimeographed).

Acharya, L.P. (1967). *Life History, Bionomics and Morphology of the Rice Hispa, Hispa armigera* Oliver. M.Sc. (Agr) Thesis, QUAT, Bhubaneswar.

Aggarwal, R., D.V. Singh, K.D. Srivastava, and P. Bahadur. (1996). The potential of antagonists for biocontrol of *Neovossia indica* causing karnal bunt of wheat. *Indian. J. Biol. Cont.* 9: 69-70.

Aggarwal, V.K., S.S. Chahal, and S.B. Mathur. (1993). Loose smut. In *Seed borne diseases and seed health testing of wheat*, eds. S.B. Mathur and B.M. Cunfer, Danish Govt. of Seed Pathology for Developing Countries, Denmark.

Aggarwal, V.K., H.S. Verma, M. Aggarwal, and R.K. Gupta. (1982). Studies on loose smut of wheat III. Effect on plant morphology and control through seed treatment with carbofuran. *Seed Res.* 10: 79-86.

Aggarwala, S.D. (1955). On the control of Paddy Hispa (*Hispa armigera*) at Pusa Bihar. *Indian J.Ent.* 17: 11-16.

Ahmad, N., P.K. Das, and Q.H. Baqri. (1984). Evaluation of yield losses in rice due to *Hirschmanniella gracilis* (de Man, 1888) Luc And Goodey 1963 (Tylenchida: Nematoda) at Hoogly (West Bengal). *Bull. Zool. Surv. India.* 5: 85-91.

Alam, K.B., M.A. Saheed, A.U. Ahmed, and P.K. Malkar. (1993). Bipolaris blight (Spot blotch) of wheat in Bangladesh. In *Wheat in heat stressed enivornments: Irrigated dry areas and rice wheat farming systems*, eds. D.A. Saunders and G.P. Hettel, Mexico, DF, CIMMYT pp. 339-342.

Amer, G.A.M. (1995). *Biocontrol of Karnal Bunt of Wheat*, Ph.D. Thesis, Division of Mycology and Plant Pathology, IARI, New Delhi.

Anjaneyulu, A. (1975). Rice stubble and self-sown rice seedlings, the reservoir hosts of Tungro during off-seasons. *Sci. Cult.* 4: 298.

Aujla, S.S., Kaur Satinder, and K.S. Gill. (1990). Efficacy of different living propagules of *Nevossia indica* in causing ear infection of wheat. *Plant Dis. Res.* 5: 112-114.

Aujla, S.S., A.S. Grewal, G.S. Nanda, and I. Sharma. (1990). Identification of stable resistance in wheat to loose smut. *Indian Phytopath.* 43: 90-91.

Aujla, S.S., I. Sharma, K.S. Gill, and A.S. Grewal. (1985). Variable resistance in wheat germplasm to *Neovossia indica. Third Nat. Seminar on Genet. and Wheat Improv., Shimla.* May 8-1 0 pp. 8-1 0 (Abst.).

Aujla, S.S., I. Sharma, K.S. Gill, and A.S. Grewal. (1986). Additional hosts of *Neovossica indica* (Mitra) Mundkur. *Indian J. Plant Path.* 4: 87-88.

Aujla, S.S., Y.R. Sharma, and A.S. Grewal. (1979). Studies on karnal bunt with particular reference to standardization technique for screening varieties. In *Proc. 17th wheat research workers workshop AICWIP*, Ranchi (Bihar).

Bansal, R., D.V. Singh, and L.M. Joshi. (1984). Effect of Karnal bunt pathogen *(Neovossia indica* (Mitra) Mundkur) on weight and viability of wheat seed. *Indian J. Agric. Sci.* 54: 663-666.

Bedi, K.S., M.R. Sikka, and B.B. Mundkur. (1949). Transmission of wheat bunt due to *Neovossia indica* (Mitra) Mundkur. *Indian Phytopath. 4: 20-26.*

Bhat, R.V., B. Roy, D.D. Roy, M. Vijayaraghavan, and P.G. Tulpule. (1981). *Toxicological evaluation of karnal bunt in monkey–A report.* Bull. Food and Drug Toxicological Research Centre, Jamaiosmania, Hyderabad, India.

Bhatia, S.K. (1978). Insect Pests. In *Wheat research in India*, ed. Ramanujam, ICAR, New Delhi, India. pp. 152-167.

Bhatti, D.S. and R.S. Dahiya. (1992). Nematode pests of Wheat and Barley (*Hetero-*

dera avenae). In *Nematode pests of crop*, eds. D.S. Bhatti and R.K. Walia, CBS Publishers and Distributors, New Delhi, India, pp. 27-42.

Bimb, H.P. and B.N. Mahto. (1992). *Wheat disease Report*. Paper presented at winter crops Technology Workshop held at WRP, Bhairahwa, Nepal.

Bindra, O.S. and H. Singh. (1970). Gujhia weevil *Tanymecus indicus* Faust (Coleoptera: Curculionidae). *Pesticides* 4: 16-18.

Birat, R.S. (1965). New records of parasitic nematodes on rice. *Sci. Cult.* 31: 494.

Bridge, J., M. Luc, and A. Plowright. (1990). Nematode parasites of rice. In *Plant parasitic nematodes in subtropical and tropical agriculture*, eds. M. Luc, R.A. Sikora, and J. Bridge, Wallingford, UK, CAB International, pp. 69-108.

Chakrabarti, N.K. (1976). Disease of rice and their control. *Pesticides Information*, 2: 11-15.

Chakrabarti, N.K., J.P. Kulshreshtha, and Y.S. Rao. (1971). Pests and diseases of new varieties and remedial measures. *Indian Fmg.* 19: 53-59.

Chatrath, M.S., N.T. Thomas, and M. Mohan. (1976). Comparative effect of loose smut infection on morphological characters of dwarf and tall wheat. *Indian Phytopath.* 29: 66-67.

Chatterjee, S.M. and P.S. Prakasa Rao. (1975). Pests of paddy and their control. In *Entomology in India*. Ent. Soc. of India.

Chatterjee, P.B. (1979). Rice leaf folder attacks in India. *Int. Rice Res. Newsletter.* 4: 21.

Chattopadhyay, S.B. (1952). Trial of different seed treating fungicides for control of primary seed borne infection of *Helminthosporium oryzae*. *Proc. 39th Indian Sci. Congr.* Part 3, p. 6.

Chaudhary, J.P. (1968). Delphacid hopper, a serious pest of paddy. *Progressive Fmg.*, Sept., 1968.

Chaudhary, J.P., M. Ramzan, and A.S. Atwal. (1968). Preliminary studies on biology of wheat aphids. *Indian J. Agric. Sci.* 39: 672-675.

Chelliah, S., and M. Bharathi. (1998). Integrated Pest Management in Rice–Opportunities and Approaches. In *IPM System in Agriculture*, eds. Rajeev K. Upadhyay, K.G. Mukerji and R.L. Rajak, Aditya Books Pvt. Ltd., New Delhi, 131-162.

Chenulu,V.V., and A. Singh. (1964). A note on estimation of losses due to leaf blight of wheat caused by *Altemaria triticina*. *Indian Phytopath*, 17: 256-257.

Chona, B.L., R.L. Munjal, K.L. Adlakha, and H.R. Sikka. (1963). An improved temperature cum humidity chamber for field inoculation of plants. *Indian Phytopath.* 16: 174-176.

CRRI. (1982). *Annual Report for 1982*. Central Rice Research Institute, Cuttack, Orissa, India. ICAR, pp. 164-168.

Das, S.R. and B. Mishra. (1990). Field evaluation of fungicides for control of sheath blight of rice. *Indian Phytopath.* 43: 94-96.

Dass, N.M., K.V. Memmon, and S.P. Christudas. (1972). Occurrence of *Nilaparvata lugens (*Stal.*) (*Delphacidae: Homoptera*)* as a serious pest of paddy in Kerala. *Agric. Res. J.* Kerala. 10: 191-192.

Dean, W.M. (1969). The effect of temperature on loose smut of wheat *(Ustilago nuda)*. *Ann. App. Biol.* 64: 75-83.

Deol, G.S. and G.S. Sandhu. (1974). Note on chemical control of brown mite. *Ind. J. Agric. Sci.* 44: 681-682.

Devadath, S. and S.Y. Padmanabhan. (1970). Apporaches to control of bacterial blight and streak diseases of rice in India. *Ind. Phytopath. Soc. Bull.* 6: 5-12.

Devkota, R.N. (1993). Wheat breeding objectives in Nepal: The National Testing System and Recent Progress. In *Wheat in heat stressed environments: Irrigated dry areas and rice wheat farming systems*, eds. D.A. Saunders and G.P. Hettel, pp. 216-223, Mexico, CIMMYT.

Dhakuwak, G.S. and Jaswant Singh. (1983). Outbreaks of whitebacked planthopper and brown planthopper in Punjab, India. *Int. Rice Commn. Newsl.* 32: 26-28.

Dhaliwal, G.S. and Ramesh Arora. (1996). *Principles of insect pest management.* National Agricultural Technology Information Centre, Ludhiana, 374 pp.

Dhaliwal, H.S. and D.V. Singh. (1989). Production and interrelationship of two types of secondary sporidia of *Neovossia indica. Curr. Sci.* 58: 614-618.

Dhaliwal, J.S. and B. Singh. (1975). Effect of simulated rain on the survival of wheat aphid, *Macrosiphum miscanthi* (Takahashi), and its syrphid predator, Eristalis tenax L. *Indian J. Ecol.* 2: 186-187.

Dhawan, S.C. and Pankaj. (1998). Nematode pests of wheat and barley. In *Nematode disease in plants*, ed. P.C. Trivedi, CBS Publishers and Distibutors, pp. 12-25.

Doval, S.L., O.P. Bhora, and S.K. Sharma. (1972). New record of *Heliothis armigera* Hubn. as a pest of wheat in India and efficacy of some insecticides against its larvae. *Indian J. Ent.* 34: 72-73.

Dubin, H.J. and M. VanGinkel. (1991). The status of wheat disease research in warmer areas, pp. 125-145. In *Wheat for the nontraditional warm areas*, ed. D.A. Saunders, Mexico, DF, CIMMYT.

Dyck, V.A and B. Thomas. (1979). The brown planthopper problem. In *Brown planthooper: Threat to rice production in Asia* IRRI, Los Banos, Phillipinnes.

Gangopadhyay, S. and N.K. Chakrabarti. (1982). Sheath blight of rice. *Commonwealth Mycological Institute.* 1645-1460.

Garg, D.K. and M.D. Jeswani. (1999). Adoptable IPM practices in rice-wheat cropping system in India. In *Training Manual-6*, eds. R.N. Singh, T.P. Trivedi, Mukesh Sehgal and Ashok Kumar, NCIPM, New Delhi, pp. 31-40.

Gargav, V.P., R.K. Patel, and O.P. Katiar. (1971). Insecticidal trial against rice leaf roller *Cnaphalocrocis medinalis* Gn. *Indian J. Agric. Sci.* 41: 47-49.

Gaur, H.S., E. Khan, and M. Sehgal. (1993). Occurrence of two species of root knot nematode infecting rice, wheat and non-host weeds in northern India. *Annals of Plant Protection Sciences*, 1: 141-142.

Gaur, H.S., M. Shah, and E. Khan. (1993). *Meloidygyne triticoryzae* sp. (Nematoda: Meloidogynidae) a root-knot nematode damaging wheat and rice in India. *Annals of Plant Protection Sciences*, 1: 18-26.

Gaur, H.S., J. Singh, J. Sharma and S.T. Chandel. (1998). *Management of root-knot nematode in rice nursery and main field by soil solarisation and nematicides.* International Conference on Integrated plant disease management for sustainable agriculture, Nov. 10-15, IARI, New-Delhi, p. 400.

Gill, H.S., K.S. Sandhu, and Tarlok Singh. (1984). Weed flora of Kharif crops of the

Ludhiana, Ferozepur and Faridkot distircts of Punjab. *Indian J. Weed Sci.* 16: 36-47.

Gill, H.S., U.S. Walia, and L.S. Brar. (1979). Chemical weed control in wheat with particular reference to *Phalaris minor* and wild oats (*Avena ludoviciana*). *Pesticides*, 13: 45-47.

Gill, K.S. (1994). Sustainability issues related to rice-wheat production system in Asia. In *Sustainability of Rice-Wheat Production Systems in Asia*, eds. R.S. Paroda, T. Woodhead, and R.B. Singh, RAPA Publication: 1994/11, FAO, Bangkok, 36-60.

Goel, L.B., S. Nagarajan, R.V. Singh, V.C. Sinha, and J. Kumar. (1999). Foliar blights of wheat: Current status in India and identification of donor lines for resistance through multilocational evaluation. *Indian Phytopath.* 52: 398-402.

Goel, L.B., D.V. Singh, K.D. Srivastava, L.M. Joshi, and S. Nagarajan. (1977). *Smuts and bunts of wheat in India*, IARI, New-Delhi, India.

Goto, M. (1964). "Kresek" and pale yellow leaf systemic symptoms of bacterial blight of rice caused by *Xanthomonas oryzae. Pl. Dis. Reptr.* 48: 858-861.

Govinda Rao, P. and D. Gopala Raju. (1955). Observations on the false smut disease of paddy caused by *Ustilaginoidea virens. Andhra Agric. J.* 2: 206-207.

Govindhu, H.C. (1977). *Plant Disease Syndromes in High Yielding Varieties with Special Reference to Cereals and Millets.* Presidential Address (Botany Sec). 65th Science Congress, pp. 29-52.

Grewal, J.S. and S.S. Bains. (1975). The role of botic and abiotic factors in the population build up of wheat aphids and extent of loss caused by them. *Indian J. Ecol.* 2: 139-145.

Gunathilagaraj, K. and M. Gopalan. (1986). Rice leaffolder (LF) complex in Madurai, Tamil Nadu, India. *Int. Rice Res. Newsl.* 11: 24.

Haai Voung, H.U. (1969). The occurrence in Madagascar of the rice nematodes *Aphlenchoides besseyi* and *Ditylenchus angustus.* In *Nematodes of tropical crops*, ed. Peachy Tech. Commn. Commonwealth. Bur. Helminth. No. 40: 270-274.

Hasan, S.F. (1973). *Wheat disease and their relevance to improvement and production in Pakistan.* Proc. Fourth FAO/Rockfeller Foundation Wheat Seminar, pp. 84-86.

Howlidar, M.A.R., M. Jaialuddin, M.B. Mealn, and L. Rahman. (1984). Effect of nitrogen and plant species on bacteria leaf blight of rice. *Madras Agric. J.* 71: 798-803.

Ichinoche, M. (1964). *A review of studies on nematodes attacking rice.* X Mtg. Intern. Rice Commn. FAO, 3-14.

Inoue, Y., and Y. Tsuda. (1959). Assessment of the disease in yield due to bacterial leaf blight of rice. *Bull. Tokai-Kinki Agr. Exp. Sta.* 6: 154-167.

Ishiyama, S. (1922). Studies on bacterial blight leaf blight. *Rep. Agr. Exp. Sta.,* 45: 223-261.

Israel, P., M.B. Kalode, and B.C. Mishra. (1968). Toxicity and duration of effectiveness of insecticides against *Sogatella furcifera (Horv), Nilaparvata lugens* (Stal.) and *Nephotettix impecticeps* Dist. on rice. *Indian J. Agric. Sci.* 38: 427-443.

Israel, P., Y.S. Rao, and P.S. Prakasa Rao. (1964). *Epidemiology of rice diseases and pests and its forecast.* Proc. 10th Mtg. IRC, Manila (Mimeograph).

Israel, P., Y.S. Rao, and P.S.P. Rao. (1966). *Major Insect pests affecting rice (a) stem borers.* FAOIRC WP on Prodn. and Prot. 11th Session, Louisina (Mimeogr.).

Israel, P., Y.S. Rao, and Y.R.V.J. Rao. (1963). Investigations on nematodes of rice and rice soil. I. *Oryza,* 3: 125-128.

Israel, P., G. Vedamurthy, and Y.S. Rao. (1959). *Review of pests of rice.* FAO, IRC WP on Plant Production and Protection, Ceylon (Mimeograph).

Jaganathan, R., K. Ramaraju, and M.S. Venugopal. (1989). Evaluation of rice culture for multiple resistance to three major pests. *Ann Plant Resis. Insects Newsl.* 15: 45-46.

Jairajpuri, M.S. and Q.H. Baqri. (1991). *Nematode pests of rice.* Oxford & IBH Publishing Co. Pvt. Ltd., New Delhi, India, pp. 66.

John, V.T. (1968). Identification and characterisation of tungro, a virus disease of rice in India. *Pl. Dis. Reptr.* 52: 871-875.

John, V.T. (1970). Yellow disease of paddy. *Indian Fmg.* 20: 27-30.

Joshi, L.M. (1978). Disease. In *Wheat Research in India,* ICAR, New-Delhi. 126-155.

Joshi, L.M. (1982). Wheat rusts management–present knowledge and future prospects. *Kavaka,* 10: 1-12.

Joshi, L.M., S.D. Gera, and E.E. Saari. (1973). Extensive cultivation of Kalyansona and disease development. *Indian Phytopath.* 26:37-73.

Joshi, L.M., D.V. Singh, and K.D. Srivastava. (1985). Status of rusts and smuts during dwarf wheat era in India. *Rachis,* 4: 10-16.

Joshi, L.M., D.V. Singh, K.D. Srivastava, and R.D. Wilcoxson. (1983). Karnal bunt–A minor disease that is now a new threat to wheat. *Bot. Rev.* 43: 308-338.

Joshi, L.M, K.D. Srivasatava, D.V. Singh, S. Nagarajan, and L.B. Goel. (1980). Wheat disease survey in India during 1975-76. *Annual Wheat Newsletter,* 24: 60-61.

Joshi, L.M., D.V. Singh, and K.D. Srivastava. (1988). *Manual of wheat diseases,* MPH, New Delhi, pp 1-75.

Kalode, M.B. and P. Israel. (1970). Persistence of insecticidal residues on the rice plant against the stem borer *Trypotyza incertulas* Walker. *Oryza,* 7: 85-90.

Kannaiyan, S. and N.N. Prasad. (1979). Seed borne nature of sheath blight pathogen *Rhizoctonia solani* in Rice. *Int. Rice Res. Newsletter* 4: 17.

Karki, P.B. (1997). Status of Nematode problems and research in Nepal. In *Diagnosis of key nematode pests of chickpea and pigeonpea and their management: Proc. Regional Training Course, 25-30 Nov. 1996, ICRISAT, India,* ed. S.B. Sharma, pp. 57-60.

Karuppuchammy, P. and M. Goplan. (1986). Influence of time of planting on the incidence of rice pests. *Madras Agric. J.* 73: 606-609.

Kashem, M.A., M.A.R. Howlider, H.A. Begum, and G.K.M.M. Rahman. (1994). Effect of fertilisers and spacing on the disease severities of bacterial leaf blight and sheath blight of rice. *Bangladesh J. of Sci. and Ind. Res.* 29: 89-95.

Khan, M.Q. (1964). *Control of paddy stem borers by cultural practices, the major insect pests of the rice plant,* IRRI Manila, The Johns Hopkins Press, Baltimore, Maryland, pp. 369-390.

Khan, M.Q., and D.V. Murthy. (1965). Rice Hispa (*Hispa armigera*). *Madras Agric. J.* 43: 158-160.

Khan, R. M., and S.K. Sharma. (1974). Shootfly on wheat. *Pl. Prot. Bull. F.A.0.* 22: 74-75.

Khanzada, A.K. and S.B. Mathur. (1984). Control of loose smut of wheat by carboxin, fenfuram and tridimenol. *Seed Sci. Tech.* 11: 947-949.

Kohli, C.K. (1966). Pathogenicity and host range studies on paddy sheath blight pathogen *Rhizoctonia solani* in rice. *J. of Res.* Punjab Agricultural University. 3: 37-40.

Kojima, K., S. Iwase, and K. Inoki. (1955). Relationship between the soil type and occurrence of bacterial leaf blight disease of rice. *Ann. Phytopath. Soc. Japan*, 19: 163.

Kulshrestha, J.P., M.B. Kalode, P.S. Prakasha Rao, B.C. Mishra, and A. Verma. (1970). High yielding varieties and the resulting change in the pattern of pests in rice in India. *Oryza* 7: 61-64.

Lingappa, S. (1972). Bionomics of the rice leaf folder, *Cnaphalocrocis medinalis*, Guenee (Lepidoptera: Pyralidae). *Mysore J. Agric. Sci.* 6: 123-134.

Mahabini, Jena, R.C. Dani, and S. Rajamani. (1992). Effectiveness of insectides against the rice leaffolder. *Indian J. Plant Prot.* 20: 43-46.

Mahto, B.N. (1999). Management of *Helminthosporium* leaf blight of wheat in Nepal. *Indian Phytopath.* 52: 408-413.

Manibhushan Rao, K. (1964). Environmental conditions and false smut incidence in rice. *Indian Pyhtopath.* 17: 110-114.

Mathur,V.K. and S.K. Prasad. (1971). Occurrence and distribution of *Hirschmanniella oryzae* in the Indian union with description of *H. mangaloriensis*. *Indian J. Nematol.* 1: 220-226.

Mehra, R., B.S. Thind, and R. Mehra. (1994). Evaluation of different seed treating agents for the control of bacterial blight of rice. *Crop Res.* Hissar. 7: 288-291.

Mehta, K.C. (1942). Control of rust epidemics of wheat and barley in India. *Indian Fmg. 3: 319-321.*

Mehta, K.C. (1940). Further studies on cereal rusts in India. *Sci. Monogr.* No. 14, Imperial Council of Agr. Res. India. pp. 1-244.

Mehrotra, R.S. (1980). *Plant pathology.* Tata McGraw-Hill Publishing Co., New Delhi. 770 pp.

Mishra, A.B. and S.P. Singh. (1969). Wheat disease situation in Madhya Pradesh, India. *PANS*, 15: 71-73.

Mishra, A.P. and A.K. Mukherjee. (1962). Effect of carbon and nitrogen nutrition on growth and sporulation of *Helminthosporium oryzae* Breda de Hann. *Indian Phytopath.* 17: 287-295.

Mishra, A.P. (1959). Incidence of *Helminthosporium oryzae* Breda de Haan on Paddy at Sabour. *Rice Newsletter*, 7: 218-221.

Mitra, M. (1931a). A comparative study of species and strains of *Helminthosporium* on certain Indian cultivated crops. *Trans. Br. Mycol. Soc.* 15: 254-293.

Mitra, M. (1931b). A new bunt of wheat in India. *Ann. Biol.* 18: 178-179.

Mitra, M. (1937). Studies on stinking smut or bunt of wheat in India. *Indian J. Agric. Sci.* 7: 469-478.

Mochida, O. (1982). *Whitebacked planthopper, Sogatellia furcifera (Horwath), a problem on rice in Asia.* Saturday Seminar, June 5,1982, IRRI, Los Banos, Philiphines.

Mondal, A.H., N.R. Sharma, M.M. Islam, M.A. Haque, M.A. Hossain, and S.A. Miah. (1992). Effect of some soil properties on the development of bacterial blight. *Bangladesh J. Pl. Path.* 8: 17-21.

Mukherjee, S.K. and B.N. Bagchi. (1965). Role of some trace elements in controlling brown leaf spot disease of paddy. *West Bengal J. Agric.* 2: 62-65.

Mukhopadhyay, S. and A.K. Chawdhury. (1973). Some epidemiological aspects of tungro virus disease of rice in West Bengal. *Int. Rice Commn. Newslet.* 22: 44-57.

Mundkur, B.B. (1949). *Fungi and plant disease.* Macmilan & Co. Ltd., London.

Munjal, R.L. (1971). *Epidemiology and control of Karnal bunt of wheat.* Proc. Sym. Epidemiology, Forecasting and control of plant diseases, INSA (Abs.) p. 42.

Munjal, R.L., and M.S. Chatrath. (1976). Studies on mode of infection of *Neovossia indica* incident of Karnal bunt of wheat. *J. Nuclear Agric. Biol.* 5: 40-41.

Nadarajan, L., and N. Rajappan. (1987). A new rice leaf folder (LF) in Kerala. *Int. Rice. Res. Newsletter,* 12: 23.

Nagarajan, S. (1991). Epidemiology of Karnal bunt of wheat incited by *Neovossica indica* and an attempt to develop a disease prediction system. CIMMYT Technical Report. International Maize and Wheat Improvement Centre (CIMMYT), Mexico.

Nagarajan, S., and L.M. Joshi. (1975). A historical account of wheat rust epidemics in India and their significance. *Cereal Rusts Bull.* 3: 29-33.

Nagarajan, S. and L.M. Joshi. (1985). Epidemiology in Indian Subcontinent. In *The cereal rusts* Vol. II, eds. A.P. Roelfs and W.R. Bushnell, Academic Press, New York, pp. 371-402.

Nagarajan, S., L.M. Joshi, K.D. Srivastava, and D.V. Singh. (1980). Epidemiology of brown and yellow rusts of wheat in north India IV. Disease management recommendations. *Cereal Rusts Bull.* 7: 15-20.

Nagarajan, S., S.K. Nayar, and S.C. Bhardwaj. (1992). Wheat rusts. In *Plant disease* Vol. I, eds. U.S. Singh, A.N. Makhopadhyay, J. Kumar and H.S. Chaube, Prentice Hall Englewood, pp. 25-40.

Nagarajan, S. and H. Singh. (1975). The Indian stem rusts rules: A concept on the spread of wheat stem rusts. *Pl. Dis. Reptr.* 59: 133-136.

Nayar, S.K. and S.C. Bhardwaj. (1998). Management of wheat rusts in India. In *IPM system in Agriculture*, eds. Rajeev K. Upadhyay, K.G. Mukerji and R.L. Rajak, Vol 3, Cereals, pp. 305-324.

Nayar, S.K. and S. Nagarajan, and P. Bahadur. (1988). Virulence distribution pattern of *Puccinia recondita F.* sp. *tritici* in India during 1983-86. *Plant Disease Research,* 3: 203-210.

Nayar, S.K., M. Prasher, J. Kumar, and S.C. Bhardwaj. (1990). Combating variation in *Puccinia recondite tritici* in India. *Pl. Dis. Res.* 10-13.

Padmanabhan, S.Y. (1983). Integrated control of bacterial blight of rice. *Oryza,* 20: 188-194.

Panda, M. and Y.S. Rao. (1969). *Evaluation of losses caused by incidence of Hirschmanniella mucronata Das in rice* (Abst.) All India Nem. Symp. New Delhi, India 1. Aug. 21-22, 1969.

Panda, S.K. and N. Shi. (1989). Carbofuran induced rice leaffolder (LF) resurgence. *Int. Rice Res. Newsl. 14: 30.*

Panwar, M.S. and Y.S. Rao. (1998). Status of Phytonematodes as pests of rice. In *Nematode disease in plants*, ed. P.C. Trivedi, CBS Publishers and Distibutors, pp. 49-81.

Paracer, C.S and D.S. Chahal. (1963). Sheath blight of rice caused by *Rhizoctonia solani*-A new record of India. *Curr. Sci.* 32: 328-329.

Parashar, M., S. Nagarajan, L.B. Goel, and J. Kumar. (1995). *Report of the coordinated experiments 1994-95.* Crop Protection (Pathology) AICWIP, DWR, KARNAL, 206 p.

Park, M. and Bertus, L.S. (1932). Sclerotial diseases of rice in Ceylon II. *Ceylon Journal of Science*, A 11: 343-359.

Paroda, R.S., T. Woodhead, and R.B. Singh. (1994). *Sustainability of rice-wheat production system in Asia.* RAPA Publication: 1994/11. Bangkok, Thailand: Regional office for Asia and Pacific (RAPA), Food and Agriculture Organization of the United Nations. 209 p.

Paruthi, I.J. (1992). Nematode Pests of Wheat and Barley–*Anguina tritici*. In *Nematode pest of crops*, eds. D.S. Bhatti and R.K. Walia, CBS Publishers and Distributors, New Delhi, India.

Patel, M.K., G.W. Dhande, and Y.S. Kulkarni. (1950). A modified treatment against loose smut of wheat. *Curr. Sci.* 19: 324-325.

Person, T.D. (1954). Destructive outbreak of loose smut in Georgia wheat fields. *Pl. Dis. Reptr.* 38: 442.

Prakasa Rao, P.S. (1972). Ecology and control of *Tryporyza incertulas* Walker and *Pachydiplosis oryzae* wood mason in rice. Ph.D. Thesis, Utkal University, Vani Vihar, Bhubaneswar.

Prakasa Rao, P.S., P. Israel, and Y.S. Rao. (1971). Epidemiology and control of the rice hispa *Dicladispa armigera* oliver. *Oryza*, 8: 345-359.

Prasad, J.S., M.S. Panwar, and Y.S. Rao. (1992). Nematode pests of rice. In *Nematode pests of crop*, eds. D.S. Bhatti and R.K. Walia, CBS Publishers and Distributors, New Delhi, India. pp. 43-61.

Prasad, J.S. and Y.S. Rao. (1984). Ectoparasitic nematode fauna of upland rice soils. *Beitrage Zur Tropischen lond Wirtschaft und Vertemar Medizin*, 22: 79-82.

Prasada, R. (1947). Discovery of the uredostage connected with the aecidia so commonly found on species of *Berberis* in Shimla hills. *Indian J. Agric. Sci.* 17: 137-151.

Prasada, R. (1951). Rusts on wild grasses. *Curr. Sci.* 20: 243.

Premalatha Dath, A. (1974). Factors influencing the development of bacterial blight of rice incited by *Xanthomonas oryzae.* Ph.D. Thesis. Ravishanker University, Raipur, India.

Prot, J.C. (1993). Monetary value estimates of nematode problems, research proposal and priorities: Rice research example in South and Southeast Asia. *Fundamental and Applied Nematology*, 16: 385-388.

Puri, S.N. (1999). Integrated Pest Management. In *Training mannual No. 5*, ed. M. Sehgal, NCIPM, New Delhi, India, pp. 1-21.

Puri, S.N. (2000). *Plant pest management curriculum development in India.* Paper

Presented at *The Expert Consultation on the Plant Pest Management Curriculum Development* at National Agricultural Universities and Related Universities, April 25-28, 2000, RAP, Bangkok.

Rahman, M.L. and I. McGeachie. (1982). Occurrences of White Tip disease in deepwater rice in Bangladesh. *Int. Rice Res. Newsl.* 7: 15.

Raj, S.M. and Y.B. Morachan. (1973). Influence of levels of fertilizers and methods of application of insecticides on 'IR-8' rice. *Madras Agric. J.* 60: 179-183.

Raju, N., M. Goplan, and G. Balasubramanian. (1990). Ovicidal action of insecticides, moult inhibitors and fungicides on the eggs of rice leaffolder and stemborer. *Indian J. Plant Prot.* 18: 5-9.

Raju, N., R. Saroja, and M. Suriachandraselvan. (1988). Compatible insecticides and fungicides to control leaf folder (LF) and sheath rot in rice. *Int. Rice Res. Newsl.* 13: 26.

Ramamoorthy, V.V. and S. Jayaraj. (1977). Efficacy of foliar and water surface application of insecticides in control of leaffolder, *Cnaphalocrocis mendinalis* Guenee. *Madras Agric. J.* 64: 252-254.

Rao, A.K. (1996). Innovative methods to manage sheath blight of rice. *J. Mycol. Res.*, 34: 13-19.

Rao, Y.S. (1985). Research on rice nematode. In *Rice research in India*, ed. S.Y. Padmanabhan, ICAR, pp. 591-642.

Rao, Y.S. and P. Israel. (1964). Recent developments and future prospects for the chemical control of the rice stem borer in India. In *The major insects pests of rice plant*, IRRI, Manila, The Johns Hopkins Press, Baltimore, Maryland, pp. 317-334.

Rao, Y.S. and J.P. Kulshreshtha. (1985). Insect Pests of Rice. In *Rice research in India*, ed. S.Y. Padmanabhan, ICAR, pp. 550-590.

Rao, Y.S. and C. Nandkumar. (1973). *Annual report of research conducted under PL 480 grant* fg-ln-470, Cutack, pp. 1-14.

Rao, Y.S. and M.S. Panda. (1970). *Study of plant parasitic nematodes affecting rice production in the vicinity of Cuttack (Orissa), India.* Final Tech. Rept. (USPL 480) Central Rice Mimphis, Tennessee.

Rao, Y.S., J.S. Prasad, and M.S. Panwar. (1986). Nematode problems in rice: Crop losses, symptomatology and management. In: *Plant parasitic nematodes of India*, eds. G. Swarup and D.R. Dasgupta, IARI, New Delhi, pp. 279-299.

Rao, Y.S. and Yadava, C.P. (1971). *Investigations on the white tip nematode Aphlenchoides besseyi of rice.* Annual Report of Central Rice Research Institute, Cuttack, 185-187.

Rao, Y.S., P. Israel, Y.R.V.J. Rao, and H. Biwas. (1971). Studies on nematodes of rice and rice soils. IV, Survey of Nematodes. *Oryza*, 8: 65-74.

Rao, Y.S., J.S. Prasad, C.P. Yadav, and C.R. Padalia. (1984). Influence of rotation crops in rice soils on the dynamics of plant parasitic nematode populations. *Biological Agriculture and Horticulture*, 2: 69-78.

Raychaudhari, S.P., M.D. Mishra, and A. Ghosh. (1967a). Preliminary note on the transmission of a virus disease resembling tungro of rice in India and other virus like symptoms. *Pl. Dis. Reptr.* 5: 300-301.

Raychaudhari, S.P., M.D. Mishra, and A. Ghosh. (1967b). Preliminary note on the occurrence and transmission of rice yellow dwarf in India. *Pl. Dis. Reptr.* 51: 1040-1041.

Razzaque, M.A. and A.B.S. Hossain. (1991). The wheat development program in Bangladesh. In *Wheat for the nontradiational warm areas*, ed. D.A. Saunders, pp. 34-73, CIMMYT, Mexico, 549 p.

Rowell, J.B. (1985). Evaluation of chemicals for rust control. In *The cereal rusts Vol II*, eds. A.P. Roelfs and W.R. Bushnell, Academic Press, New York, pp. 371-402.

Royer, M.H. and J. Rytter. (1988). Comparison of host ranges of *Tilletia indica* and I. Berclayans. *Plant Dis.* 72: 133-136.

Saari, E.E. and J.M. Prescott. (1985). World Disribution in Relation to Economic Losses. In: *The Cereal Rusts* Vol. II, eds. A.P. Roelfs and W.R. Buchnell, Academic Press, New York, pp. 259-298.

Sadasivan, T.S., S. Suryanaryanan, and L. Ramakrishnan. (1965). *Influence of temperature on rice blast disease.* Proc. Symp. IRRI, July 1963, Johns Hopkins Press, Baltimore, Maryland, 163-171.

Sahni, V.M. and D.K. Bhutani. (1966). A note on control of termites damaging Wheat. *Sci. Cult.* 32: 426-427.

Sasser, J.N. and D.W. Freckman. (1987). A world perspective on nematology: The role of society. In *Vistas on nematology. A commemoration of the twenty-fifth anniversary of the Society of Nematologists*, eds. J.A. Veech and D.W. Dickson, Hyattsville, Maryland, USA, Society of Nematologists, Inc., pp. 7-14.

Savary, S., N.P. Castilia, F.A. Elazequi, C.G. McLaren, M.A. Ynalvez, and P.S. Teng. (1995). Direct and indirect eeffects of nitrogen supply and disease source structure on rice sheath blight spread. *Phytopath.* 85: 959-965.

Sehgal, M. (1991). *Studies on moisture stress induced morphological and morphometrical changes and survival in nematodes.* Ph.D. Thesis, IARI, 114 pp.

Sehgal, M. (1999). *Surveillance of nematodes and their management in rice-wheat in Northern India*, Annual Report NCIPM, 1998-99, pp. 50-51.

Sehgal, M. and H.S. Gaur. (1999). *Important nematode problems of India.* Tech. Bull. NCIPM, New Delhi, India, 16 p.

Sehgal, M., N. Kalra, and H.S. Gaur. (2000). Rice root nematode zonation in India. *Annal. Pl. Prot. Sci.* (In Press).

Sekhon, K.S., K.S. Gill, S.K. Randhawa, and S.K. Saxena. (1980). Effect of Karnal bunt on quality characteristics of wheat. *Proc. 19th all India Wheat Res. Workers Workshop.* AICWIP, Anand, Gujarat.

Sen, P., and S. Chakravorty. (1970). Biology of hispa (*Dicladispa armigera*) Oliv. (Coleoptera: Chrysomelidae). *Indian J. Ent.* 32: 123-126.

Sharma, S.B., C. Johansen, and S.K. Midha. (1998). *Nematode pests in rice-wheat-legume cropping systems*, Proc. of a Regional Training Course 15 Sept. 1997, Facilitation Unit Rice-Wheat Consortium for Indo-Gangetic Plains, IARI, Pusa, New Delhi, India, p. 105.

Sharma, S.B., N.S. Price, and J. Bridge. (1997). *The past, present and future of plant nematology in International Agricultural Research Centres*, Nematological Abstracts, 66: 119-142.

Sharma, S.K. and R. Prasada. (1967). *Origin of physiologic races in wheat rusts through mutations.* Proc. First Int. Symp. Plant Pathology, IARI, New Delhi.

Singh, A. and R. Prasad. (1978). Date of sowing and meterological factors in relation to occurrence to Karnal bunt of wheat in U.P. Tarai. *Indian J. Mycol. Pl. Pathol.* 8: 2.

Singh, A. and R.C. Rai. (1986). Chemical control of wheat diseases in India. In *Problem and progress of wheat pathology in South Asia*, eds. L.M. Joshi, D.V. Singh and K.D. Srivastava, Malhotra Publishing House, New Delhi, pp. 374-397.

Singh, A., A.N. Tiwari, and R.C. Rai. (1985). Control of Karnal bunt of wheat by spray of fungicides. *Indian Phytopath.* 38: 104-108.

Singh, A., Y.L. Yene, and H.S. Chaube. (1974). Detection of loose smut infected plants before ear emergence in wheat cultivar Sonalika (RR21). *Curr. Sci.* 43: 295.

Singh, D.V. (1986). Bunts of wheat in India. In *Problem and progress of wheat pathology in South Asia*, eds. L.M. Joshi, D.V. Singh and K.D. Srivastava, Malhotra Publishing House, New Delhi. pp. 124-161.

Singh, D.V., R. Aggarwal, J.K. Shreshtha, B.R. Thapa, and H.J. Dubin. (1989). First report of *Tilletia indica* on wheat in Nepal. *Plant Dis.* 73: 273.

Singh, D.V., H.S. Dhaliwal, and R.J. Metzgar. (1988). Innoculum and time for screening against Karnal bunt disease of wheat. *Indian Phytopath.* 41: 632-633.

Singh, D.V., L.M. Joshi, and K.D. Srivastava. (1983). Karnal bunt, new threat to wheat in India. In *Recent advances in plant pathology*, eds. A. Hussain, Kishan Singh, B.P. Singh and V.P. Agnihotri, Print House (India), Lucknow, pp. 121-135.

Singh, D.V., K.D. Srivastava, and L.M. Joshi. (1979). Prevent losses from Karnal bunt. *Indian Farming.* Feb. 1979.

Singh, D.V., K.D. Srivastava, and L.M. Joshi. (1980). *Karnal bunt of wheat: Factors associated with its spread.* Proc. Indian National Seminar on Genetics and Wheat improvement, Hisar, Feb. 8-10, p. 29.

Singh, D.V., K.D. Srivastava, R. Aggarwal, S. Nagarajan, and B.R. Verma. (1999). Mulching as a means of Karnal bunt management. *Plant Dis. Res.* 6: 115-116.

Singh, P.J., H.S. Dhaliwal, and K.S. Gill. (1989). Chemical control of Karnal bunt (*Neovossia indica*) of wheat by single spray of fungicides at heading. *Indian J. Agric. Sci.* 59: 131-133.

Singh, R.B. and R.S. Paroda. (1994). Sustainability and productivity of rice-wheat systems in the Asia-Pacific region: Research & technology development needs. In: *Sustainability of rice-wheat production systems in Asia*, eds. R.S. Paroda, T. Woodhead, and R.B. Singh. RAPA Publication: 1994/11, FAO, Bangkok 1-35.

Singh, S.L. and P.P. Singh. (1985). Cultural practices and Karnal bunt (*Neovossia indica*) incidence. *Indian Phytopath.*, 38: 594.

Singh, V.S. (1999). Pest management strategies in wheat. In *Training mannual No. 6*, eds. R. N. Singh, T.P. Trivedi, M. Sehgal and Ashok Kumar, NCIPM, New-Delhi, India. pp. 40-47.

Srivastava, A.S. (1966). Occurrence of *Aphlenchoides* sp. causing white tip of rice on "Taichung (Native)-1" variety of paddy for first time in Uttar Pradesh, India. *Labdev J. Sci. Tech.*, 4: 281-282.

Srivastava, D.N. (1969). Bacterial blight and streak diseases of rice in India. *Indian Phytopath.* 22: 163-164.

Srivastava, D.N. and Y. Rao. (1963). Epidemic of bacterial blight of rice in North India. *Indian Phytopath.* 16: 393.

Srivastava, D.N. and Y.P. Rao. (1966). Symptoms and diagnosis of bacterial blight of rice. *Curr. Sci.* 35: 60-61.

Srivastava, D.N., Y.P. Rao, and J.C. Durgapal. (1966). Can "Taichung (Native 1)" stand up to bacterial blight. *Indian Fmg.* 16: 15.

Srivastava, K.D., D.V. Singh, and R. Aggarwal. (1998). Loose smut of wheat. In *IPM system in agriculture*, eds. Rajeev K. Upadhyay, K.G. Mukherji and R.L. Rajak, Aditya Books Pvt. Ltd., New Delhi, India Vol 3. pp. 291-293.

Srivastava, K.D., D.V. Singh, R. Aggarwal, P. Bahadur, and S. Nagarajan. (1992). Occurrence of loose smut and its source of resistance in wheat. *Indian Phytopath.* 45: 111-112.

Stakman, E.C. and J.G. Harrar. (1957). *Principles of plant pathology.* Ronald Press. New York.

Stubbs, R.W. (1985). Stripe rusts. In *The cereal rusts* Vol. II, eds. A.P. Roelfs and W.R. Bushnell, Academic Press, New York, pp. 61-91.

Subramaniam, T.R., T.S. Muthukrishnan, C.V. Sivakumar, P. Rajagoplan, Meerzianudeen, J. Chandrasekhran, and G. Rajendran. (1973). *Current work and recent findings in Tamil Nadu nematodes.* Proc. All India Nematology Workshop, Mar. 1973, New Delhi.

Subramanian, A. and V.K.R. Sathiyanadam. (1983). Studies on the relative efficacy of pesticide application methods on the control of rice pests. In *Pest management in rice*, eds. S. Chekkuah and Balasubramanian, Tamil Nadu Agricultural University, Coimbatore, pp. 89-94.

Sundararaman, S. (1922). *Helminthosporium* disease of rice. *Bull. Agric. Res. Inst. Pusa.* 128 pp.

Suryanaranyanan, S. (1958). Role of nitrogen in host suspectibility to *Pyricularia oryzae. Curr. Sci.* 27: 447-448.

Suryanaranyanan, S. (1966). Enivornment and the blast disease. *Bull. Indian Phytopath. Soc.* 3: 110-114.

Suzuki, H. (1975). Meteorological factors in the epidemiology of rice blast. *Ann. Rev. Phytopath.* 13: 239-256.

Swaminathan, M.S., S.P. Raychaudhuri, L.M. Joshi, and S.D. Gera. (1971). Incidence of black point and Karnal bunt in north-western plains zone during 1969-70. *Wheat Disease Newsletter.* 4: 1-4.

Tandon, J.P. and A.P. Sethi. (1991). *Wheat Production Technology*, Directorate of Wheat Research, Karnal, Haryana, India.

Taylor, A.L. (1969). Nematode parasites. In *Nematodes of tropical crops*, ed. Peachy, Tech. Commn. Commonwealth. Bur. Helminth. No. 40: 264-268.

Thind, B.S., and R.K. Mehra. (1992). Chemical control of bacterial blight of rice. *Pl. Dis. Res.* 7: 226-234.

Uthamasamy, S. (1985). Problems and priorities in the management of rice leaf folder. In *Integrated pest and disease management*, ed. S. Jayaraj, Tamil Nadu Agricultural University, Coimbatore pp. 54-61.

Warham, E.J. (1984). A comparison of inoculation methods for Karnal bunt *Neovossia indica. Phytopathology*, 74: 856-857.

Warham, E.J. (1990a). Effect of *Tilletia indica* infection on viability, germination and vigour of wheat seed. *Plant Dis.* 74: 130-132.

Warham, E.J. (1990b). A comparison of inoculation techniques for assessment of germplasm suspectibility to Karnal bunt (*Tilletia indica*) disease of wheat. *Ann. App. Biol.* 116: 43-60.

Warham, E.J., J.M. Prescott, and E. Griffiths. (1989). Effectiveness of chemical seed treatment in controlling Karnal bunt disease of wheat. *Plant Dis.* 73: 585-589.

Yadava, C.P., G. Santaram, P. Israel, and M.B. Kalode. (1972). Life history of rice leaf roller *Cnaphalocrocis medinalis* (Guenee) (Lepidoptera: Pyraladae) and its reaction of rice varieties and grasses. *Indian J. Agric. Sci.* 42: 520-523.

Plant Parasitic Nematodes
in Rice-Wheat Based Cropping Systems
in South Asia:
Present Status and Future Research Thrust

S. B. Sharma

SUMMARY. Rice-wheat based cropping systems in South Asia are among the most highly evolved production systems in the world. The productivity growth of these systems in South Asia is declining due to several factors including the biotic stresses of plant parasitic nematodes. This article reviews the research on rice and wheat nematodes in a cropping systems perspective and identifies nematodes that have wide host ranges and are greatly influenced by the crop rotations and sequences. These polyphagous nematodes can cause significant damage to rice and (or) wheat crops, either alone or in combination with other microorganisms. The research projects on pest management in the region lack inter-disciplinarity and it is important for nematologists to become integral members of interdisciplinary teams on improving the productivity and sustainability of the rice-wheat cropping systems. *[Article copies available for a fee from The Haworth Document Delivery Service: 1-800-342-9678. E-mail address: <getinfo@haworthpressinc.com> Website: <http://www.HaworthPress.com> © 2001 by The Haworth Press, Inc. All rights reserved.]*

KEYWORDS. Management, nematodes, research thrust, rice, South Asia, wheat

S. B. Sharma is affiliated with Crop Improvement Institute, 3 Baron Hay Court, South Perth, WA 6151, Australia (E-mail: sbs_nema@hotmail.com; ssharma@agric.wa.gov.au).

[Haworth co-indexing entry note]: "Plant Parasitic Nematodes in Rice-Wheat Based Cropping Systems in South Asia: Present Status and Future Research Thrust." Sharma, S. B. Co-published simultaneously in *Journal of Crop Production* (Food Products Press, an imprint of The Haworth Press, Inc.) Vol. 4, No. 1 (#7), 2001, pp. 227-247; and: *The Rice-Wheat Cropping System of South Asia: Efficient Production Management* (ed: Palit K. Kataki) Food Products Press, an imprint of The Haworth Press, Inc., 2001, pp. 227-247. Single or multiple copies of this article are available for a fee from The Haworth Document Delivery Service [1-800-342-9678, 9:00 a.m. - 5:00 p.m. (EST). E-mail address: getinfo@haworthpressinc.com].

227

INTRODUCTION

The rice (*Oryza sativa*)-wheat (*Triticum aestivum*) cropping systems in South Asia cover a vast area of about 12 million hectares in Bangladesh, India, Nepal and Pakistan (Table 1). About 32% of total rice area and 42% of total wheat area in the region are under rice-wheat rotations (Singh and Paroda, 1994; Hobbs and Morris, 1996). The rice and wheat together account for 90% of the cereal production in South Asia. Other crops commonly cultivated in the rice-wheat cropping system rotation are: mustard (*Brassica juncea*), rapeseed (*B. juncea*), groundnut (*Arachis hypogaea*), sunflower (*Helianthus annuus*), mungbean (*Vigna mungo*), chickpea (*Cicer arietinum*), blackgram (*Vigna radiata*), lentil (*Lens culinaris*), linseed (*Linum usitatissimum*), berseem (*Trifolium alexandrium*), sesbania (*Sesbania aegyptica*), fodder sorghum (*Sorghum vulgare)* and pearl millet (*Pennisetum glaucum*), barley (*Hordeum vulgare*), potato (*Solanum tuberosum*), sugarcane *(Saccharum offcinarum)*, jute (*Corchorus capsularis*) and vegetables including onion (*Allium cepa*) and garlic (*A. sativum*) (Hobbs and Morris, 1996).

The productivity of rice and wheat in South Asia, particularly in the intensively cultivated states of Punjab in India is either declining or has become stagnant. A deficit in cereal production of 20-25 million t/year is envisaged by the year 2020–and to circumvent this deficit a production growth rate of 2.5% per year is needed (Hobbs and Morris, 1996). The causes of productivity decline are multifold and complex. Plant parasitic nematodes are among important causes for productivity decline globally and they are estimated to cause 10% yield loss in rice and 7% in wheat (Sasser and Freckman, 1987). Economically, nematode-induced losses to rice in South and South East Asia are estimated at U.S. \$5.2 billion (Prot, 1993a). In parts of Orissa in eastern India, losses due to the root-knot nematode (*Meloidogyne graminicola*) and lesion nematode (*Pratylenchus indicus*) on rice are estimated between 18% and 64% (Ali, 1997). More than 144 species of nema-

TABLE 1. Area of rice-wheat systems in the South Asian rice-wheat countries.

Country	Total rice area (Mha)	Total wheat area (Mha)	Rice-wheat rotation area as % of total rice area	Rice-wheat rotation area as % of total wheat area	Total rice-wheat area (Mha)
Bangladesh	10.20	0.59	5	85	0.50
India	42.10	23.5	23	40	9.6
Nepal	1.30	0.58	35	84	0.5
Pakistan	2.0	7.9	72	19	1.5

Table adapted from Paroda and Singh, 1994.

todes are reported on rice and 90 species on wheat (Dhawan and Pankaj, 1998; Panwar and Rao, 1998). The losses caused by these nematodes depend on many factors and nematode densities at the time of planting are crucial in estimating the likely damage to a crop. The cropping cycles greatly influence the densities of polyphagous nematodes. For example, if a preceding crop is a poor host of a given nematode, the nematode-induced damage and yield loss to a succeeding susceptible crop may be unimportant. Changes in cropping patterns, and economic and policy changes affect the status of nematode pests of a crop (Prot, 1994a).

This chapter examines the work done on rice and wheat nematodes in a cropping systems perspective, assesses the influence of other crops in the cropping system on nematode populations, discusses interactions of nematode species with other soil microorganisms, and proposes future research thrust. It is envisaged that by judging the nematode problems within a cropping systems perspective, a more sustainable approach to pest management can be developed. This approach represents a step forward towards integration of crop specific research into a more holistic vision.

NEMATODE PESTS SPECIFIC TO RICE IN RICE-WHEAT BASED CROPPING SYSTEMS

Rice is a low-value crop and nematode problems of low-value crops are not fully recognized by farmers and policy makers (Prot, 1994b). Important genera of plant parasitic nematodes that are associated with rice in South Asia are *Aphelenchoides* (in all the ecosystems), *Ditylenchus angustus* (in deepwater rice), *Hirschmanniella* spp. (in permanently or intermittently flooded ecosystems), *Pratylenchus* spp. (in upland rice), and *Meloidogyne graminicola* (in all the ecosystems).

White Tip

Aphelenchoides besseyi causes white tip of rice. In parts of Bangladesh, more than 50% fields in deep water rice are infested with this nematode (Rahman and McGeachie, 1982; Rahman and Taylor, 1983). The white tip has also been reported from Pakistan, Myanmar and Nepal (Karki, 1997; Swe, 1997). When infested seeds are sown, the dormant adult females rapidly become active and feed ectoparasitically around the apical meristem during early plant growth. The new leaves have chlorotic tips. The infected plants mature late and have sterile panicles borne on tillers produced from upper nodes. The nematode population increases rapidly at late tillering and the reproductive phase of plant growth. Annual loss due to improper grain filling and reduction in chlorophyll content is estimated at 4% in deepwater rice in Bangladesh (Catling et al., 1979), and 21-46% in India (Nandkumar et al., 1975). The yield loss data are not available from Nepal and Pakistan. The economic damage threshold level of *A. besseyi* in parts of Tamil Nadu in southern India is 3-nematodes/grain (Muthukrishnan, Rajendran and Chandrashekaran, 1974). The nematode infection favors invasion by *Trichoconis padwickii*, *Corticium sasaki* and *Acrocylindrium oryzae* that increase the severity of stem rot and leaf blast diseases (McGawley, Rush and Hollis, 1984; Table 2). Saprophytic and pathogenic fungi are also good hosts of the nematode, therefore, chances of recurrence in nematode infested fields cannot be ruled out even when nematode free seeds are used (Sivakumar, 1987). Options for management of white tip nematodes have been developed (see Sharma and Rahaman, 1998).

Ufra

The ufra disease of rice, caused by *D. angustus*, is called "dak pora" in Bangladesh (Rahman and Taylor, 1983) and causes severe damage in very wet years (Bridge, Luc and Plowright,1990). In localized areas, severe infestation can even cause complete crop loss (Ou, 1972). In Bangladesh, up to 80% damage has been reported (Rahman, Sharma and Miah, 1981). Yield losses range from 10% in West Bengal (Rao, Jayaprakash and Mhoanty, 1986) to 50% in Uttar Pradesh (Singh, 1953).

The residues of the preceding crop and irrigation water are the sources of nematode infestation. The nematode infection produces white patches, or

TABLE 2. List of pathogenic fungi and bacteria with which plant parasitic nematodes interaction rice-wheat based cropping systems.

Crop	Nematode	Other pathogens
Rice	Aphelenchoides besseyi	Trichoconis padwickii, Sclerotium oryzae, Pyricularia oryzae, Corticum sasaki, Acrocylindricum oryzae, Enterobacter agglomerans
	Ditylenchus angustus	Pyricularia oryzae, Fusarium sp., Cladosporium sp.
	Hoplolaimus indicus	Sclerotium rolfsii, Pseudomonas sp., Fusarium oxysporum, Rhizoctonia solani
	Hirschmanniella oryzae	Rhizoctonia solani
	Heterodera oryzicola	Sclerotium rolfsii
Wheat	Heterodera avenae	Rhizoctonia solani
	Anguina tritici	Clavibacter tritici, Dilophospora alopecuri, Ustilago tritici, Tilletia foetida, Neovossia indica

speckles at the bases of young leaves. Infected panicles are usually crinkled with empty, shriveled glumes and flag leaf are twisted and distorted. In Bangladesh and eastern India, the rainy season crop suffers severe damage and post rainy season crops suffer moderate to insignificant damage. The feeding of nematode on young foliar tissues facilitates invasion of *Fusarium* and *Cladosporium* sp. (Vuong, 1969). The nematode infected plants become more susceptible to *Pyricularia oryzae* (Mondal et al., 1986). The host range of *D. angustus* is confined to wild and cultivated species of rice and *Leeria hexandra*, *Echinocloa colona* and *Oryza* spp. (Panwar and Rao, 1998). Options for management of this nematode have been developed.

POLYPHAGOUS NEMATODES
IN RICE-BASED CROPPING SYSTEMS

Wide ranges of polyphagous nematode species have been reported from rice-based cropping systems (Bridge, Luc and Plowright, 1990). *Hirschmanniella* spp., *Hoplolaimus* spp., *Heterodera* spp., *Meloidogyne* spp., *Pratylenchus* spp., *Tylenchorhynchus* spp., *Helicotylenchus* spp., *Xiphinema* spp., *Paralongidorus* spp., *Longidorus* spp., and *Criconemoides* are important polyphagous nematodes. These nematodes do not produce specific aboveground symptoms of their attack and stunting of plants is a symptom commonly associated with infection of these nematodes. Crops that are generally grown in rotation with rice are susceptible to these species (Table 3).

Stunting

The lance nematodes (*Hoplolaimus galeatus*, *H. indicus*, *H. columbus*) are commonly associated with rice in Bangladesh, India, Nepal and Pakistan. These nematodes are commonly found in the states of Uttar Pradesh, Madhya Pradesh, Bihar and Andhra Pradesh in India where sorghum and finger millet are grown in rotation with rice (Rao, Prasad and Rao, 1984). At-planting time population densities of 100-10,000 nematodes/plant can reduce number of tillers by 21-36% and grain yields by 10-20% (Ramana and Rao, 1977). The economic damage threshold level of lance nematodes is 65-nematodes/100 g of soil at sowing time. The lance nematodes enhance the chances of root infection by *Sclerotium rolfsii*, *Pseudomonas* sp., *Fusarium oxysporum* and *Rhizoctonia solani* (Ramana and Rao, 1976).

The rice root nematode species (*Hirschmanniella* spp.) are important parasites of rice. Four species, *H. gracilis*, *H. mucronata*, *H. oryzae and H. spinicaudata*, cause damage to rice crops in irrigated, lowland and deep water rice (Rahman and Taylor, 1983; Prasad and Rao, 1984a, 1984b). During early

TABLE 3. Host status of crops that are commonly grown in rice-wheat based cropping systems to associated plant-parasitic nematodes.

Nematode	Crop											
	1	2	3	4	5	6	7	8	9	10	11	12
Aphelenchoides besseyi	++	−	−	−	−	−	−	−	−	−	−	−
Ditylenchus angustus	++	−	−	−	−	−	−	−	−	−	−	−
Hoplolaimus columbus	++	++	+	+	*	+	+	+	*	++	+	+
H. indicus	++	++	+	+	+	+	+	+	+	+	+	+
H. galeatus	++	+	+	+	*	+	+	+	*	*	+	*
Hirschmanniella gracilis	++	−	−	−	−	−	−	−	*	−	+	−
H. mucronata	++	−	−	−	−	−	−	−	*	−	+	−
H. imamuri	++	+	−	−	−	−	−	−	*	−	+	−
H. oryzae	++	−	−	−	−	−	−	−	*	−	+	−
H. spinicaudata	++	+	−	−	−	−	−	−	*	−	+	−
Helicotylenchus abunaamai	++	+	+	+	*	*	+	*	*	*	+	+
H. mucronatus	+	+	+	*	*	*	+	*	*	*	+	+
H. erythrinae	+	+	+	*	*	*	+	*	*	+	+	+
H. dihystera	+	+	+	+	*	*	+	*	+	+	+	+
H. multicinctus	+	+	+	*	*	*	+	*	*	*	+	+
Tylenchorhynchus annulatus	++	+	+	+	*	*	+	*	*	+	+	+
T. martini	+	+	+	*	*	*	+	*	*	+	+	+
T. brassicae	+	+	+	*	*	*	+	*	*	+	+	+
T. vulgaris	−	++	+	+	*	+	+	*	*	+	+	+
Pratylenchus brachyurus	++	+	++	+	*	+	*	+	*	++	+	+
P. indicus	++	++	+	+	*	+	+	*	+	+	+	+
P. thornei	++	++	+	+	*	*	*	+	+	++	+	+
P. zeae	++	+	+	+	*	+	+	*	*	*	+	+
Meloidogyne graminicola	++	+	+	−	−	+	−	+	−	+	+	−
M. javanica	*	++	++	+	+	−	+	−	+	++	++	++
Heterodera avenae	−	++	−	−	−	−	−	−	−	−	−	−
H. oryzicola	++	−	−	−	−	−	−	−	−	−	−	−
Anguina tritici	−	++	−	−	−	−	−	−	−	−	−	−

1. rice, 2. wheat, 3. legumes, 4. groundnut, 5. berseem , 6. sorghum, 7. pearl millet, 8. sesbania, 9. potato, 10. sugarcane, 11. vegetables, 12. jute. ++ good host; + host ; * likely host; − non-host (not reported).

seedling stage, the symptoms are similar to that of nitrogen deficiency. Leaves of infected seedlings are initially yellow and later turn brown and brittle with ash colored tips. *Hirschmanniella oryzae* is ubiquitous in rice growing areas; more frequent and abundant in clay soils and less frequent in dry and alkaline soils (Prasad et al., 1987), whereas *H. mucronata* occurs in sandy soils (Israel, Rao and Rao, 1966). Sympatric occurrence of more than one species of *Hirschmanniella* is common. Aboveground symptoms of damage caused by *Hirschmanniella* spp. are non-specific. Affected plants are stunted with poor root growth, fewer tillers, and reduced yield (Edward, Sharma and Agnihotrudu, 1985). Yield losses of up to 25% occurs due to *H. mucronata* and up to 70% due to *H. oryzae* (Mathur and Prasad, 1972). Economic damage threshold levels are 120 and 100 nematodes/100 g of soil for *H. oryzae* and *H. mucronata*, respectively (Rao, Prasad and Panwar, 1985; Panda and Rao, 1971). The nematodes favor secondary infection by microorganisms and development of fungi such as *Rhizoctonia solani* (Rao, 1970; Gokulpalan and Nair, 1983). Some other hosts of the *H. oryzae* include *Abelmoschus esculentum, Gossypium hirsutum, Lycopersicon esculentum, Pennisetum typhoides, Saccharum offcinarum, Triticum aestivum* and *Zea mays*.

The spiral nematodes (*Helicotylenchus* spp.) are common on upland rice. In some situations, spiral nematodes may cause up to 90% yield losses (Rao, Prasad and Rao, 1984). *Helicotylenchus abunaamai*, an ectoparasite polyphagous species prefers upland and well drained rice soils in India. It feeds on roots, rootlets and root hair, but juveniles may become endoparasitic when roots are heavily infected (Padhi and Das, 1985). *Helicotylenchus mucronatus* and *H. erythrinae* have been reported frequently on rice in Pakistan (Malik and Yasmeen, 1978).

The stunt nematodes (*Tylenchorhynchus* spp.) are common in upland, lowland, and deep-water rice; *Tylenchorhynchus annulatus* is most widely distributed. These nematodes feed on epidermis and cortex causing stunting of foliage, but without any tangible retardation of the root system (Israel, Rao and Rao, 1966). Economic injury level of stunt nematodes is 32 nematodes/100 g of soil (Rao, 1990). *Tylenchorhynchus martini* causes significant yield reduction in deepwater rice in Bangladesh (Page and Bridge, 1979) while *T. brassicae* damages rice nurseries in Punjab in northern India (Chhabra, Sajan and Singh, 1974). Other less commonly associated species are *T. claytoni, T. elegans, T. indicus, T. mashhoodi, T. vulgaris* and *T. zeae*. These species have wide host ranges including legumes and cereals (Table 3).

The sheath nematode (*Caloosia heterocephala*) damages rice crop in upland and flooded conditions. The symptoms of nematode attack are not pronounced and in young plants the leaf tips may become dry (Rao, 1984).

The nematode feeds ectoparasitically and can arrest the apical growth of roots (Rao and Mohandas, 1976).

Root-Lesion

Ten species of root-lesion nematodes (*Pratylenchus* spp.) have been reported from rice-based cropping systems; *P. zeae* and *P. indicus* cause damage to upland rice in India and Pakistan. Losses caused by *P. zeae* may range from 13% to 29% (Bridge, Luc and Plowright, 1990) and by *P. indicus* from 18% to 55% (Prasad and Rao, 1978b). *Pratylenchus brachyurus* and *P. pratensis* are also potentially important on upland rice. The damage threshold level of *P. indicus* is 40 nematodes/100 g of soil at the time of planting (Prasad and Rao, 1978b). The soil moisture regime of 28-50% is ideal for reproduction and development of *P. indicus* which indicates that irrigated upland ecosystems are suitable habitats (Prasad and Rao, 1980). *Oryza glaberrima, O. breviligulata, Eleusina coracana, Sorghum bicolor, Zea mays, Pennisetum pedicellatum, Triticum aestivum, Sesamum indicum, Avena sativa, Hordeum vulgare, Secale cereale, Vigna radiata, Amaranthus* sp., *Lycopersicon esculentum, Ipomea batatus, Glycine max, Arachis hypogaea, Saccharum* spp., *Solanum tuberosum, Brassica juncea, Allium cepa, Lactuca sativa, Nicotiana tabacum, Gossypium* spp. and many weed plants are hosts of *P. indicus.*

Root-Knot

The root-knot nematodes are important parasites of rice. *Meloidogyne graminicola* is the most damaging species in rainfed upland rice in India (Orissa, Madhya Pradesh, West Bengal, Assam and Tripura states) and lowland and deepwater rice in Bangladesh (Bridge, Luc and Plowright, 1990). Yellowing of foliage and delay in flowering by 2-3 weeks are common symptoms associated with nematode infection while swollen and hooked root tips are characteristic of *M. graminicola* (Israel, Rao and Rao, 1963). The nematode completes its life cycle in 24-32 days depending on growing season (Rao and Israel, 1973). It causes 10% loss in yield in upland direct seeded rice (Rao, Jayaprakash and Mhoanty, 1986). In flooded rice, the seedling damage in nurseries before transplantation is high in the infested soils; the tolerance limit in the nursery is less than one juvenile/cm^3 soil (Table 2) and economic threshold level for *M. graminicola* is 125 juveniles/seedling of rice (Rao, 1984). This nematode is a problem in Bangladesh rice probably as a consequence of poor water management, as flooding reduces the incidence of the nematode. Many crops that are commonly grown in rotation with rice are hosts of *M. graminicola* and some of the host plants are *Alopercurus* sp.,

Avena sativa, Beta vulgaris, Brachiara mutica, Brassica juncea, B. oleracea, Colocasia esculenta, Cyperus procerus, C. pulcherrimus, C. rotundus, Echinochloa colona, Eleusine indica, Fimbristylis miliacea, Fiurena sp.*, Glycine max, Lactua sativa, Lycopersicon esculentum, Monochloria vaginalis, Oryza sativa, Panicum miliaceum, P. repens, Paspalum scrobiculatum, Pennisetum typhoides, Phaseolus vulgaris, Poa annua, Ranunculus* sp.*, Saccharum officinarum, Sorghum bicolor, Spaeranthus* sp.*, Sphenoclea zeylandica, Spinacia oleracea, Triticum aestivum, Vicia faba* (Rao, Prasad and Rao, 1984). Options for management primarily based on chemicals have been developed (see Sharma and Rahaman, 1998). *Meloidogyne incognita* and *M. javanica*–the most widely distributed root-knot nematodes in South Asia–have not been reported as important parasites of rice or wheat. Recently, infection of *M. javanica* on rice was observed in Punjab and Andhra Pradesh (Sharma and Rahaman, 1998). The nematode population in Punjab attacked chickpea, rice, and wheat while Andhra Pradesh population attacked groundnut, finger millet, and rice.

Cyst Nematodes

The rice cyst nematode (*Heterodera oryzicola*) is potentially an important parasite of rice in upland and well drained soils in Orissa, Madhya Pradesh, West Bengal and Goa in India (Evans and Rowe, 1998). Browning and chlorosis of leaves, retarded functioning of roots, early flowering of the plants by 10-13 days and partial filling of grains are symptoms associated with *H. oryzicola* infection (Rao and Jayaprakash, 1977). Presence of white pearl-like nematode females on 5-6 week old seedlings is characteristic sign of nematode infection. Damage threshold level is 28-eggs/g soil and yield losses of 21-42% have been attributed to *H. oryzicola* (Kumari and Kurian, 1981). Some other hosts are *Cynodon dactylon, Brachaira* sp., *Musa* spp. The nematode is an endoparasite and interacts with *Sclerotium rolfsii. Heterodera oryzae*, another species of cyst nematodes is present in Bangladesh (Page and Bridge, 1978) and it is adapted to flooding. It completes 2-3 generations in a crop season (Netscher, 1969).

NEMATODE PESTS SPECIFIC TO WHEAT
IN RICE-WHEAT BASED CROPPING SYSTEMS

The cereal cyst nematode (*Heterodera avenae*) and seed gall nematode (*Anguina tritici*) are important nematode pests of wheat. The symptoms of *H. avenae* infection are commonly known as "Molya" in India. These two species also interact with fungal and bacterial pathogens (Table 2) eventually augmenting the yield losses.

Molya

Heterodera avenae is one of the important nematode pests of wheat in India and Pakistan (Dhawan and Pankaj, 1998). The roots show a characteristic elongation of the main root with pearly root (due to presence of white females of the nematodes) character. The tips of the rootlets appear bunchy. The economic damage threshold level is 20 eggs/g soil and application of nematicides in the infested fields results in 9-95% greater yield than that in untreated controls depending on the initial nematode density. In Pakistan, yield losses attributed to *H. avenae* range from 15-20% (Maqbool, 1981, 1988). *Heterodera avenae* has limited host range and cultivation of rice in the rainy season has greatly suppressed the nematode population in the rice-wheat system than in cotton-wheat and maize-wheat systems (Sakhuja et al., 1998).

Seed Gall

The seed gall nematode, *Anguina tritici*, is widespread and occurs in nearly all the wheat growing areas in Pakistan (Maqbool, 1988), central and eastern submontane region in Nepal (Manandhar and Amatya, 1987), and major wheat growing region of India (Swarup, 1986). Crop losses of more than 60% were reported due to *A. tritici* infestation (Paruthi and Bhatti, 1985). Second-stage juveniles attack young floral primordia and form galls in place of grains. The nematode in association with a bacterium, *Clavibacter tritici* causes ear rot of wheat. Attempts are being made in India for eradication of this nematode from the country.

POLYPHAGOUS NEMATODES IN WHEAT-BASED CROPPING SYSTEMS

Many multi-host nematode species have been reported from wheat-based cropping sequences. *Meloidogyne* species are pathogenic to almost all the crops in rotation with wheat. These nematodes have wide host preferences and attack most of the rotational crops in wheat-based cropping systems. These nematodes are root feeders and their infections do not produce specific symptoms on the aerial part of the plants. Stunted plant growth, as explained previously (see polyphagous nematodes in rice-based cropping system) is commonly associated with infection of these nematodes. The root-knot nematodes (*Meloidogyne* spp.) attack wheat in India, however, data on the extent of damage are not available. The importance of *Meloidogyne* spp. (e.g., *M. graminicola*) that are parasites of rice, wheat and other crops in the rice-wheat based cropping systems has not been sufficiently examined (Gaur, Saha and Khan, 1993). The lesion nematodes (*Pratylenchus zeae, P. thornei, P. pratensis, P. indicus* and *Pratylenchus* spp.) attack wheat in many parts of the wheat growing regions. The economic importance of lesion nematodes on wheat is not clear in South Asia. The lesion nematodes (*P. thornei* in particular) are important pests of wheat in Australia. *Vigna radiata, V. mungo, V. unguiculata* and *Sesamum indicum* are poor hosts of *P. zeae*.

The stunt nematodes cause poor growth and sometimes economic loss in specific wheat growing regions, some of the frequently encountered and important species are *Tylenchorhynchus vulgaris, T. annulatus, T. brassicae, T. brevilineatus, T. mashhoodi, T. persicus*, and *Merlinius brevidens*. At-planting time inoculum of 1000-*T. vulgaris*/plant causes significant reduction in root, shoot lengths and fresh weights (Patel and Thakar, 1989).

HOST STATUS OF CROPS THAT ARE COMMONLY ROTATED WITH RICE AND (OR) WHEAT

Many plant parasitic nematodes which parasitizes rice and (or) wheat also attack plant species that are commonly rotated with these cereals in South Asia (Table 3).

Legumes

Pratylenchus, Tylenchorhynchus, Hoplolaimus species attack legumes as well as rice and wheat in rice-wheat cropping systems. *Pratylenchus thornei* is an important nematode on chickpea and mungbean. It causes significant plant growth reduction and yield loss in chickpea even at low population densities of > 0.1/g soil (Walia and Seshadri, 1985). Species of *Pratylenchus, Tylenchorhynchus*, and *Criconemella* cause damage to groundnut (Sharma and McDonald, 1990). Plant parasitic nematodes besides causing direct damage, suppresses root nodulation by *Rhizobium* and nitrogen fixing ability of legumes and enhances the aggressiveness of soil-borne pathogenic fungi such as *Fusarium* spp., *Meloidogyne javanica*, and *Helicotylenchus* sp., which are occasionally found associated with lentil in India (Mulk and Jairajpuri, 1974; Prakash, 1981). Mung bean is a good host of *M. incognita* (Hussaini and Seshadri, 1975), and simultaneous infection of *M. incognita* and *M. javanica* reduces seed yield of lentil by 39%. *Meloidogyne javanica* is an important parasite of groundnut in India. These species are well known for their crop damaging potential, and emergence of populations that can feed on cereals as well as on legumes is of great concern. Heavy infestation of stunt nematode (*Tylenchorhynchus vulgaris*) reduces root and shoot growth of several legumes (Gill and Swarup, 1977). *Hoplolaimus indicus* at an initial inoculum level of 0.2-juveniles/g soil adversely affects germination of green gram seeds (Haidar, Nath and Prasad, 1978) and *H. seinhorsti* affects growth of pigeonpea (Sharma and Nene, 1988). *Sesbania rostrata* reduces field populations of *Hirschmanniella* spp., but it is a very good host for *M. graminicola* when grown in non-flooded soils. *Sesbania rostrata* should be used with caution particularly under rainfed conditions and other alternative leguminous crops that are resistant to *M. graminicola* should be used (Prot, 1993b).

Oilseeds

Groundnut, sunflower, rapeseed, mustard and sesame are important oil yielding crops in the rice-wheat cropping systems. Sunflower is susceptible to the root-knot nematodes; however, sesame and mustard are not good hosts of the root-knot nematodes and these crops are expected to suppress the population densities of the root-knot nematodes (Sharma and Sharma, 1993).

Berseem

Species of *Tylenchorhynchus* and *Helicotylenchus* damage berseem. *Tylenchorhynchus mashhoodi* causes reduction in fodder and seed yield of berseem. Although large numbers of nematodes have been recorded associated with this crop, little information is available on nematode-induced specific problems.

Sorghum

Tylenchorhynchus vulgaris, Pratylenchus zeae are important parasites in India and *P. thornei, T. mashoodi* in Pakistan. The threshold level for sorghum is 2-5 nematodes of *Pratylenchus* sp./g of soil in India (Sharma and McDonald, 1990).

Pearl Millet

Pearl millet is a host of both *M. incognita* and *M. javanica* in northwestern India (Handa, Mathur and Bhargava, 1971). *Helicotylenchus* sp. and *Tylenchorhynchus* sp. are also important parasites on pearl millet in India and Pakistan. Pearl millet is a good host for *T. vulgaris* (Upadhyay and Swarup, 1972). A population of 1 nematode/g of soil is the threshold level of *Tylenchorhynchus* sp. on pearl millet (Sharma and McDonald, 1990).

Potato

Meloidogyne incognita and *M. javanica* are important nematodes found in all the potato growing areas. Several other plant parasitic nematodes such as lesion, spiral, and stunt nematodes are frequently found in potato fields. Stunt and spiral nematodes cause 9-29% loss in tuber yield in the potato growing tracts in the hills of northern India (Varaprasad, 1994).

Sugarcane

Meloidogyne incognita and *M. javanica* are commonly occurring species in sugarcane growing regions. Lesion (*P. zeae, P. brachyurus* and *P. pratensis*); spiral (*Helicotylechus dihystera* and *H. erythrinae*), stunt (*Tylenchorhynchus martini* and *T. elegans*) and lance (*H. indicus* and *H. columbus*) nematodes are widespread in India and Pakistan.

Vegetables

The root-knot nematodes are very damaging nematode pests of vegetable crops such as okra, tomato, egg plant, onion, cauliflower, turnip, cucumber, pumpkins, squash, bottle gourd, sponge gourd, carrot, sweet potato, pea, radish, spinach and beans. *Meloidogyne incognita* and *M. javanica* are the most important species on vegetables in the region (Sharma and Sharma, 1993). *Tylenchorhynchus* spp., *Hoplolaimus* spp., *Trichodorus* spp., *Paratrichodorus* spp., *Longidorus* spp., *Hemicycliophora* spp., *Helicotylenchus* spp., *Rotylenchus* spp., and *Xiphinema* spp. attack vegetables but estimates of crop losses caused by these species are lacking.

Jute

The most important nematode pests (*M. incognita* and *M. javanica*) of jute are the root-knot nematode species (Saikia and Phukan, 1986). Other frequently associated and possibly economically important species are *Hoplolaimus indicus, Helicotylenchus* sp. Weeding and removal of jute stubbles combined with rotation with paddy rice, followed by crop of wheat increases yield of jute.

FUTURE RESEARCH THRUST

Research efforts during the last 30 years have captured the diversity in plant parasitic nematode fauna associated with rice- and wheat-based cropping systems. A large number of species that are capable of damaging the rice, wheat and associated crops have been identified. It is apparent from the available data that nematodes adversely affects the productivity of rice-wheat based production systems. As sustainability and stability of rice-wheat based cropping systems are the major research and developmental goals of national and international agricultural centers, it is crucial to consider the contributions plant parasitic nematodes make to yield decline. Increase in the population build up of specific polyphagous nematodes in the rice-wheat based cropping system and their active interactions with other soil microbes (pathogens as well as beneficial), does, in all likelihood, influences the soil health (Prot, 1993b). Many scientists view the build up of plant parasitic nematode populations, rather than a loss of soil fertility, as a major component of the phenomenon of 'soil sickness' (e.g., Ferris and Ferris, 1974). The research projects on pest management in rice-wheat based systems in the region lack interdisciplinarity and nematologists are still busy in drawing the attention of others to the pest potential of nematodes (Sharma, Nigel and Bridge, 1997). The first need is to involve nematologists as integral members of any interdisciplinary teams on improving the productivity and sustainability of the rice-wheat cropping systems. This is essential to broadly identify and rate the pest problems at selected sites in South Asia, to identify opportunities for research and extension on major nematode pests in diffrerent cropping domains, and to develop interdisciplinary approach to pest management. The nematologists on priority basis, using previous or new data, must characterize the rice-wheat nematode problems in relation to locations and farmers' management practices, quantify the relative importance of nematode pests in contributing to gaps between attainable and actual yields, and identify the nematode pests that require integrated management at system level. This will require an attempt to describe the nematode pest problems in relation to crop or cropping system management. The nematologists also must work on the concept of 'accumulative pathogenicity' of the associated plant parasitic nematode fauna. The population densities of some commonly associated nematodes (e.g., polyphagous ectoparasites) generally do not increase to high levels but in polyspecific community structure, the cumulative damage by all such species put together could be significant enough to warrant management options. Application of non-parametric multi-variate data analysis techniques such as correspondence analysis would assist in linking of qualitative attributes (e.g., crop cultivar) with quantitative attributes (e.g., nematode population density, yield) (Prot and Savary, 1993). Use of geographic information system would help in mapping the nematode endemic areas (Pande et al., 1998). On-farm experiments would be needed to develop crop loss databases from which the contribution of yield losses caused by each nematode pest or pest group can be quantified and their relative contribution to the yield gaps determined. Monetary values can be assigned, depending on the attainable yield levels, to the value of losses caused by each nematode pest.

There is also a need to assess the impact of new and anticipated production technologies on changes in the nematode community structure and their natural enemies. This can be done at locations where new emerging technologies are being tested in comparison with existing farmers' practices. Some of the new production technologies that may be assessed for effect on nematodes include: new tillage and crop establishment practices such as reduced and zero-tillage, direct seeding of rice, new crop rotation and diversification of rice-wheat systems, crop residue management, introduction of new varieties and germplasm such as hybrid rice, effect of weedicide applications on nematode populations, relationship between nematode population build up with soil fertility status. The nematologists should hypothesize changes, based on knowledge and perception that can be made in the system for better nematode management. There is a distinct need for development of management technologies for the major nematode pests, assessment of these technologies for their complementarity with other crop production technologies, and ease of adoption of the management technologies by the farmers. It is essential to identify the research as well as developmental gaps in the present nematode management systems for nematode pests.

The renewed research thrust are expected to describe:

- Linkages between the cropping systems management practices and nematode problems,
- Losses caused and economic importance of each nematode pests in relation to cropping system management,
- A model for projecting trends in nematode pest scenario in relation to water, nutrient, and crop management,
- Determination of changes in nematode population dynamics in relation to introduction of new production technologies in the system,
- Recommendations for farmers for adoption of nematode pest management,
- Methodologies for developing farmer participatory nematode management strategies,
- Development of extension publication and video for wider dissemination of practical aspects of nematode management, and
- Development of simple diagnostic tools for identification of nematode pests by farmers and extension workers.

A greater research and development input on nematode pests will benefit the farmers and will enhance the productivity and sustainability of the rice-wheat cropping system of South Asia.

REFERENCES

Ali, S.S. (1997). Status of nematode problems and research in India. In *Diagnosis of key nematode pests of chickpea and pigeonpea and their management*, ed. S.B. Sharma, Proceedings of a regional training course, Andhra Pradesh, ICRISAT, pp. 74-82.

Bridge, J., Luc, M. and Plowright, A. (1990). Nematode parasites of rice. In *Plant parasitic nematodes in subtropical and tropical agriculture*, eds. M. Luc, R.A. Sikora, and J. Bridge, Wallingford, UK, CAB International, pp. 69-108.

Catling, H.D., Cox, P.G., Islam, Z. and Rahman, M.L. (1979). Two destructive pests of deepwater rice-yellow stem borer and ufra. *ADAB News* 6: 16-21.

Chhabra, H.K., Sajan, S.S. and Singh, J. (1974). Occurrence and control of the stunt nematode (*Tylenchorhynchus brassicae* Siddiqi, 1961) infection rice nursery in Punjab. *Current Science* 43: 632.

Dhawan, S.C. and Pankaj. (1998). Nematode pests of wheat and barley. In *Nematode diseases in plants*, ed. P.C. Trivedi, CBS Publishers and Distributors, pp. 13-25.

Edward, J.C., Sharma, N.N. and Agnihotrudu, V. (1985). Rice root nematode (*Hirschmanniella* spp.)–A review of the work done in India. *Current Science* 54: 179-182.

Evans, K. and Rowe, A. (1998). Distribution and economic importance. In *The Cyst Nematodes*, ed. S.B. Sharma, Kluwer Acadmic Publishers, pp. 1-30.

Gaur, H.S., Saha, M. and Khan, E. (1993). *Meloidogyne triticoryzae* sp. n. (Nematoda: Meloidogynidae) a root knot nematodedamaging wheat and rice in India. *Annals of Plant Protection* 1: 18-26.

Gill, J.S. and Swarup, G. (1977). Pathogenic effect of *Tylenchorhynchus vulgaris* on gram. *Indian Journal of Nematology* 7: 155-156.

Gokulpalan, C. and Nair, N.C. (1983). Field screening for sheath blight and rice root nematode resistance. *International Rice Research Newsletter* 8: 4.

Ferris, V.R. and Ferris, J.M. (1974). Inter-relationship between nematodes and plant communities in agricultural ecosystems. *Agro-Ecosystems* 1: 275-299.

Haider, M.G., Nath, and Prasad, S.S. (1978). Studies on the lance nematode *Hoplolaimus indicus* 1–Pathogenicity and histopathogenesis on maize. *Indian Journal of Nematology* 8: 9-12.

Handa, D.K., Mathur, B.N. and Bhargava, L.P. (1971). Occurrence of root-knot on pearl millet. *Indian Journal of Nematology* 1: 244.

Hobbs, P., and Morris, M. (1996). *Meeting South Asia's future food requirements from rice-wheat cropping systems: Priority issues facing researchers in the post-green revolution era.* NRG paper 96-01. Mexico, DF, CIMMYT.

Hussaini, S.S. and Seshadri, A.R. (1975). Interrelationships between *Meloidogyne incognita* and *Rhizoctonia* sp. *Indian Journal of Nematology* 5: 188-199.

Israel, P., Rao, Y.S. and Rao, V.N. (1966). *Survey of nematodes in rice fields and evaluation of their damage.* July 23-28, 11th meeting. FAO, Rome, International Rice Commission.

Israel, P., Rao, Y.S. and Rao, Y.R.V.J. (1963). Investigations on nematodes of rice and rice soils. I. *Oryza* 3: 125-128.

Karki, P.B. (1997). Status of nematode problems and research in Nepal. In *Diagnosis of key nematode pests of chickpea and pigeonpea and their management, Pro-*

ceedings of a regional training course, ed. S.B. Sharma, ICRISAT: Andhra Pradesh, India, pp. 57-60.

Kumari, U. and Kuriyan, K.J. (1981). Cyst nematode. *Heterodera oryzicola*, on rice in Kerala. I. Estimation of loss in rice due to *H. oryzicola* infestation in field conditions. *Indian Journal of Nematology* 11: 106.

Malik, R. and Yasmeen, Z. (1978). Nematodes in paddy field of Pakistan. *International Rice Research Newsletter* 3: 16-17.

Manandhar, H.K. and Amatya, P. (1987). Plant parasitic nematodes in Nepal. *International Nematology Network Newsletter* 4 :30-34.

Maqbool, M.A. (1981). Occurrence of root-knot nematodes in Pakistan. *Nematologia Meditteranea* 9: 211-212.

Maqbool, M.A. (1988). Present status of research on plant parasitic nematodes in cereals and food and forage legumes in Pakistan. In *Nematodes parasitic to cereals and legumes in temperate semi-arid regions*, eds. M.C. Sexena, R.A. Sikora, and J.P. Srivastava, ICARDA, Aleppo, Syria, pp. 173-180.

Mathur, V.K. and Prasad, S.K. (1972). Role of the rice root nematode *Hirschmanniella oryzae* in rice culture. *Indian Journal of Nematology* 2: 158-168.

McGrawley, E.C., Rush, M.C. and Hollis, J.P. (1984). Occurrence of *Aphelenchoides besseyi* in Louisiana rice seed and its interaction with *Sclerotium oryzae* in selected cultivars. *Journal of Nematology* 16: 65-68.

Mondal, A.H., Rahman, L., Ahmed, H.J. and Miah, S.A. (1986). The cause of increasing blast susceptibility of ufra infected rice plants. *Bangladesh Journal of Agriculture* 111: 77-79.

Mulk, M.M. and Jairajpuri, M.S. (1974). Nematodes of leguminous crops in India. II. Five new species of *Helicotylenchus* Steiner, 1945 (Hoplolaimidae). *Indian Journal of Nematology* 4: 212-221.

Muthukrishnan, T.S., Rajendran, G. and Chandrashekaran, J. (1974). Studies on the white-tip nematode of rice, *Aphelenchoides besseyi* in Tamil Nadu. *Indian Journal of Nematology* 4: 188-193.

Nandakumar, C., Prasad, J.S., Rao, Y.S. and Rao, J. (1975). Investigations on the white-tip nematode (*Aphelenchoides besseyi* Christie, 1942) in rice (*Oryza sativa* L.). *Indian Journal of Nematology* 5: 62-69.

Netscher, C. (1969). L'ovogense et la reproduction chez *Heterodera oryzae* et *H. sacchari* (Nematoda: Heteroderidae). *Nematologica* 15: 10.

Ou, S.H. (1972). *Rice disease*. Commonwealth Mycological Institute. England.

Padhi, N.N., Das, S.N. (1985). Observation of *Helicotylenchus abunaamai* in host tissue. *Nematologia Meditteranea* 13: 115-116.

Page, S.C. and Bridge, J. (1979). Root and parasitic nematode of deep water rice area in Bangladesh. *International Rice Research. Newsletter* 4: 10.

Page, S.L.J. and Bridge, J. (1978). Plant nematodes on deep water rice in Bangladesh. *Report on visit to Bangladesh*. Ministry of Overseas Development, UK, pp. 48.

Panda, M. and Rao, Y.S. (1971). Evaluation of losses caused by the rice root nematode (*Hirschmanniella mucronata* Das) in rice (*Oryza sativa* L.). *Indian Journal of Agriculture Science* 41: 611-614.

Pande, S. Asokan, M., Sharma, S.B. and Joshi, P.K. (1998). Use of geographic information system in plant parasitic nematode research in the rice-wheat-legumes cropping systems. In *Nematode pests in rice-wheat cropping systems. Proceedings of a regional training course*, eds. S.B. Sharma, C. Johansen and S.K. Midha. Haryana, Haryana Agricultural University and Patancheru, Andhra Pradesh, ICRISAT (in press).

Panwar, M.S. and Rao, Y.S. (1998). Status of phytonematodes as pests of rice. In *Nematode diseases in plants*, ed. P.C.Trivedi, CBS Publishers and Distributors, pp. 49-81

Paruthi, I.J. and Bhatti, D.S. (1985). Estimation of loss in yield and incidence of *Anguina tritici* in wheat in Haryana (India). *International Nematology Network Newsletter* 2:3,13-16.

Patel, P.N. and Thakar, N.A. (1989). Damaging threshold level of the stunt nematode, *Tylenchorhynchus vulgaris* on wheat variety J-24. *Indian Journal of Nematology* 19: 78.

Prakash, A. (1981). *Lens culinaris*–A new host for root-knot nematode, *Meloidogyne javanica* in India. *National Academy of Science Lettter* 4: 459.

Prasad, J.S., Panwar, M.S., Rao, Y.S. and Rajamani, S. (1987). Control of root lesion nematode, *Pratylenchus indicus* Das 1990 in rainfed upland rice. *Rice Research Newsletter* 8: 3.

Prasad, J.S. and Rao, Y.S. (1978b). Potentiality *of Pratylenchus indicus* Das, 1960, the root lesion nematode as a new pest of upland rice. *Annals of Zoologie and Ecologie Animale* 10: 634-640.

Prasad, J.S. and Rao, Y.S. (1980). Influence of edaphic factors in the build up of the root lesion nematode *Pratylenchus indicus* Das, 1960 in rice. Effect of type, texture, porosity and moisture content of soil. *Revue de Ecologie Et De Biologie Du Sol.* 17: 173-179.

Prasad, J.S. and Rao, Y.S. (1984a). Ectoparasitic nematode pests of upland rice soils. *Beitrage Zur Tropischen Landwirtschaft und Veterinaumedzin* 22: 79-82.

Prasad, J.S and Rao, Y.S. (1984b). The status of root nematodes (*Hirschmanniella* sp.) as pests of rice (*Oryza sativa*). *Beitrage Zur Tropischen Landwirtschaft und Veterinaumedzin* 22: 281-284.

Prot, J.C. (1993a). Monetary value estimates of nematode problems research proposal and priorities, The rice research example in south and southeast Asia. *Fundamental and Applied Plant Nematology* 16: 385-388.

Prot, J.C. (1993b). Biological and genetic basic of fungus-nematode interactions. In *Nematode interactions*, ed. M.W. Khan, London, UK, Chapman and Hall, pp. 288-301.

Prot, J.C. (1994a). Effects of economic and policy changes on status of nematode pests of rice in Vietnam and Philippines. *Fundamental and Applied Plant Nematology* 17: 195-198.

Prot, J.C. (1994b). Nematode constraints to crop production and the role of international agricultural research centers in constraints alleviation: role of the International Rice Rsearch Institute. In *International agricultural research on diseases caused by nematodes–Needs and constraints*. Summary and recommendations of a satellite meeting, eds. S.B. Sharma, and D. McDonald, Montreal Canada, International Congress of Plant Pathology, pp. 10-13.

Prot, J.C. and Savary, S. (1993). Interpreting upland rice yield and *Pratylenchus zeae* relationships: Correspondece analyses. *Journal of Nematology* 25: 277-285.

Rahman, M.L. and McGeachie, I. (1982). Occurrence of white tip disease in deepwater rice in Bangladesh. *International Rice Research Newsletter* 7: 15.

Rahman, M.L. and Taylor, B. (1983). Nematode pests associated with deepwater rice in Bangladesh. *International Rice Research Newsletter* 8: 20-21.

Rahman, M., Sharma, B.R. and Miah, S.A. (1981). Incidence and chemical control of ufra in boro fields. *International Rice Research Newsletter* 6: 12.

Ramana, K.V. and Rao, Y.S. (1976). On the inter-relationship between the lance nematode *Hoplolaimus indicus* and soil pathogen interaction in rice. *Indian Journal of Plant Protection* 4: 125-127.

Ramana, K.V. and Rao, Y.S. (1977). Evaluation of damage and yield losses due to the lance nematode (*Hoplolaimus indicus* Sher) in rice. *Andhra Agriculture Journal* 24: 121-125.

Rao, Y.S. (1970). *Rice nematodes*. PANS manual No. 3. Rice, London, HMSO, pp. 99-107.

Rao, Y.S. (1984). Nematode problems of rice in submerged soils and their management. *Panel discussion on Nematological problems of North Zone*, Hisar, Haryana, India.

Rao, Y.S. (1990). Nematode problems of rice and their control. In *Progress in plant nematology*, eds. S.K. Saxena, M.W. Khan, A. Rashid, and R.M. Khan, Professor Abrar, M. Khan, Festschrift volume, New Delhi, India, CBS Publisher, pp. 475-502.

Rao, Y.S. and Israel, P. (1973). Life history and bionomics of *Meloidogyne graminicola*, the rice root-knot nematode. *Indian Phytopathology* 26: 333-340.

Rao, Y.S. and Jayaprakash, A. (1977). Leaf chlorosis in rice due to a new cyst nematode. *International Rice Research Newsletter* 2: 5.

Rao, Y.S., Jayaprakash, A. and Mhoanty, J. (1986). Nutritional perturbations in rice due to root infection by sedentary endoparasitic nematodes. *Rice Research Newsletter* 7: 6-7.

Rao, Y.S. and Mohandas, C. (1976). Occurrence of *Caloosia heterocephala* n. sp. (Nematoda: Hemicycliohporidae) on roots of rice (*Oryza sativa*). *Nematologica* 22: 227-234.

Rao, Y.S., Prasad, J.S. and Panwar, M.S. (1985). Nematode pests of rice in India. In *Non-insect pests and predators*. All India Scientific Writers Society, New Delhi, India, pp. 65-71.

Rao, Y.S., Prasad, J.S. and Rao, A.V.S. (1984). Interaction of cyst and root-knot nematodes in roots of rice. *Revue de Nematologie* 7: 117-120.

Rao, Y.S., Prasad, J.S., Yadav, C.P. and Padalia, C.R. (1984). Influence of rotation crops in rice soils on the dynamics of parasitic nematodes. *Biological Agriculture and Horticulture* 2: 69-78.

Saikia, D.K. and Phukan, P.N. (1986). Estimation of loss in jute due to root-knot nematode, *Meloidogyne incognita*. *Indian Journal of Nematology* 16: 108-109.

Sakhuja, P.K., Singh, I., Kaur, D.J. and Randhawa, N. (1998). Nematode problems in rice-wheat-legume cropping systems in Punjab. In *Nematode pests in rice-wheat cropping systems. Proceedings of a regional training course* eds. S.B. Sharma, C. Johansen, and S.K. Midha, Haryana Agricultural University and Patancheru, Andhra

Pradesh, International Crops Research Institute for the Semi-Arid Tropics (in press).

Sasser, J.N. and Freckman, D.W. (1987). A world perspective on nematology; The role of the society. In *Vistas on nematology*, eds. J.A. Veech, D. Dickson, A Commemoration of the Twenty-Fifth Anniversary of the Society of Nematologists, Society of Nematologists, Inc. Hyattsville, Maryland, USA. pp. 7-14.

Sharma, S.B. and McDonald, D. (1990). Global status of nematode problems of peanut, pigeonpea, chickpea, sorghum, and pearl millet and suggestions for future work. *Crop Protection* 9: 453-458.

Sharma, S.B. and Nene, Y.L. (1988). Effect of *Heterodera cajani, Rotylenchulus reniformis, and Hoploloaimus seinhorsti* on pigeonpea biomass. *Indian Journal of Nematology* 18:273-278.

Sharma, S.B., Nigel, N.S. and Bridge, J. (1997). The past, present and future of plant nematology in international agricultural research centres. *Nematological Abstracts* 66:119-142.

Sharma, S.B. and Rahaman, P.F. (1998). Nematode pests in rice and wheat cropping systems in the Indo-Gangatic Plain. In *Nematode pests in rice-wheat-legume cropping system*, Proceedings of a regional training course, eds. S.B. sharma, C. Johanson and S.K. Midha, Rice wheat consortium for Indo-Gangatic Plain, IARI, New Delhi, India.

Sharma, S.B. and Sharma, R. (1993). Changing scenario in the status of nematode pests of crops. In *Changing scenario in the pests and pest management in India*, eds. H.C. Sharma, and M.V. Rao, Plant Protection Association of India, 193-202.

Singh, U.B. (1953). Some important diseases of paddy. *Agriculture and Animal Husbandry* 30: 27-30.

Singh, R.B. and Paroda, R.S. (1994). Sustainability of rice-wheat production systems in Asia-pacific region: Research and technology development needs. In *Sustainability of rice-wheat production systems in Asia*, eds. R.S. Paroda, T. Woodhead, and R.B. Singh, Food and Agriculture Organization of the United Nations, Bangkok, Thailand, pp. 1-36

Sivakumar, C.V. (1987). The white tip nematode in Kanyakumari district, Tamil Nadu, India. *Indian Journal Nematology* 17: 72-75.

Swarup, G. (1986). Investigation on wheat nematodes. In *Twenty five years of coordinated wheat research* 1961-86, eds. J.P. Tandan and A.P. Sethi, ICAR, Wheat Project Directorate, New Delhi, India, pp. 189-206.

Swe, A. (1997). Status of nematode problems and research in Myanmar. In *Diagnosis of key nematode pests of chickpea and pigeonpea and their management: Proceedings of a regional training course.* ed. S.B. Sharma, ICRISAT, Andhra Pradesh, India, pp. 47-52.

Upadhyay, K.D. and Swarup, G. (1972). Culturing, host range and factors affecting multiplication of *Tylenchrohynchus vulgaris* on maize. *Indian Journal of Nematology* 2: 139-145.

Varaprasad, K.S. (1994). Nematode constraints to crop production in South Asian countries and the role of international agricultural research centers in constraint alleviation. In *International agricultural research on diseases caused by nematodes–Needs and constraints*, eds. S.B. Sharma, and D. McDonald, Summary and recommendations of a satellite meeting. Montreal, Canada, International Congress of Plant Pathology, pp. 10-13.

Vuong, H.H. (1969). The occurrence in Madagascar of the rice nematode. *Aphelenchoides besseyi* and *Ditylenchus angustus*. In *Nematodes of tropical crop*, ed. J.E. Peachy, *Technical Communication No. 40*. St. Albans, England, Commonwealth Bureau of Helminthology, pp. 274-288.

Walia, R.K. and Seshadri, A.R. (1985). Pathogenicity of the root-lesion nematode *Pratylenchus thornei* on chickpea. *International Chickpea Newsletter* 12: 31.

Water Management
in the Rice-Wheat Cropping Zone
of Sindh, Pakistan:
A Case Study

M. Aslam
S. A. Prathapar

SUMMARY. Rice and wheat are Sindh's most important cereal crops, extensively grown in rotation. Rice-wheat rotations occur most widely in the Upper Sindh. In Sindh, rice is grown on 0.75 million hectares (mha) with an average yield of 2 tons/ha, whereas the area under wheat is over one mha with average yield of 2.1 tons/ha. The current yields of rice and wheat in Sindh are far below their yield potentials. This is partly due to improper water management practices. During the rice season, the major problem of water management is an early season water shortage followed by an excessive amount of water in the region with the onset of the monsoon season. During the wheat season, excess soil moisture in the rice fields could delay wheat planting, and secondly, later in the wheat season, there could be a shortage of irrigation water. This paper presents an overview of current water management practices in the Sindh rice-wheat zone as a case study, reviews water management practices in rice-wheat zones around the world, and identifies measures, which may result in real water savings in the rice-wheat zone of Sindh. Identified measures are at farm level. These measures include water efficient methods of rice establishment, irrigation sched-

---

M. Aslam is Senior Irrigation Engineer and S. A. Prathapar is Director, International Water Management Institute, IWMI 12 KM, Multan Road, Chowk Thokar Niaz Baig, Lahore–53700, Pakistan (E-mail: s.prathapar@cgiar.org).

[Haworth co-indexing entry note]: "Water Management in the Rice-Wheat Cropping Zone of Sindh, Pakistan: A Case Study." Aslam, M., and S. A. Prathapar. Co-published simultaneously in *Journal of Crop Production* (Food Products Press, an imprint of The Haworth Press, Inc.) Vol. 4, No. 1 (#7), 2001, pp. 249-272; and: *The Rice-Wheat Cropping System of South Asia: Efficient Production Management* (ed: Palit K. Kataki) Food Products Press, an imprint of The Haworth Press, Inc., 2001, pp. 249-272. Single or multiple copies of this article are available for a fee from The Haworth Document Delivery Service [1-800-342-9678, 9:00 a.m. - 5:00 p.m. (EST). E-mail address: getinfo@haworthpressinc.com].

uling, water-saving irrigation regimes, land leveling, improved layout of irrigation ditches and fields, discontinuation of pancho system and improved drainage and farmer's participation. *[Article copies available for a fee from The Haworth Document Delivery Service: 1-800-342-9678. E-mail address: <getinfo@haworthpressinc.com> Website: <http://www. HaworthPress.com> © 2001 by The Haworth Press, Inc. All rights reserved.]*

KEYWORDS. Cropping system, drainage, Indo-Gangetic Plains, Pakistan, rice, Sindh, water-logging and salinity, water management practices, wheat

OVERVIEW OF THE RICE-WHEAT CROPPING SYSTEM AND ITS EXISTING WATER MANAGEMENT PRACTICES IN SINDH

A cropping sequence in which wheat is grown after rice and then, rice after wheat in the same field is known as the rice-wheat cropping system. Pakistan has 21 million hectares (mha) of farmland, of which rice-wheat cropping sequence is adopted in 2.2 mha. There are four rice-wheat cropping zones (Figure 1) in various Agro-Ecological Regions (AERs) of Pakistan: Zone 1 (Northern Zone), Zone 2 (Punjab Rice-Wheat Zone or PRWZ), Zone 3 (Upper Sindh Zone: Sindh Rice-Wheat North or SRWN) and Zone 4 (Lower Sindh Zone: Sindh Rice-Wheat South or SRWS). The SRWN and SRWS constitute the Sindh Rice-Wheat cropping zone where rice is the dominant crop. Wheat yield in this zone are the lowest in the Basin because wheat is mostly sown after rice when irrigation water availability in most of the area is scarce. The SRWN and SRWS zones are irrigated through nine canals off taking from the Guddu, Sukkur and Kotri Barrages on the River Indus (Table 1).

Area and Yield

In Sindh, total rice area is 0.75 mha with a total rice production of about 1.5 million tons (m tons) at an average rice yield of 2 tons/ha. The area under wheat cultivation is over one mha with an average yield of 2.1 tons/ha for irrigated wheat. Therefore the areas of rice and wheat in Sindh are 38% and 12% of the country's rice (about 2 mha) and wheat (about 8 mha) areas, respectively. The rice and wheat production of Sindh is 50% and 13% of the national rice (3 m tons) and wheat (16 m tons) production, respectively.

The 0.75 mha of rice and over 95% of the one million hectare of wheat grown in Sindh is irrigated. Rice dominates on the Right Bank of the Indus River partly because of high water tables, which makes the land unsuitable

FIGURE 1. Schematic map showing the rice-wheat growing zones (shaded areas) of Pakistan.

Zone 1: Rice grown in mountain environment. High rainfall and cold temperature region.
Zone 2: Subtropical Northern Irrigated Plains. Rice (Basmati and IRRI rice)-Wheat Cropping system practiced.
Zone 3: Southern Irrigated Plains. High temperature region. Rice-Wheat cropping system practiced.
Zone 4: Southern Irrigated Plains and Indus Delta Tropical Region. Poor drainage. Rice-Wheat cropping system practiced.

TABLE 1. Irrigation system of the Sindh rice-wheat zones

Zone	Canal Command	CCA (m ha)	% CCA
SRWN			
	Desert Feeder (NP)	0.2	100
	Begari Feeder (NP)	0.3	100
	North West (P)	0.3	100
	Rice Canal (NP)	0.2	100
	Dadu Canal (P)	0.2	100
SRWS			
	Kalri Begar (NP)	0.3	
	Lined Channel (NP)	0.2	100
	Fuleli Canal (NP)	0.4	100
	Pinyari Canal (NP)	0.3	100

P = Perennial; NP = Non-Perennial

for the other crops in the kharif (monsoon) season. In contrast, most of the wheat in Sindh is grown on the Left Bank of the Indus. Rice-wheat rotations occur most widely in the Upper Sindh region. It is estimated that 32% of the rice area is in Lower Sindh whereas 68% is in Upper Sindh region. The IRRI rice varieties are cultivated on about 90% of the total rice area in Sindh, and on the remaining 10% area, traditional varieties are grown. IRRI6 has a yield potential of over 9 tons/ha when grown under farmers' conditions, using recommended practices. Yet, average yield of IRRI6 across the Sindh province is about 4 tons/ha, less than half the yield potential. Yields are substantially higher in the Upper (4.3 tons/ha) than in the Lower (2.2 tons/ha) Sindh region.

The total wheat area increased sharply in the mid 1970s to over 1 mha with the introduction of short-duration wheat varieties. Wheat yields increased steadily from a provincial average of less than 1 ton/ha in the mid-sixties to 2.1 ton/ha in the mid-eighties. Wheat yields in Upper Sindh (2 tons/ha) are slightly, but consistently below those of Lower Sindh (2.2 tons/ha) region. Mean wheat yields in irrigated areas (2.1 tons/ha) are double those of the non-irrigated wheat tracts (1 ton/ha). Approximately 98% of wheat production in Sindh is from irrigated areas.

However, the present rice and wheat yields are much below the potential yield of this region (Table 2). This yield gap can be attributed to inadequate levels of vital inputs and cultural practices (seedbed preparation, seeding rate and depth, planting dates, time, method and amount of fertilizers and irrigation water use, pesticides use, etc.) and environmental constraints including soils (soil salinity/sodicity), irrigation water supplies and drainage problem.

Water Requirements

Rice is a semi-aquatic plant that yields best under flooded conditions. The average water requirement is 620 mm and 720 mm for early maturing and late maturing rice varieties, respectively. It is a normal practice to supply irrigation water as a continuous flow to a rice field to maintain constant flooding at a depth of 50-75 mm during most of its growing season. Water requirement of the rice crop is the highest during: (1) the initial seedling period, up to 10 days after transplanting; (2) the pre-flowering and flowering period, spanning about 25 days; and (3) 5 to 7 days of grain formation period. Rice crop is adversely affected in case of any deficiency of water during these

TABLE 2. Yields (kg/ha) of rice and wheat with their potential

Crop	Present National Yield	Potential Yield	Yield Crop
Wheat	1881	6400	4519
Rice (basmati)	1567	5200	3633

critical periods. On the basis of its water requirement of 620-720 mm and the acreage under rice, total annual requirement of irrigation water for rice in Sindh is 12340 million m^3.

The water requirement varies for different stages of the rice crop establishment process. Very little amount of water is required for raising nursery seedling as the nursery:transplanted area ratio is 1:40, i.e., seedlings grown on one ha of nursery is sufficient for transplanting 40 ha. The nursery period is about one month during which 25 mm of water is sufficient for the growth of seedlings in the nursery. However, at the time of sowing the soil is kept saturated for one week.

For puddling the field prior to transplanting of seedlings, 60-80 mm of water is required. The optimum water level at transplanting stage is about 25 mm deep, which is to be maintained even after transplanting. The total number of productive tillers is a varietal characteristic of rice, but is very much affected by the field water level during tillering stage. It is essential that the field water level at this stage is shallow and light irrigations are required to allow the plant to produce maximum number of tillers. During the reproductive phase (panicle initiation, booting, heading and flowering) spanning 25-30 days, water depth may vary between 75-100 mm until seed formation and development. More than 60% of total water requirement for growing rice is during the reproductive and grain-filling phase. The water is drained out 10-12 days before maturity.

Wheat crop has a growing season of about 130-150 days, during which water requirement is about 325 to 450 mm. Gross application depth of water per irrigation is 75 mm with irrigation interval of 25-30 days. Crop growth stages sensitive to water shortages are the tillering, flowering and grain filling period. Water shortage and uncertain irrigation schedules are severe problems being faced by the majority of wheat farmers. Water availability in adequate and reliable quantity also affects use of other inputs in proper amounts. In Sindh rice-wheat zone, majority of the farmers apply 5-6 irrigations.

Water Logging and Salinity

In the Sindh rice-wheat areas, presence of extensive depressional areas, lack of adequate drainage and improper water management practices result in widespread problems of water logging and salinity, which cause considerable reduction in rice and wheat yields. Tables 3 and 4 present the water logging and soil salinity situation in the Sindh rice-wheat zone. These tables indicate that 32% land in the Sindh rice-wheat zone is waterlogged and salinized, which is not favorable for optimum growth of most crops. Water logging is a major constraint for wheat production. Wheat cultivation can be done successfully when water table is below 1.5 m, but yields decrease drastically when water table is within 100 cm depth (Table 5). In Sindh, water table is

TABLE 3. Canal command wise area under less than 1.5 m watertable depth during 1989

Canal Command	Gross Area (m ha)	Waterlogged Area (m ha)	
		April	October
Desert	0.1	0.04	0.1
Begari	0.4	0.04	0.4
North West	0.5	0.3	0.5
Rice	0.2	0.02	0.2
Dadu	0.3	0.08	0.2
Kalri	0.3	0.04	0.2
Lined Channel	0.2	0.06	0.2
Fuleli	0.4	0.2	0.4
Pinyari	0.4	0.1	0.3

Adapted from MacDonald and Nespak, 1993.

TABLE 4. Extent of soil salinity in various canal commands

Canal Command	Gross Area (ha)	Survey Area (%)	Salt-Affected Area*	
			Ha	(%)
Desert	129853	100	24672	19
Begari	441663	100	77733	17.6
North West	437608	100	89272	20.4
Dadu	235246	100	65046	27.7
Rice	233082	100	46616	20
Kalri	273899	–	–	–
Lined Channel	187263	77	70889	49
Fuleli	389926	100	220698	56.6
Pinyari	444693	63	137252	49

*Salt-affected area out of surveyed area.
Adapted from Rehman et al., 1998.

TABLE 5. Effect of different watertable depths on wheat yield

Watertable Depth (cm)	% Yield Reduction
0-25	79
25-50	49
50-75	28
75-100	13
100-125	5
125-150	1
150-175	0

Adapted from MacDonald et al. 1990.

shallow in October during the rice season and declines to its lowest in June after wheat harvest.

Irrigation Supplies

In Sindh, majority of the farmers indicates a general water shortage and uncertain irrigation supplies. The shortage appears to be particularly acute in late rabi and early kharif season coinciding with the latter half of the wheat season and the early part of the rice season, respectively. Prolonged periods of canal closure are detrimental to wheat yield. Table 6 presents the effect of water shortage at critical stages on wheat yield. The yield is lowest when the water shortage occurs both during November and February.

Groundwater Salinity

The groundwater in Sindh at most places has been found to be saline (TDS > 3000 ppm) and unfit for irrigation. Table 7 presents the canal command wise proportions of fresh (TDS < 3000 ppm) and saline (TDS > 3000 ppm) groundwater areas for the Sindh rice-wheat zone. This table shows that about 19 percent of the total CCA of Sindh rice-wheat zone has fresh groundwater and about 81 percent has saline groundwater. In SRWN, about 63 percent of the CCA (1.6 mha) has saline groundwater whereas in SRWS, about 100 percent of CCA (1.2 mha) has saline groundwater with no fresh water in that zone.

Drainage

In the Sindh rice-wheat zone, where rice is grown in kharif followed by wheat in Rabi, surface drainage is required for:

- Removal of excess water from land currently flooded to excessive depths in the rice season, thus bringing more land under cultivation during the winter season.
- Removal of storm water runoff, thus protecting the crop from flood damage and reducing the risk of losing the entire crop.
- Improved time lines for wheat and other winter crop planting due to rapid drying of fields at the end of the monsoon season due to improved water control and reduced flooding. This increases the bearing capacity of the soil for mechanical harvesting of rice and land preparation for the following wheat crop.

Present Drainage Schemes

In the Sindh rice-wheat zone, four surface drainage projects, i.e., Hairdin, Larkana-Shikarpur, North Dadu and Kotri surface drainage projects, and five

TABLE 6. Effect of water shortage on wheat yield

Water Shortage Months		Average Yield (kg/ha)
November	February	
Yes	No	2855
Yes	Yes	2585
No	No	2977

Source: Revised action program for irrigated agriculture, 1979.

TABLE 7. Culturable command area by groundwater quality in thousand hectares

Canal Command	Fresh	Saline	Total CCA
Desert	157.8	–	157.8
Begari	170.4	170.4	340.8
North West	46.4	263	309.4
Rice	46.2	163.8	210
Dadu	39.2	205.7	244.8
Kalri	–	257.4	257.4
Lined Channel	–	220.1	220.1
Fuleli	–	360.6	360.6
Pinyari	–	323.3	323.3

Adapted from Rehman and Rehman, 1998.

subsurface (tube well) drainage projects, i.e., Kandh Kot, Shikarpur, Sukkur and Larkana Tube Well Pilot Projects, and Sukkur Right Bank Fresh Ground Water (FGW) Project, are operational. Table 8 presents the annual volumes of drainage effluent generated from the various surface and subsurface (tube well) drainage schemes in the Sindh rice-wheat zone.

The drainage effluent from Hairdin Project is pumped into the Kirther Branch off taking from the Northwest canal. Effluent from more than 30% of the Larkana-Shikarpur Surface Drainage Project is recycled into the Northwest and Rice canals or pumped into the river. The salinity concentrations of effluent from rice areas were found to be less than 1000 ppm. The drainage effluent of 2000 ppm from North Dadu project is partly disposed off into the Main Nara Valley drain and the rest is recycled into the Ghar Branch off taking from the Rice canal. The surface drainage effluent generated from the Kotri command is drained directly into the Arabian Sea.

For Kandh Kot Tube Well Pilot Project, water quality analysis of 20 tube wells in 1987-89 showed that 15 tube wells were pumping usable quality water, 2 were pumping marginal quality water and 3 were pumping hazard-

TABLE 8. Annual volume of drainage effluent from the Sindh rice-wheat zone

Drainage Project	Year	Volume (BCM/yr)	Source
Hairdin	1995	0.4	MacDonald and Nespak, 1995a
Larkana-Shikarpur	1998-99	0.4	Nespak, 1999
North Dadu	1998-99	0.3	Nesspak, 1999
Kotri	1995	7.2	MacDonald and Nespak, 1995a
Kandh Kot Pilot	1988-89	0.01	WAPDA, 1990a
Shikarpur Pilot	1988-89	0.03	WAPDA, 1990b
Sukkar Pilot	1988-89	0.01	WAPDA, 1990c
Larkana Pilot	1988-89	0.01	WAPDA, 1990d
Sukkar Right Bank FGW	1982-89	0.1	WAPDA, 1990

Adapted from MacDonald et al. 1990.

ous waters. In case of Shikarpur Pilot Project, water quality analysis done in 1979-80 showed that 47 of the 50 tube wells were in the usable category, and analysis done again between 1987-89 showed no significant changes in the quality of tube well drainage effluent. For Sukkur Pilot Project, water quality analysis of 1979-80 showed that 14 tube wells were discharging water of usable quality and the remaining were of marginal quality. Subsequent analysis in 1987-89 showed no significant changes in water quality. In case of Larkana Pilot Project, water quality analysis of 34 tube wells during 1979-80 showed that 40 tube wells were pumping usable groundwater and the rest were of marginal quality. Analysis of all wells during 1987-89 showed that 19 tube wells were pumping usable quality water. Overall there is no indication of any worsening trend. For Sukkur Right Bank FGW Tube Well Project, water quality data for 360 of the 400 tube wells monitored during 1987-89 reveals that about 94 percent of the tube wells had usable waters (WAPDA, 1990e).

Proposed Right Bank Master Plan

Presently, surface and tubewell drainage projects have been implemented on 30% of the Guddu and Sukkur Right Bank Command area based on the Lower Indus Plan (LIP) recommendations made in the 1960s. Recently, WAPDA has prepared a regional development plan known as the Right Bank Master Plan (RBMP) based on the studies carried out over a period of two years (1989-91). This plan is supplemental to the surface and subsurface drainage projects implemented by the Government of Pakistan under the recommendations of the LIP, covering an area of 1.8 mha in the commands of the Guddu and Sukkur barrages. The central feature of the plan is the enlargement and extension of the existing Main Nara Valley Drain to become the Right Bank Outfall Drain (RBOD), mainly to dispose of excess surface waters from the rice growing areas in the commands of Desert, Begari, Northwest, Rice and Dadu canals. The anticipated annual volumes of drainage effluent to be generated during the four stages of implementation of RBOD are given in Table 9.

PROSPECTS FOR IMPROVED WATER MANAGEMENT PRACTICES IN THE RICE-WHEAT ZONES

Poor water management practices at the farm level cause wastage of limited irrigation water, whereas sustained rice-wheat production systems need improved irrigation management practices at the farm level. The appropriate and promising irrigation water management practices for the rice-wheat system at the farm level being practiced in other countries and in the Sindh of Pakistan are discussed below.

Water Efficient Methods of Rice Establishment: Direct Seeding of Rice

There are two systems of direct seeding of rice, i.e., the wet and the dry system. Under the wet direct seeding system, pre-germinated seeds are sown on the saturated field that has been prepared under wet condition. In the dry direct seeding method, field preparation is done under dry condition, immediately followed by seed sowing either before irrigation water is applied or before rainfall occurs, to enable germination and seedling establishment. Under this system, the total cropping season could be reduced by about two weeks avoiding the nursery preparation and transplanting phase and the overall irrigation requirement is reduced, resulting in significant water savings of up to 25%. It has been recorded that transplanting rice requires more water and results in more wastage than the dry direct seeding system.

Bhuiyan (1992) and Bhuiyan, Sattar and Khan (1995) reported that wet-seeded rice, in which pre-germinated seeds are sown directly on puddled fields in lieu of transplanting seedlings to establish a rice crop, offers a

TABLE 9. Anticipated annual volume of drainage effluent from RBOD

Stage	Volume (BCM/yr)
I	0.5
II	0.9
III	1.1
IV	1.7

Adapted from Rehman, 1996.

significant opportunity for improved management of irrigation water. Land preparation is more water-efficient for wet-seeded rice than for transplanted rice. In the case of wet-seeding technique, preparation of pre-germinated seeds by seed soaking takes 24-36 hours, and land preparation can be completed in about a week, avoiding long periods of water losses occurring in transplanted rice. Research has shown that about 30% less water is required in order to prepare a typical field up to the same puddled condition for wet-seeded rice than for transplanting rice. Wet-seeded rice yields more in both water-sufficient and water-short situations, requires less labor and produces a better return on investment than transplanted rice.

De Datta (1986) reported that farmers in the Philippines, Malaysia and Thailand are switching from transplanted to direct-seeded wet system. Under this system, during the first four to five days after seeding, the field should be kept moist, but not flooded. Thereafter, 2-3 cm of water is allowed to flood the rice field. About 10 days after seeding, the water level is increased to 5 cm and maintained at that depth until crop maturity. Direct seeding is usually faster and easier than transplanting rice seedlings, and grain yields are similar or occasionally higher than transplanted rice. This technique requires better leveling of the field for good water control and crop establishment. A recommended seed rate of 100 kg/ha pre-germinated rice seeds gave the best results in terms of adequate stand establishment and weed control. The direct-seeded flooded rice technique has the advantages of reduced labor cost because it eliminates nursery preparation, care of seedlings in seedbed, pulling of seedlings, hauling and transportation, shorter cropping cycle because of the absence of transplanting shock and the improved water control and hence its better management. Further, direct seeding would require shallow water depth to be maintained in the rice field compared to transplanted rice fields. In Sindh, the traditional practice of rice establishment is the transplanting (TPR) of young rice seedlings from rice nurseries. Adoption of direct seeded rice (DSR) method for rice crop establishment would offer opportunities for improved management of irrigation water.

Irrigation Scheduling

Prihar and Grewal (1985) addressed the issue of improving water use efficiency of rice, specifically for the rice-wheat cropping system. They showed that improved irrigation scheduling developed at the research stations promise considerable water savings and based on soil conditions, irrigation to rice can be delayed for variable time periods after infiltration of water from rain or previous irrigation.

Gunawardena (1992) developed a rotational irrigation schedule for the Mahaweli System B project in Sri Lanka for improved water management practices for paddy. The adoption of this rotational scheduling of frequent irrigation could significantly reduce water losses through seepage and percolation from the paddy fields. This practice will also encourage the farmer to stick to his rótation since the farmer is assured of the next supply within a short interval instead of waiting for several days. This assurance will help in reducing the excess water abstraction from the field canal, thereby reducing inequity in water distribution.

Efficient irrigation schedules have been developed for both rice and wheat in the high productivity rice-wheat zone of India (Gill 1994). It has been shown that one week of submergence and keeping soil wet for the remaining of the growing period can save about 33% irrigation water for rice without significant yield losses. Adoption of irrigation schedules for wheat could save 20 to 25% of irrigation water.

In the Sindh rice-wheat zone, the entire rice crop is grown in flooded fields, which are initially irrigated individually. Later in the crop season, with the increase in water depth, a largely uncontrolled water flow takes place from field to field, thereby resulting in continuous flow irrigation. The necessary irrigation and drainage control facilities, and separate inlets for individual fields for intermittent irrigation are not usually present. Paddy farmers in Sindh practice continuous irrigation of field because of the uncertainty of water availability, resulting in over-flooding of rice fields. At other places, particularly at the tail end of water channels, the fields are irrigated whenever water is available. Generally, fields are not properly leveled. Consequently, in order to irrigate the high spots, the fields are over-irrigated resulting in prolonged periods of deep water at the lower spots.

The average amount of water applied per hectare of rice grown in various canal commands and the minimum and maximum applications recorded on individual study areas for a total assessed area of 530 ha are presented in Table 10.

Clearly, more water is applied to rice in the Kotri Barrage Command (Kalri, Fuleli, Pinyari) than in the Sukkur Right Bank (Rice, Northwest,

TABLE 10. Average, minimum and maximum water amounts applied to rice fields during Kharif, 1964

Canal Command	Average (mm)	Minimum (mm)	Maximum (mm)
Rice Canal	945	731	1432
North West Canal	823	640	2560
Dadu Canal	1707	1707	1707
Kalri, Fuleli, Pinyari	2316	1432	2926
Lined Channel	1372	1341	1753
All Commands	1433	640	2926

Adapted from Lower Indus Report, 1965.

Dadu) rice growing areas. This is a result of more water being available per hectare in the Kotri Commands.

Studies have related evapotranspiration (ET) and irrigation requirements for wheat to the class A pan evaporation (E) (Kijne 1994). ET averaged about 80% of E pan after full ground cover. Irrigation application based on 75 to 80% of E pan provided adequate water for wheat. In northwestern India and Pakistan, where commonly five irrigations are applied to wheat based on its developmental stage, a scheduling technique based on water application at 75% of E pan could reduce the number of early wheat season irrigation and hence irrigation requirement for the growing season.

In Sindh, wheat is grown as dubari (growing of crops in the rabi season on residual moisture from kharif irrigation) and bosi (growing of crops in the rabi season on water supplied at the end of the kharif season) crop mainly in the non-perennial irrigation areas. In the perennial irrigation areas, it is usually grown under the basin flood irrigation method. Table 11 presents the average wheat yield per hectare grown under dubari, bosi and basin flood irrigation methods. The average yield from flood-irrigated fields is higher than that of the other two methods. In general dubari wheat yields more than the bosi wheat.

Under basin flooded irrigation areas, more than 75 percent of the wheat receives a pre-sowing irrigation. The average interval between sowing and the first irrigation is approximately 4 weeks in the perennial irrigated areas, and is between 5 and 7 weeks in the non-perennial areas. Generally, more irrigation is applied in the perennial than in the non-perennial areas (Table 12). The average maximum interval between successive irrigations is 6 weeks in the perennial areas where there is a canal closure during January. The interval is little shorter in the non-perennial irrigated areas.

Based on a field survey of 100 farmers who actually grew irrigated rice and wheat on the same field during 1986-87 (Bhatti et al. 1987) in the

TABLE 11. Irrigation methods vs. wheat yields (kg/ha)

Canal Command	Method of Irrigation		
	Normal	Dubari	Bosi
	Average Yield (kg/ha)		
Lined Channel	731	548	–
Kalri, Fuleli, Pinyari	183	274	183
Rice Canal	–	640	–
Dadu Canal	1005	–	822
North West Canal	640	457	91

Adapted from Lower Indus Report, 1965.

command of Dadu Canal, rice was continuously flooded. During the wheat season unlike the common practice in Punjab, only two of the sampled farmers gave pre-sowing irrigation to their wheat fields. All farmers irrigated their wheat crop post-sowing: 28 percent irrigated twice; 60 percent irrigated thrice; and 12 percent of the sampled farmers irrigated their wheat crop four times. Thus, the most frequent number of post-sowing irrigations for wheat was three, with range from two to four. In general, farmers who seed their wheat crop late irrigated their crop more often.

Yaseen, Rao and Sipio (1995) reported that the total water requirement for wheat is 375 mm for Lower Sindh areas whereas farmers apply about 450 mm of irrigation water to wheat. Thus, farmers are applying about 17 percent more water than the crop water requirement. Water use efficiency was therefore low (8.11 kg/mm/ha) in case of farmers' irrigation practices against the recommended scheduling of wheat (9.97 kg/mm/ha). This reflects more water used than required, with reduced yield (3.65 ton/ha) against yield (3.74 t/ha) obtained by applying 375 mm of irrigation. By better management of irrigation water, water use efficiencies could be increased which would result in water savings coupled with reduced drainage effluent.

Irrigation scheduling techniques designed to understand when and how much to irrigate are very important to assist in efficient on-farm water use, especially in the areas of limited water supply. By managing other production inputs too, the maximum crop yield could be achieved. The improved irrigation scheduling and irrigation practices like land leveling by laser technology and appropriate irrigation method could result in reduced seasonal irrigation water requirements and consequently, help in controlling water logging and salinity conditions in irrigated areas.

Thus, adoption of efficient irrigation schedules developed by the research institutions, like DRIP, LIM, etc., offers more opportunities for improving water use efficiency of the rice-wheat farming systems in Sindh. However, over-irrigation is done by farmers because of their lack of awareness about crop water requirements and fixed warabandi system of water distribution. Over-irrigation causes not only water logging and salinity problems, but also reduces crop yield. Optimum irrigation scheduling can prevent water logging

TABLE 12. Number of irrigations applied to wheat

Canal Command	Number of Irrigation
Lined Channel	3
Kalri, Fuleli, Pinyari	1
Rice Canal	0
Dadu Canal	4
North West Canal	2

and soil salinity, will increase crop yields, and, save irrigation water. By adopting irrigation scheduling of wheat, about 17% of irrigation water could be saved compared to farmers' practices. The irrigation scheduling adjusts water application to climatic evaporative demand and allowable soil water depletion, particularly for wheat grown on high water-storage soils, thereby causing considerable reduction in irrigation requirements for the growing season; and also, improving productivity of water.

Water Saving Irrigation Regimes

Sandhu et al. (1980) evaluated the effect of various irrigation practices on the irrigation requirement of rice on a sandy loam soil. They found that the irrigation needs of rice were highest (170-204 cm) with continuous submergence (Table 13). The mean increase (average of 4 years) in irrigation water efficiency with 1-day, 2-day, 3-day and 5-day drainage was 30, 54, 57 and 88% over the mean irrigation water efficiency of 28.9 kg/ha-cm with continuous submergence (Table 13). This study revealed that continuous submergence of soil is not necessary in order to obtain high rice yields. Once the transplanted seedlings are well established, irrigation could be delayed for some period after complete infiltration of ponded water without any yield loss. The potential saving of 20 to 50% in irrigation water primarily results from the reduction in percolation losses.

Mishra, Rathore and Pant (1990), and Mishra, Rathore and Pant (1997) reported that an optimum rice yield with high water use efficiency of 3.21-3.67 kg/ha-mm could be obtained by adopting an intermittent irrigation schedule of 3-5 days after the disappearance of ponded water under shallow water tables of 7-92 cm deep. Similarly, intermittent irrigation of 1-3 days under medium water table (13-126 cm deep) conditions, in place of continuous submergence would be beneficial.

Tabbal, Lampayan and Bhuiyan (1992) evaluated alternative water management strategies on farmers' rice fields on clay loam soils with average

TABLE 13. Various irrigation regimes for rice

Irrigation Regime	Irrigation Water Applied (cm)					Mean Irrigation Water Efficiency (kg/ha-cm)
	1974	1975	1976	1977	Mean	
Continuous Submergence	204	195	170	192	190	28.9
1-day drainage	151	138	130	160	145	37.6
2-day drainage	–	117	121	136	125	44.4
3-day drainage	114	92	107	128	113	45.3
5-day drainage	96	94	81	112	96	54.4

Adapted from Sandhu et al., 1980.

water table depth of 95 cm for four consecutive dry seasons (1988-1991). This study showed that: (1) water use under saturated soil regime was about 40% less than continuous flooded conditions without any yield loss. In the alternate wet and dry regime, water use was 60% less than the traditional practice, but it resulted in about 28% yield loss; (2) high rice yields could be obtained under continuous saturated soil regime maintained in the field to allow ET to take place at the potential rate. However, if weed growth is a problem, continuous submergence up to panicle initiation stage of crop growth and, then continuous saturation could be an effective technique of water-efficient irrigation without yield reduction; and (3) percolation rate could be significantly reduced by maintaining continuous saturated soil conditions. But, if soil cracks develop, percolation losses increase, and water losses could even be greater than those in the continuously submerged field conditions.

Kijne (1994) suggested three alternatives to continuous flooding, i.e.,

- Intermittent irrigation, where the field is irrigated as soon as the soil water content falls slightly below saturation, in an amount sufficient to maintain shallow submergence of the field.
- Heading stage submergence method, in which the soil is kept at saturation or is lightly submerged during most of the growing period, followed by field submergence to a depth of 10 cm for a period of 25 days after heading.
- Water saving method, in which after puddling and transplanting, the field is supplied with sufficient water to keep the soil water content in the root zone at not less than 75% of saturation during the growing period, with moderate submergence only during a period of 30 days starting at head initiation till the end of the flowering. Intermittent irrigation could save 20% water compared to continuous flooding and leads to only 50% of the potential yield; heading stage submergence could save 40% water and yields are reduced by 25%, whereas water saving technique requires 25% less water without yield loss.

The practice of continuous submergence of rice for initial 2 weeks after transplanting followed by irrigation at 2-day drainage saved, on average, 73% irrigation water compared to continuous submergence (Singh et al. 1996). Mishra, Rathore and Tomar (1995) evaluated the effect of different irrigation regimes for wheat on groundwater contribution, irrigation requirement and water use efficiency under fluctuating shallow water table (SWT) at a depth of 40 to 90 cm and medium water table (MWT) at a depth of 80 to 130 cm. They found that irrigation given only at crown root initiation and milk stages under shallow water table conditions, and at crown root initiation, maximum tillering and milk stages under medium water table conditions, could give optimum wheat yields, resulting in higher water use efficiency.

In Sindh, one or more pre-irrigations are provided in order to have some soil moisture before rice transplanting. This is done when the seedlings are in the nursery. An additional irrigation is given just prior to transplanting. Once the seedlings are established, normally a few days after transplanting, water is again applied to flood the field and to maintain the water level just below the growing tips of the rice plants. Once a depth of about 30 cm is attained, further water applications are provided to maintain this depth. The levelness of the fields is rarely, if ever, up to the standard required for efficient water control. Irrigation is discontinued about 15-20 days before harvesting in order to hasten maturity and also, to have fields free of standing water at harvest. In order to improve water-use efficiency of rice crop, farmers in Sindh should adopt water-saving practices of maintaining a thin layer of standing water in the rice field, saturated or alternate wet and dry soil regimes in place of the traditional practice of continuous submergence. These irrigation regimes can save 20-70% of irrigation water without significant yield loss. In clay loam soils, high rice yields can be obtained by maintaining a saturated soil moisture regime, thereby allowing potential ET to occur. However, if weed growth is a problem, submergence up to panicle initiation stage and continuous saturation thereafter, can increase water-use efficiency without significant rice yield reductions. Therefore, adopting these moisture regimes provide opportunities for improving water-use efficiency in rice.

Land Leveling

Farmers should be encouraged to carry out land leveling in order to improve water conservation. Land leveling effectively facilitates on-farm water control and management. It is a basic requirement at field level to avoid over- or under-irrigation due to the micro-ground surface undulations. In order to facilitate this activity, the local government agencies should provide subsidies, technical assistance and training to the farmers. As the scope of land leveling by bullock and manual labor is very limited in Sindh due to clayey nature of the soils, farmers can benefit from laser technology for land leveling. Laser equipment is now being locally manufactured by the private sector. In general, farmers have a tendency to over-irrigate to water undulating fields. Poorly leveled fields are considered to be a major factor in causing over-irrigation and giving rise to waterlogging and salinity problems.

Discontinuation of Pancho System

In Sindh, some of the rice farmers, especially on the Right Bank of the Indus, practice "Pancho System" of irrigation. This system involves draining of standing water from the field at intervals of 4 to 5 days and re-irrigating the same rice fields. The standing water in the rice fields is drained out to adjoining low-lying areas and fresh water is applied. Though, some of "Pancho water" let out of rice fields is lost through evaporation and percolation, this practice ultimately results in water logging, thereby creating acute drainage problems in the rice growing areas. In most of the rice fields, the groundwater table rises to such an extent that in some areas it reaches the surface layer of soil. Under these conditions, percolation either slows down or does not take place. The only way to continue this practice is through lateral drainage. There is an artificial drainage system prevailing in the rice growing area of Sindh. The farmers' practice of the Pancho System seems to be helpful in increasing rice production, but in the absence of an efficient drainage system, the water is drained to low-lying areas resulting in water logging.

The factors that encourage the Pancho System of irrigation are:

- Long watercourses and continuous flow of water in them. The farmers are forced to use their irrigation turns irrespective of their irrigation needs.
- Irrigation by block of fields. There is only one water inlet for a number of fields from the irrigation ditch; water is let into one field and through it to the next and so on. Often water is kept moving slowly through the paddy fields; and
- Control of salinity. In newly reclaimed saline land, water is let out of the field for washing the salts from the land.

In order to improve the drainage of rice areas, the practice of Pancho System should be stopped and this could be achieved by reducing the length of the watercourses and farmers should irrigate each field separately from the irrigation ditch. For achieving these objectives, there is a need to construct additional minor channels by the Irrigation Department and additional irrigation ditches by the farmers, so that each field has a separate water inlet from a watercourse for separate and controlled irrigation. There should also be an efficient drainage system to drain out water from the marshes.

Improved Layout of Irrigation Ditches and Fields

In Sindh, significant amounts of water are wasted due to negligence by farmers, bad layout of water channels and weak bunds of rice fields, resulting in low irrigation and application efficiencies. These channels follow a zigzag pattern, which slow down water speed and cause siltation necessitating frequent cleaning of water channels. Clark and Aniq (1993) reported that the rice fields on the right bank of the Indus receive a total of about 1.2 m to 1.7

m of water, of which 0.9 m is for consumptive use, therefore resulting in an application efficiency of 53 to 75% only.

Problem of wastage of water, over-irrigation and shortage of water could be solved by improved layout of water channels and fields and keeping the water channels straight and clear. Also, in order to improve the irrigation efficiency, leveling of individual fields and making the bunds of water channels and fields strong is very important. Farmers should be advised to construct proper field bunds to efficiently retain and control water to the required standing depth, which prevents or minimizes water losses through reduced drainage effluent.

Improved Drainage

In China, major constraints to rice-wheat cropping are either from too much water causing water logging of wheat in the south or too little water causing drought stress to wheat in the north. Thus, in the middle and lower reaches of the Yangtze River, wheat season rainfall is 500-700 mm, which in the absence of drainage results in water logging to wheat seedlings. In northern China, wheat season rainfall is only 150 mm, which is insufficient to meet wheat water requirements and, therefore, three irrigations are applied to wheat. In the lower and middle reaches of the Changjiang River Basin, most paddy fields are situated in depressions with high ground water table and majority of the soils are clay to loamy clay in texture. After rice harvest, the paddy fields are wet and the excessive soil moisture reduces the growth and yield of wheat after rice. Constructing a perfect drainage system was effective in minimizing the excessive moisture damage. Open-ditch drainage systems are more common throughout the central and southern China. This drainage system consists of ditches (0.4-0.6 m deep) running parallel to or transverse to the length of the rice-wheat fields and each ditch is connected to drainage canals (0.8-1.0 m deep) surrounding a block of fields (Yixia, 1989; Lianzheng and Yixian, 1994).

In Japan, Ogino and Murashima (1993), and Ogino and Murashima (1996) reported that in order to grow non-paddy crops like wheat and to obtain appropriate soil and working conditions for farm machinery in large-scale paddy plots, subsurface drainage system be installed. To realize this national policy, land consolidation projects have been promoted and the installation of subsurface drainage has been intensified, supported by central and local governments' subsidies. Agricultural drainage systems consisting of open drainage canals, farm ditches and surface drains have been effectively combined with irrigation canal systems to reduce diversion requirements and prevent water logging. Each plot has its own inlet for irrigation and outlet for surface drainage and a farmer can control water in each plot independently of other fields.

Reuse of Drainage Effluent

Recycling of drainage water (Ngion 1994) is being adopted in Malaysia for water saving and conservation and for sustaining agricultural development. In a number of existing irrigation schemes, drainage water is being pumped from the drains into nearby canals or directly onto the adjoining paddy fields for recycled use, which directly raises the cropping intensity.

Zulu, Toyota and Misawa (1996) evaluated the potential of agricultural drainage water reuse for improving irrigation water management in rice land of a water shortage area in Niigata Prefecture, Japan. They found that for three years (1991-1993), the average water reuse component was about 14.5% of the total irrigation water supply for the whole Kaliyada area. Apart from meeting the water needs at peak demand periods, water reuse was a quick response water supply solution during dry spells, increasing both the water reliability and rice crop security. By using drainage water for irrigation, the system fresh water requirements were reduced by about 14.5% of the irrigation water supply for the whole Kaliyada. This study proved that coupled with efficient water use methods, water reuse has a high potential for improving irrigation water management for the rice-wheat system.

In Sindh, the water from canals and rice fields coupled with Pancho water either percolates into the groundwater or stands in the depressions. These pools of water remain filled throughout the year because only a portion of it seeps down to groundwater or is lost due to evaporation. These pools of water go on expanding every year as a result of which valuable agricultural land is being lost. In the kharif rice-growing season, water table is generally at or near the surface with negligible deep percolation. Excess water removal by evaporation leaves behind the concentrated salts in the area. In the absence of drainage, surface water accumulation during the kharif season is not drained off in time resulting in reduced area for the rabi crop.

In Sindh, to grow wheat after rice and to obtain appropriate soil and working conditions for farm machinery in the paddy fields, installation of subsurface drainage system should be intensified and be supported by government subsides. There should be participatory drainage management schemes in which farmers should share the capital cost and be responsible for the operation and maintenance costs. Subsurface drainage systems operated and maintained by the government perform poorly due to management and financial resource constraints. Therefore, collaborative tile drainage systems with capital, operational and maintenance cost sharing by the beneficiaries (farmers) can be established to sustain on-farm drainage systems.

The reuse of drainage water offers another opportunity for improving irrigation water management for rice-wheat cropping system in water shortage areas of Sindh. Depending upon the quality of drainage effluent, the water reuse component can contribute significantly to the total irrigation water supply for the rice-wheat irrigated areas. Apart from meeting the water needs at peak demand periods, drainage water reuse would be a quick response water supply solution during water shortage periods, increasing both the water reliability and rice-wheat crop security. For Sindh, the possibility of reuse of drainage effluent should be considered as a means of water saving and conservation.

Farmers Participation

Magliano et al. (1990) reported that in the Philippines active farmer participation resulted in improved equity and reliability of water distribution, which in turn resulted in earlier planting of rice and an increase in the irrigated area. Khan (1992) also stressed on the importance of farmers' participation in irrigation programs for efficient water management of paddy fields.

In Sindh, farmers should be organized in order to participate in the decision-making process and the Irrigation Department should consider them as active partners in this process. This would improve equity and reliability of water distribution, which in turn would result in timely planting of rice and wheat.

In conclusion, several options exist for improving water use efficiency of the rice-wheat cropping system in the Sindh zone of Pakistan and these examples can be extrapolated to other areas of the Indo-Gangetic Plains of South Asia. Adoption and spread of improved water management practices will sustain the high rice and wheat yields and should therefore be an integral part of the cropping system.

REFERENCES

Bhatti, I.M., B.B. Khokar, G.H. Brohi, K.B. Chana, J.C. Flinn, and L.E. Velasco. (1987). *Rice-wheat rotations in upper Sindh, Pakistan*. Rice Research Institute, Dokri, Department of Agriculture, Livestock and Fisheries, Government of Sindh, Pakistan.

Bhuiyan, S.I. (1992). Water management in relation to crop production: Case study on rice. *Outlook on Agriculture* 21(4): pp. 293-299.

Bhuiyan, S.I., M.A. Sattar, and M.A.K. Khan. (1995). Improving water use efficiency in rice irrigation through wet seeding. *Irrigation Science* 16(1): pp. 1-8.

Clark, A.K., and M. Aniq. (1993). Canal irrigation and development opportunities for the Indus Right Bank in Sindh and Baluchistan. *ICID Bulletin* 42(1): p. 11.

De Datta, S.K. (1986). Technology development and the spread of direct-seeded flooded rice in South-East Asia. *Experimental Agriculture* 22: pp. 417-426.

Gill, K.S. (1994). *Sustainability issues related to rice-wheat production system in Asia. Sustainability of rice-wheat production systems in Asia.* Regional office for Asia and the Pacific (RAPA) publication 1994: pp. 36-60.

Gunawardena E.R.N. (1992). Improved water management for paddy irrigation: A case study from the Mahawali System B in Sri Lanka. In *Soil and water engineering for paddy field management*, eds. V.V.N. Murty and K. Koga, Proceedings of the International Workshop on Soil and Water Engineering for Paddy Field Manageemnt, January 28-30, AIT, Bangkok, Thailand, pp. 307-316.

Khan, L.R. (1992). Water management issues for paddy fields in the low-lying areas of Bangladesh. In *Soil and water engineering for paddy field management*, eds. V.V.N. Murty and K. Koga. Proceedings of the International Workshop on Soil and Water Engineering for Paddy Field Management, January 28-30, 1992, AIT, Bangkok, Thailand, pp. 339-345.

Kijne, J.W. (1994). *Irrigation and drainage for the rice-wheat system: Management aspects. Sustainability of rice-wheat production system in Asia.* Regional office for Asia and Pacific (RAPA) publication, 1994: pp. 112-125.

Lianzheng, W., and G. Yixian. (1994). *Rice-Wheat systems and their development in China. Sustainability of rice-wheat production systems in Asia.* Regional office for Asia and the Pacific (RAPA) publication, 1994, FAO of Limited Nations Bangkok: pp 160-171.

Lower Indus Report. (1965). Vol. 10.

Mac Donald Sir, M., and Partners Ltd., Harza Engineering Co., International LP, Nespak (Pvt.) Ltd. and ACE (Pvt.) Ltd. (1990). *Water sector investment planning study*, Vol. 1: Main report, federal planning cell, Lahore.

Mac Donald Sir, M. and Nespak (Pvt.) Ltd. (1993). *Pakistan Drainage sector Environmental Assessment, National Drainage Program I*, Volume 4-Data (Water, soil and agriculture), WAPDA, Lahore.

Mac Donald Sir, M. and Nespak (Pvt.) Ltd. (1995). *Feasibilty study, National Drainage Program I*, Volume 2-Annexes, WAPDA, Lahore.

Magliano, A.R., D.M. Cablayan, R.C. Undan, T.B. Moya, and C.M. Pascual. (1990). *Technical consideration for rice-based farming systems: Main irrigation system management.* In inter-country workshop on irrigation management for rice-based farming systems, Colombo, Sri Lanka, November 1990. International Irrigation Management Institute (IIMI), Colombo, Sri Lanka, pp. 15.

Mishra, H.S., T.R. Rathore, and R.C. Pant. (1990). Effect of intermittent irrigation on groundwater table contribution, irrigation requirement and yield of rice in Mollisols of the Terai Region of India. *Agricultural Water Management* 18: pp. 231-241.

Mishra, H.S., T.R. Rathore, and V.S. Tomar. (1995). Water use efficiency of irrigated wheat in the Terai Region of India. *Irrigation Science* 16: pp. 75-80.

Mishra, H.S., T.R. Rathore, and R.C. Pant. (1997). Root growth, water potential and yield of irrigated rice. *Irrigation Science* 17: pp. 69-75.

Nespak (Pvt.) Ltd. (1999). *Data Review Report No. 13*, September 1999.

Ngion, K.C. (1994). Improved irrigation water management for sustainable agricultural development in Malaysia. In *JICA; FAO, RAPA, Irrigation performance and evaluation for sustainable agricultural development:* Report of the expert consultation of the Asian Network on Irrigation/Water Management, Bangkok, Thailand, 16-20 May 1994. Bankok, Thailand: FAO, RAPA: pp. 151-160.

Ogino, Y., and K. Murashima. (1993). *Subsurface drainage system of large size paddies for crop diversification in Japan.* International commission on irrigation and drainage. Fifteenth congress, The Hague, 1993.

Ogino, Y., and K. Murashima. (1996). *Drainage in Asia (I) rice-based farming in the humid tropics.* Special contribution. Rural and environmental engineering No. 31 (1996. 8): pp. 4-12.

Prihar, S.S., and S.S. Grewal. (1985). *Improving irrigation water use efficiency in rice-wheat cropping sequence–technical and policy issues.* In *National Seminar on Water Management: The key to developing agriculture.* Agricole Publishing Academy, New Delhi, India.

Rehman, C.A.U. (1996). Salinity management in the Indus Basin. Proceedings of the Workshop on Groundwater and Irrigation Management Issues. Centre of Excellence in Water Resources Engineering, University of Engineering and Technology, Lahore, Pakistan. April 21-22, 1996.

Rehman, G., Hussain, A., Hamid, A., Almas, A.S., Tabassum, M., Nomani, M.A., and Yousaf, K. (1998). *The Irrigated Landscape: Resource Availability Across the Hydrological Divides. Waterlogging and Salinity Management in the Sindh Province, Pakistan.* Volume one, IIMI Report No. R-70.1.

Rehman, A., and Rehman, G. (1998). *Strategy for Resource Allocation Across the Hydrological Divides. Waterlogging and Salinity Management in the Sindh Province, Pakistan* Volume Three, IIMI Report No. R-70.3.

Revised Action Programme for Irrigated Agriculture. (1979). *Agronomy: Supporting report. Master planning and review division,* Water and Power Development Authority (WAPDA), Lahore, Pakistan.

Sandhu, B.S., K.L. Khera, S.S. Prihar, and Baldev Singh. (1980). Irrigation need and yield of rice on a sandy-loam soil as affected by continuous and intermittent submergence. *Indian Journal of Agricultural Sciences* 50(6): pp. 492-496.

Singh, C.B., T.S. Aujla, B.S. Sandhu, and K.L. Khera. (1996). Effect of transplanting date and irrigation regime on growth, yield and water use in rice (*Oryza sativa*) in Northern India. *India Journal of Agricultural Sciences* 66(3): pp. 137-141.

Tabbal, D.F., R.M. Lampayan, and S.I. Bhuiyan. (1992). Water-efficient irrigation technique for rice. In *Soil and water engineering for paddy field management,* eds. V.V.N. Murty and K. Koga. Proceedings of the International Workshop on Soil and Water Engineering for Paddy Field Management, January 28-30, AIT, Bangkok, Thailand, pp. 146-159.

WAPDA. (1990a). *Tubewell Performance Report of Kandh Kot Pilot Project,* 1987-89, SCARP Monitoring, SM (South).

WAPDA. (1990b). *Tubewell Performance Report of Shikarpur Pilot Project,* 1987-89, SCARP Monitoring, SM (South).

WAPDA. (1990c). *Tubewell Performance Report of Sukkur Pilot Project,* 1987-89, SCARP Monitoring, SM (South).

WAPDA. (1990d). *Tubewell Performance Report of Larkana Pilot Project,* 1987-89, SCARP Monitoring, SM (South).

WAPDA. (1990e). *Water Quality of Tubewells in Sukkur Right Bank FGW Project*, 1987-89 SCARP Monitoring, SM (South).

WAPDA. (1996). *Hydrological Monitoring and Tubewell Performance Report of Sukkur Right Bank FGW Project*, SCARP Monitoring, SM (South).

Yaseen, S.M., and M.C. Haroon. (1990). *Drainage and reclamation in Pakistan–an overview*. Paper presented at Indo-Pak Workshop on Soil Salinity and Water Management, held on February 10-14, PARC, Islamabad, Pakistan.

Yaseen, S.M., M.I. Rao, and Q.A. Sipio. (1995). *Some preventive measures to control waterlogging and salinity on farm level*. Paper presented at National Workshop on Draiange System Performance in Indus Plain and Future Strategies, held at DRIP, Tando Jam, Sindh, Pakistan. January 28-29, 1995.

Yixian, G. (1989). *Rice-wheat cropping systems in China*. Proceedings of the 20th Asian Rice Farming Systems Working Group Meeting. International Rice Research Institute (IIRI), Los Banos, Philippines, pp. 201-266.

Zulu, G., M. Toyota, and S. Misawa. (1996). Characteristics of water reuse and its effects on paddy irrigation system water balance and the rice land ecosystem. *Agricultural Water Management* 31(3): pp. 269-283.

Use and Management
of Poor Quality Waters
for the Rice-Wheat Based
Production System

P. S. Minhas
M. S. Bajwa

SUMMARY. India's major improvements in food production have been made possible via a shift towards rice-wheat systems as a consequence of enhanced utilization of ground waters. Rice-wheat rotation covering 10.5 million ha contribute about 75 percent of total food production. But it is being observed now that yield increases in rice and wheat has slowed down and there is rather a decline in factor productivity. One of the major reasons for this decline is indiscriminate use of alkali waters constituting about 25-42 percent of ground waters surveyed especially in the northwestern states of the Indo-Gangetic Plains (IGP). Because of high water requirements of the system, sodication rates of soil being irrigated and their steady state pH and sodicity values are much more (about 1.8 times) than that of the low water requiring rotations like millet/maize-wheat. So there has been a dilemma on whether or not rice-wheat system should be advocated with alkali irrigation waters. Consistent research efforts have lead to the guidelines for irrigation with such waters with respect to their amendment needs (gypsum requirements, frequency and mode of application), conjunc-

P. S. Minhas is Project Coordinator, AICRP on Management of Salt-Affected Soils and Use of Saline Water in Agriculture, Central Soil Salinity Research Institute, Karnal–132001, India. M. S. Bajwa is Director of Research, Punjab Agricultural University, Ludhiana–141004, India.

[Haworth co-indexing entry note]: "Use and Management of Poor Quality Waters for the Rice-Wheat Based Production System." Minhas, P. S., and M. S. Bajwa. Co-published simultaneously in *Journal of Crop Production* (Food Products Press, an imprint of The Haworth Press, Inc.) Vol. 4, No. 1 (#7), 2001, pp. 273-306; and: *The Rice-Wheat Cropping System of South Asia: Efficient Production Management* (ed: Palit K. Kataki) Food Products Press, an imprint of The Haworth Press, Inc., 2001, pp. 273-306. Single or multiple copies of this article are available for a fee from The Haworth Document Delivery Service [1-800-342-9678, 9:00 a.m. - 5:00 p.m. (EST). E-mail address: getinfo@haworthpressinc.com].

tive use with canal waters and use of organic materials and chemical fertilizers, etc. These results do show that subject to the following of specific soil-water-crop management systems, it is possible to control the build up of sodicity in soils and sustain crop yields. Options available in terms of management practices and some of the researchable issue are discussed in this paper. *[Article copies available for a fee from The Haworth Document Delivery Service: 1-800-342-9678. E-mail address: <getinfo@haworthpressinc.com> Website: <http://www.HaworthPress.com> © 2001 by The Haworth Press, Inc. All rights reserved.]*

KEYWORDS. Alkaline soils, amendment needs, conjunctive use of waters, sodicity control, sodic water, soil health, sustainability of rice-wheat

INTRODUCTION

Large-scale development of surface and ground water resources during post-independence period has reduced the susceptibility of Indian agriculture to the vagaries of monsoons. Recent estimates show that 37.6 percent of the net sown area of 142.2 million hectares (M ha) has provisions for irrigation facility, particularly in arid and semiarid regions. Increased water availability along with high yielding fertilizer responsive varieties particularly of rice and wheat led to ushering in of the 'Green Revolution.' Due to its stable economic returns, the last three decades have witnessed a shift in cropping towards the rice-wheat system (10.2 M ha), particularly in the northwest Indian states of Punjab, Haryana, western Uttar Pradesh and Rajasthan (Abrol and Gill, 1994). About 73% of the national food grain requirement in India is now being met by the rice and wheat crops which is likely to increase to about 77% by the year 2010. More recently, however, yield stagnation and reduced factor productivity or even a declining tendency in yields with time, are being reported from many parts of the Indo-Gangetic Plains. This has led to increasing concerns about long-term sustainability of the rice-wheat system, which is a key to the national food security system. At the same time, excessive withdrawal of ground water in regions having good quality ground waters is leading to decline in ground water levels at alarming rates. While in some other regions the water table has risen resulting in waterlogging and salinity. The latter problem is particularly acute in areas underlain with marginal and poor quality ground waters. Remedies to tackle the problems for the two situations would be different. In many areas, the good quality water supplies are inadequate and farmers are left with no option but to use brackish ground waters for irrigation of the rice-wheat system. Long-term use of these poor quality waters in the absence of proper soil-water-crop management practices, however, can pose grave risks to soil health and the environment. Experimental evidences show that the build up of salinity/sodicity in soils receiving brackish water irrigation can be controlled by following specific system of management (Minhas and Gupta, 1992). Nevertheless, researchers still face a dilemma whether or not rice-wheat system is sustainable under brackish water irrigated conditions. This compilation has been developed to describe some of these aspects and also reports possible options to sustain rice-wheat yields under conditions where poor quality waters have to be used.

WATER RESOURCES AND THEIR QUALITY IN RICE-WHEAT GROWING AREAS

The major rice-wheat growing areas of north-west India encompasses mostly semi-arid parts of the states of Punjab, Haryana, western districts of Uttar Pradesh adjoining Haryana and also some parts of Rajasthan (Figure 1) where the rainfall (Table 1) is inadequate to meet the water requirements of the system. The rice-wheat growing areas are, however, relatively well placed with respect to ensured supplemental water supplies with 95-98% of net sown area being irrigated through surface and ground water resources, both conjunctive or in isolation.

Surface Waters

Upper Bari Doab canal, Bist Doab canal, Sirhand canal, Sirhand Feeder canal, Bhakra Main line and Eastern canal are the major canal water distribution systems in Punjab. The total annual diversion of canal water is 1.53 M ha-m, out of which the Ravi sub-system, Sutlej sub-system, Beas + Sutlej system and the Shah Nahar contribute 0.315, 0.811, 0.35 and 0.050 M ha-m, respectively. About 53% of the total canal water supplies in the state are used in southwestern parts. Net area irrigated through canals is about 38% of the total irrigated area. Similarly, in Haryana, the canal water is carried from Bhakra reservoir through Nangal hydel channels-Bhakra main line, Narwana branch, NBK link, Barwala link. The second important canal irrigation system of the state, one of the oldest in India, is Western Yamuna Canal. However, in the absence of any storage across the Yamuna, the supply of Western Yamuna Canal is dependent upon runoff of the river. The net area irrigated through canals in Haryana is about 1.35 M ha which forms about 49% of the state's irrigated area.

As most of the canals receive their water supplies from river flows, their quality has been observed to be of high order. The salinity (electrical conductivity), ECw is usually below 0.5 dS/m. The predominant anions are HCO_3^- and SO_4^{2-} and the major cations are Ca^{2+} (Ca and Mg on equivalent basis)

FIGURE 1. Schematic map showing distribution of groundwater quality for irrigation in India.

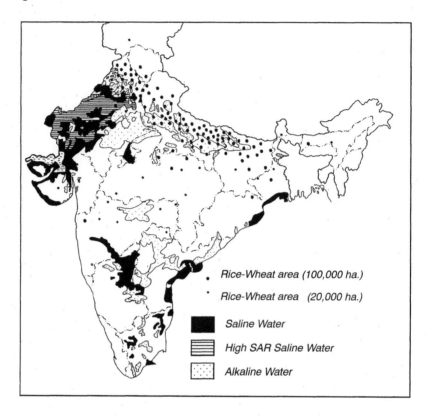

and Na⁺. The residual alkalinity (RSC) is invariably < 1 meq/l. Although the northwest parts of India have an impressive network of canals, these regions have extensive problems of waterlogging and salinity in their canal command areas. The affected areas mostly exist adjacent to canals especially where drainage facilities are poor, canal water levels are higher than ground level and where ground water is not harvested at rates sufficient enough to arrest the rise in ground water table due to seepage from canals. These problems get aggravated in some areas because of poor soil-water-crop management practices followed by the farmers.

Ground Waters

Major areas under rice-wheat system have come to stay in the regions with a relatively better quantity and quality of ground waters. In fact, the develop-

TABLE 1. Some basic, land-use, water resources and their quality and yield statistics of major paddy-wheat growing states of northwest India

Characteristic	Punjab	Haryana	Uttar Pradesh	Rajasthan
Geographical area (M ha)	5.0	4.4	29.4	34.2
Net sown area	4.2	3.5	17.2	16.1
Cropping intensity	176	167	147	117
Fertilizer use (kg)				
(N + P_2O_5 + K)/yr	168	106	89	23
Wheat production				
(m Mg)/yield (Mg/ha)				
1965-66	1.92 (1.24)	0.87 (1.28)	4.25 (0.96)	NA
1997-98	13.38 (4.01)	7.23 (3.62)	20.82 (2.31)	3.46 (1.72)
Rice production				
(m Mg)/yield (Mg/ha)				
1965-66	0.29 (1.00)	0.20 (1.06)	2.06 (0.46)	NA
1997-98	7.90 (3.40)	2.54 (2.97)	12.16 (2.12)	0.19 (NA)
Arid area (%)	26.8	29.3	Nil	57.4
Semi-arid area (%)	60.2	59.7	70	36.7
Mean annual rainfall (mm)	611	556	836(W)	700(W), 300(E)
Net irrigated area (%)	94	73	59	29
Canals	38.3	48.9	30.1	35.1
Wells	61.6	50.8	65.5	23
Irrigation benefit from major & medium schemes by the end of 1984-85 (M ha)				
Potential	1.568	1.929	6.813	2.462
Utilization	1.508	1.745	5.513	2.444
Ground water resources (M ha-m/yr)				
Utilizable	1.31	0.88	9.27	1.83
Net draught	0.93	0.61	2.68	0.46
Potential available	0.36	0.27	6.59	1.37
Use of poor quality waters	0.38	0.38	1.28	0.39
Rating of ground water quality				
Good (%)	59	33	37	16
Marginal (%)	22	8	20	16
Poor (%)	19	55	43	68
Characteristic features of poor quality waters				
Saline (%)	22	24	NA	16
Sodic (%)	54	30	NA	35
Saline sodic (%)	24	46	NA	49

ment of ground water resources through shallow tube wells has played a pivotal role in enhancing the overall agricultural production. Where as, the increase in area irrigated through canal irrigation during the last three decades was about 19%, this was of the order of 160 and 196% through tube wells in Haryana and Punjab, respectively. The irrigation through ground waters accounts for 60-65% of the total irrigation requirement and the remaining is met through other sources. As a consequence of over-exploitation of ground waters, ground water table is declining at an alarming rate. According to latest water balance studies, out of 118 blocks in Punjab, 70 blocks are 'dark' (their ground water exploitation is > 85% of the annual recharge) out of which 6 blocks are overexploited. Similarly, out of 16 districts in Haryana, 11 districts (mostly rice-wheat growing area) have declining water table.

Ground water surveys have shown that about 41-84% of the well waters in different states of the Indo-Gangetic Plains is brackish (Figure 1, Table 1). Vertical as well as lateral variations in ground water quality are encountered, even at short distances. High salinity waters are mostly encountered in arid parts of northwest states, of course where rice-wheat is not a common crop rotation. Ground waters of rice-wheat areas are largely of good quality except for the incidence of high residual alkalinity in many parts. Ground waters with higher RSC are common in central and south western parts of Punjab covering about 25% of the total area of the sate (Bajwa et al., 1975). These include parts of Amritsar (Khara Majha), Bhatinda (Mansa and Phul), Ferozepur (Zira and Dharamkot), Moga (Bagha Purana, Nihalsighwala), Ropar (Kharar), Sangrur (Malerkotla and Sangrur) and southern Ludhiana. In Haryana, these waters are encountered in the districts of Jind (Rajaund, Narwana), Karnal (Nilokheri, Nissang), Kaithal (Gulha Cheeka, Pundari, Dand, Kaithal), Panipat (Assandh), Bhiwani, Mahendergarh (Narnaul, Dadri, Pataudi), Gurgaon (Bawal), Fridabad (Ballabargh and Sohna) and Sirsa covering almost 21% of the total area of the state. Alkali waters are also common in Agra, Mathura, Aligarh, Mainpuri, Etah, Unnao, Fatehpur, Ballia and several other districts of Uttar Pradesh and to the east of Aravalli range in Rajasthan including parts of Jaipur, Kota, Udaipur, Tonk, Nagaur, Sikar and Jhunjhunu districts. Associated with salinity, the ground waters in some pockets may contain toxic levels of boron, fluoride, nitrate, selenium, silica, etc.

EFFECTS OF IRRIGATION WITH ALKALI WATERS ON SOILS AND CROPS

Irrigation with alkali waters containing high Na^+ relative to Ca^{2+} and Mg^{2+} and high carbonates (CO_3^{2-} and HCO_3^-) leads to increase in alkalinity and sodium saturation in soils. The increase in exchangeable sodium percentage (ESP) adversely affects soil physical properties including water infiltration and soil aeration. On drying, soils become very hard and on wetting, the soil particles get dispersed and clog the pores that affects root respiration and development. Due to poor respiration, the young seedlings of arable crop, e.g., wheat, exhibit yellowish appearance after the first irrigation. The waters with low calcium ($Ca^{2+} < 2$ meq/l) and high amounts of carbonates result in specific toxicity symptoms on plants particularly scorching and leaf burning at the early seedling development stage of crops. The parameters generally analyzed for knowing their potential to create sodicity/alkalinity hazards are as follows.

Sodium Adsorption Ratio

The sodium adsorption ratio (SAR) is determined by using the equation:

$$SAR = \frac{Na^+}{\sqrt{(Ca^{2+} + Mg^{2+})/2}} \tag{1}$$

where, all concentrations are expressed in meq/l.

Residual Sodium Carbonate

Residual sodium carbonate (RSC) is another empirical approach proposed by Eaton (1950) to assess the sodicity hazard of carbonate and bicarbonate rich waters. RSC is expressed as:

$$RSC = (CO_3^{2-} + HCO_3^-) - (Ca^{2+} + Mg^{2+}) \tag{2}$$

where all the concentrations are expressed in meq/l. The complete precipitation of CO_3^{2-} and HCO_3^- as is the basis of this concept does not hold good for field situations.

Adjusted SAR (adj.SAR)

When water is applied for irrigating the crops, salt concentrates in the soil because of only pure water being extracted by the crop roots. If no precipitation of salts takes place then $SAR_{(soil\ water)}$ which is in equilibrium with the exchange, is expected to increase by a factor 1.4 ($\sqrt{2}$) and vice versa. Since in soils, precipitation and dissolution reactions (mainly involving Ca^{2+} ions) do occur and SAR alone can not be used accurately to predict ESP build up in soils, the concept of SAR was modified. The new parameter, i.e., the "adj.SAR" (Bower, Ogata and Tucker, 1968), is calculated using the following equation:

$$\text{adj.SAR} = \frac{Na^+}{\sqrt{(Ca^{2+} + Mg^{2+})/2}} [1 + (8.4 - pHc)] \qquad (3)$$

where Na^+, Ca^{2+} and Mg^{2+} are concentrations of these ions in irrigation water (meq/l), the value 8.4 is approximate pH of a non-sodic soil in equilibrium with $CaCO_3$. Theoretically, pHc is the calculated pH of water in contact with lime and in equilibrium with soil CO_2:

$$pHc = (pK'_2 - pK'c) + p(Ca + Mg) + p(Alk) \qquad (4)$$

The first, second and third terms on the RHS of the equation are obtained using the sum of Ca + Mg + Na, Ca + Mg and CO_3 + HCO_3 (meq/l) in water analysis, respectively. At pH > 8.4, there is no precipitation of $CaCO_3$ and at pH < 8.4, there is a linear relationship between amount of $CaCO_3$ precipitated and saturation index (SI = 8.4 − pHc). As leaching fraction (LF) increases, the precipitation of $CaCO_3$ decreases. Thus to accurately predict the ESP, LF has been included in the equation as:

$$ESP = SARsoil = \frac{1}{\sqrt{LF}} SARiw [1 + (8.4 - pHc)] \qquad (5)$$

Rhoades (1968) found that weathering of soil minerals and consequent addition of Ca and Mg to soil solution affect the calculated values of ESP and modified the above equation to:

$$ESP = \frac{y^{(1+2LF)}}{\sqrt{LF}} SARiw [1 + (8.4 - pHc)] \qquad (6)$$

where 'y' is the weathering constant usually taken as 0.70. Suarez (1981) observed that pH_{eq}, i.e., pH for soil solutions in equilibrium with $CaCO_3$ is not constant at 8.4 but depends upon solution composition, leaching fraction and P_{CO_2} in the root zone. He then proposed the following derivation for better prediction of SAR_{dw} and in turn ESP.

$$SAR_{dw} = \frac{Na_{iw}/LF}{[Mg_{iw}/LF + X(P_{CO_2})^{1/3}]^{1/2}} \qquad (7)$$

The 'X' represents the Ca in applied water modified due to salinity (ionic strength) and HCO_3^-/Ca^{2+} ratio (concentration in meq/l at the estimated P_{CO_2} of 7×10^4 Mpa in surface few millimeters of soil). Modified value of

Ca (Ca_x) was then used by Ayers and Westcot (1985) to correct calcium concentration vis-a-vis the SAR of waters and used the term adj.R_{Na}.

$$\text{adj.}R_{Na} = \frac{Na}{\sqrt{(Ca_x + Mg)/2}} \quad (8)$$

Sodication/Alkalization of Soils

In the early stages of irrigation with alkali waters, large amounts of divalent cations are released into the soil solution from exchange sites and minerals. Alternating irrigation with alkali waters and rainy season water can induce a cycle of precipitation and dissolution of salts in soils. The sodicity build up in soils, thus is the outcome of equilibrium between all these processes. Several reports on the sodication behaviour of soils upon irrigation with waters having high residual alkalinity have come up from northwest parts of India (Bajwa, Hira and Singh, 1983; 1986, Bajwa and Josan 1989a & b, Bajwa, Josan and Chaudhary, 1993). The general conclusions drawn with respect to sodicity (ESP) build up in soils (Table 2) from a series of long-term experiments on soils varying in texture (loamy sand to silt loam) under wheat based rotations are as follows:

- Whenever the residual alkalinity increased in irrigation waters of nearly same adj.SAR, the alkalization in soils was quicker.
- When waters of high SAR and high RSC contain sufficient Ca^{2+} (> 2 meq/l), the rate of sodication vis-a-vis alkalization of soils is retarded.
- The build up of ESP and pH is sharp under rice based cropping system especially in the upper soil layers. This was obviously due to larger number and greater depth of applied irrigation water than those observed under other upland crops like cotton, maize and pearl millet in rotation with wheat.
- Under field conditions, the sodication of soils irrigated with alkali waters is a continuous process over the years though the pace slows down during the later years. With reduction in infiltration (RIR = 0.3 at an ESP of > 20), the opportunity for alkali waters to penetrate deeper is reduced, thereby, alkali solutions induce sodicity in the upper layers when concentrated through loss of water due to evapo-transpiration. Such conditions do not allow for the achievement of steady state conditions that has been the basis for the development of various indices of sodicity (adj.SAR/RNa). Due to this, the results are contradictory to the theoretical indices like adj.SAR/RNa that predict ESP build up will be lower in upper layers where leaching fractions (LF) are expected to be higher than those from lower layers.

TABLE 2. Effect of long-term use of alkali waters on sodicity (ESP), pH and infiltration rate of soils and yield of wheat

ECiw (dS/m)	RSC (meq/l)	SAR	adj.SAR (1)	(2)	(3)	Ca²⁺ (meq/l)	Soil texture	Kharif crop	No. of years	pH	ESP	RIR*	Wheat yield
3.2	4.0	21.4	38.2	ND	ND	3.5	Loamy sand	Maize/millet	10	8.9	36	46	91**
1.5	14.8	19.5	41.0	33.1	19.5	0.5	Silt loam	Millet	9	10.0	43	–	20
1.4	10.0	15.8	31.6	25.3	17.4	0.6	Silt loam	Millet	9	9.6	32	–	60
1.4	10.1	13.5	26.7	19.6	16.2	0.4	Sandy loam	Rice	6	9.6	46	14	45
1.4	10.0	15.8	31.6	25.3	15.8	0.6	Silt loam	Rice	9	9.7	46	–	42
1.5	8.0	11.6	25.6	17.4	20.9	2.2	Sandy loam	Maize	5	9.1	20	71	96
3.0	8.0	25.0	52.6	35.0	42.5	2.2	Loam	Maize	5	9.0	28	21	92
1.2	10.6	10.1	25.5	ND	ND	1.1	Sandy loam	Cotton	8	9.2	24	–	74

Data compiled from Bajwa and associates (1983, 1986 and 1989a,b,c, 1993), (1), (2) and (3) represent adj.SAR (using pCa + Mg), adj.SAR (using pCa) and latter corrected for rainfall, respectively. *RIR represent relative infiltration rate referenced to canal water. **Relative yields averaged over years.

Several researchers have evaluated the sodication/alkalization of soils by relating these changes with various indices of irrigation waters. Singh, Rana and Bajwa (1977) observed that ESP of the soils upon use of alkali waters was better related with their concentrations of CO_3^{2-} + HCO_3^- ions (r = 0.60), followed by RSC (r = 0.54) and SAR (r = 0.45) for 139 soil samples collected from Bhatinda district in Punjab. A comparison between adj.SAR and RSC as the sodicity indices made by Sharma and Mondal (1981) showed that root zone ESP could be better correlated with adj.SAR (r = 0.857) than RSC (r = 0.598). Similarly, Bajwa, Hira and Singh (1983) reported that ESP was better related with adj.SAR (r = 0.90) than with RSC (r = 0.49). Long-term field experiments on sandy loam soils, that were mono-cropped (wheat/barley-fallow) and irrigated with alkali waters (RSC upto 15 meq/l; SAR [22-37]) showed much higher sodication of the plow layer. Upto 50% of additional ESP build up could be ascribed to the direct effect of RSC in irrigated waters (Dhir, Sharma and Singh, 1980; Manchanda, Gupta and Jain, 1989; and Gupta, 1980). However, with the inclusion of a *kharif* crop in rotation, the build up in ESP was much higher. Bajwa and Josan (1989) observed that ESP of the surface 60 cm soil was significantly related to SARiw (r = 0.937), adj.SARiw (r = 0.949), RNa (r = 0.941), RSCiw (r = 0.921) and SARe (r = 0.980). The small differences in the coefficients of determination for the various indices tested indicated their almost similar effectiveness. The build up of ESP at the bottom of the root zone was much less than predicted by adj.SARdw. Actually the adj.SARdw defines the adj.SAR/ESP at the bottom of the root zone after the steady state conditions have been attained so it should not be confused with average ESP of that zone. Nevertheless, several field observations have shown that though the steady state is not achieved, quasi-stable salt balance conditions are attained after 4-5 years of sustained irrigation with alkali waters under the monsoonal climatic conditions, and that further rise in pH and ESP is very slow. Predictions by Bajwa, Choudhary and Josan (1992) with cotton-wheat system showed that ESP of surface 60 cm (where maximum build up of ESP took place) was significantly related to SARdw as predicted by Bower (r = 0.981), Rhoades (r = 0.985) and modified Ayers and Westcot (r = 0.941) equations. Minhas and Gupta (1992) also concluded that though there may be some discrepancies in the use of various sodicity parameters in vogue (SAR, adj.SAR, RNa) under the monsoonal climatic conditions, where most of the rain water is received in a shorter span of time (July-September) that do not allow for the attainment of steady state conditions. Yet these indices can be effectively employed for predicting sodicity hazards of alkali waters for the most important surface layer (0-30 cm). The averaged values of coefficients for these parameters as derived from several experiments in northwest India are presented in Table 3. It seems that adj.RNa can serve as a useful index to

TABLE 3. Predicting ESP (0-30 cm) from the parameters of alkali waters

| Crop rotation | Value of empirical constant | | | | |
| | SAR | | | adj.SAR | adj.RNa |
	(1)	(2)	(3)	(2)	
Rice-wheat	2.9	1.5	1.8	2.6	2.6
Millet/maize-wheat	2.2	0.7	0.8	1.3	1.0

(1), (2) and (3) represent adj.SAR calculated using pCa + Mg, pCa and latter corrected for rainfall. Data compiled by Minhas and Gupta (1992).

predict the ESP build up upon irrigation with alkali waters in millet/maize-wheat rotation, particularly because it does not require the use of any empirical constant as in case of adj.SAR predictions. But for the rice based cropping sequences, development of empirical constant is needed for all the indices or a value of 2.6 adj.RNa seems to be reliable. Manchanda (1993) also reported that for wheat-fallow system, SAR and adj.RNa under-predicted ESP (0-30 cm soil) by 1.5 and 1.3 times whereas adj.SAR over-predicted it by 2 times. For wheat-rice, SAR, adj.RNa and adj.SAR under-predicted it by 3.4, 1.9 and 1.5 times, respectively, whereas none of these fitted well for wheat-millet crop sequence. For wheat-fallow and wheat-rice, ESP equaled 1.5 SAR and 1.5 adj.SAR, respectively.

Infiltration Problems

Long-term use of alkali waters has been known to induce permeability problems in soils. Poor intake of the irrigation or rainwater often reduces the replenishment of the soil water storage and consequently decreases water supply to the plant roots. Low permeability of soils to water also leads to poor soil aeration and gas exchange problems. Influence of the major ion chemistry of irrigation waters on the hydraulic conductivity (K) of the alluvial and vertisol soils has been evaluated by several workers under laboratory conditions (Pal, Singh and Poonia, 1980; Girdhar and Yadav, 1980; Minhas and Sharma, 1986; Verma, Gupta and Sharma, 1987). In general, it is observed that the K-values of soils decrease with increase in SAR and the effects are more pronounced at low electrolyte concentration. Relative contributions of SAR and EC of the water in decreasing permeability were observed to vary with texture, Ca:Mg ratio and mineralogy of the soils. Sharma and Minhas (1986) studied the influence of ECiw, SAR/adj.SAR of water of two distinct chemical compositions on infiltration rates of an illitic sandy loam soil. The influence of the two types of waters on the relative infiltration rate (RIR) of the illitic calcareous sandy loam soil could be presented as:

I. $Cl^- - SO_4^{2-}$ dominant waters having no residual alkalinity

$$RIR = 0.887 + 0.037 \, EC - 0.015 \, SAR \quad (R^2 = 0.74^{**})$$

II. HCO_3^- dominant waters with residual alkalinity

$$RIR = 0.690 + 0.077 \, EC - 0.015 \, adj.SAR \quad (R^2 = 0.80^{**})$$

Here for computing RIR, the IR (3.0 mm/hr) determined with a water of ECiw = 0.3 dS/m, SAR < 1.0 for the normal soil was taken as reference. For the two kinds of salts in water, the effect of SAR/adj.SAR was negative and similar in magnitude and that salinity of water mitigated the adverse effect of high SAR. The above relation also points out that for similar EC and SAR, waters having alkalinity prove more harmful in reducing the infiltration rates of the soils. It may be pointed that the soils in the above studies were subjected to a sequence of infiltrating waters with reduced electrolytic concentrations while maintaining their SAR/RSC and drying in between the cycles was not allowed. Nevertheless, reductions in infiltration under natural field conditions are the result of the greater opportunities for the clay movement as well as the re-arrangements of the resultant sub-aggregates during the separated successive wetting and drying cycles. So the long-term use of alkali waters are expected to cause greater impact on infiltration. Manchanda, Sharma and Singh (1985), reported that long-term use of an alkali water (ECiw 4.0 dS/m, SAR > 30 and RSC 10 meq/l) reduced the infiltration rates to 5 mm/d in a sandy loam soil. The infiltration rate of a sandy loam (11% clay) and a loam (24% clay) soil determined after 5 years of irrigation with alkali waters to maize-wheat rotation (Table 3), decreased progressively with an increase in adj.SAR of alkali waters (Bajwa, Hira and Singh 1983). Reductions in RIR were more in case of loam soil as well as when ECiw increased at the same level of RSC. Similarly, infiltration rate of a loamy sand (5-8.8% clay) soil was reduced to 0.46 and 0.29 (referenced to canal water) after 5 years of irrigation with an alkali water (ECiw 3.2 dS/m, RSC 4 meq/l and adj.SAR 38.5) under rice-wheat and maize/millet-wheat rotations, respectively. The values of RIR for the two rotations after 10 years were 0.25 and 0.18, respectively. When 5 years data for the above two experiments were pooled to develop relationship between RIR and adj.SARiw, it could be described by:

$$RIR = 1.104 - (0.0147 \times adj.SARiw) \quad r = 0.87 \quad n = 12$$

This relation indicates that RIR decreases at a rate of about 1.5 percent per unit increase in adj.SAR beyond its value of 7.1, which by sheer chance are almost similar to that reported by Sharma and Minhas (1986) for alkali waters. In a light textured soil irrigated with waters having RSC of 2.4, 11.0 and 16.0 meq/l, Dhankar et al. (1990) reported the respective steady infiltration rates of 7.50, 0.70 and 0.60 mm/h under fallow-wheat rotation. The counter values were 6.40, 0.63 and 0.06 mm/h under millet-wheat rotation. Minhas et al. (1994) reported that steady infiltration rates of a sandy loam soil after irrigation with high-SAR (30-40 mmol/l) saline waters for a period of eight years were reduced to 5-10% of the normal soil. These reductions were irreversible as even though saline water (120 meq/l) with much higher electrolytic concentrations than the floc values of soil clays (30-40 meq/l) were used, the recovery was only 22-28%. It was concluded that the deeper layers ingressed with clays, i.e., below the plough layer where re-mixing of moved-in clay did not occur, it controlled the steady IR of soils irrigated with high SAR-saline waters. This contradicted the in vogue concept of surface layer controlling the infiltration in brackish water irrigated soils and that the permeability problems upon irrigation with saline-alkali waters are combated either when EC of the incoming solution or the salt released from soils maintain solution concentrations higher than their floc values. However, under continental monsoon climates, the irrigation season is usually followed by rainy season when salts are partially leached from the surface soil layers and with the alterations in the balance between EC and SAR, the clays get dispersed and move downwards with the traction of solution. Thus, the dynamic equilibrium between ESP of the soil, inherent infiltration characteristics (soil texture and mineralogy) and salt release in relation to rainfall pattern of the site will determine the amount and depth to which colloidal clays can migrate and consequently cause permeability problems (Minhas et al., 1999). Therefore, a rethinking is required on water quality guidelines such that structural changes in soils could be accurately described on the basis of quality parameters of waters as well as the soil, climate and other management parameters. Nevertheless, Minhas et al. (1999) have proposed that for evaluating infiltration hazards upon irrigation with saline-sodic waters, measurements of K-values after consecutive cycles of saline and simulated rain water can serve as a better diagnostic criteria.

Crop Responses to the Use of Alkali Waters

Several reports on the use of alkali waters for irrigation by the farmers in Jalore district (Saksena et al., 1966), Alwar and Jaipur district of Rajasthan (Singh and Sharma, 1971) and in Gurgaon district in Haryana (Verma 1973) have pointed out that irrigation of wheat-fallow system with waters having RSC upto 7.5 meq/l is a common practice. In studies conducted at Agra (AICRP Saline Water 1989) and at Karnal (Gupta, 1980), wheat yields were not reduced due to use of the waters having RSC value upto 10 meq/l (ECiw 2 dS/m, SAR < 10) on sandy loam soils. In both these studies, wheat was grown during *rabi* season and the micro-plots remained fallow during the monsoon season. The average rainfall of about 500-550 mm during the monsoon season mitigated the adverse effect of the sodicity/alkalinity build up during irrigation to wheat. In fallow-wheat rotation, sodicity/alkalinity build did not achieve the levels required to manifest its effect on wheat, which has been reported to be quite tolerant to sodicity (Abrol and Bhumbla, 1979). As mentioned earlier, use of alkali water in both the seasons leads to faster deterioration of soils. Many of the waters used at farmer's fields have residual alkalinity much above the limits prescribed for safe use (Singh and Singh 1971). The results suggest that when waters having $M^{2+}/\Sigma M^{n+}$ ratio more than 0.25 (or concentration of $Ca^{2+} > 2.0$ meq/l) used for prolonged periods did not pose any notable deterioration in soil properties. Based on the experimental data on the deterioration of soil properties and also on experiences of the Indian farmers, Manchanda, Verma and Khanna (1982) concluded that economic use of high RSC waters requires that soils be well drained and coarser textured (loamy sand to sandy loam), only semi-tolerant *rabi* crops (wheat, barley) should be grown in winter season with the lands left fallow in *kharif* and rainfall during the fallow *kharif* period exceeds 400 mm, which serves to re-dissolve the precipitated $CaCO_3$. Later, Manchanda, Sharma and Singh (1985) also reported that the wheat yields started declining during each successive year on a sandy loam soil and the crop failed to germinate during the eighth year because of high ESP (92) and pH (10.0) when irrigated with alkali water (ECiw 1.6 dS/m, RSC 10 meq/l). Bajwa, Hira and Singh (1983) used waters of ECiw (1.15-4.5 dS/m), RSC (2 and 8 meq/l) and SAR (11.6-38.5) in maize-wheat system for five years on a sandy loam and a loam. The reduction in maize yield (Y) due to build-up of sodicity (ESP) in surface 75 cm soil could be described according to relation: Y = 109.36-1.948 ESP, whereas the yields of wheat remained unaffected as the highest ESP encountered (30.5) was well below its sodicity tolerance limit. In a similar study on the use of different RSC/adj.SAR (calculated using pCa + Mg) water in rice-wheat and pearl millet-wheat systems, Bajwa et al. (1989) reported that millet yields were not significantly affected by sustained alkali water irrigation during the initial two years. On the basis of the quadratic relationships developed between adj.SAR and wheat yields, it was predicted that for the maintenance of 90, 75 and 50 percent of the yield obtained under canal water, irrigation waters having adj.SAR less than 23, 28 and 32, respectively, should be used for the rice-wheat system. The corresponding values of adj.SAR for the pearl millet-wheat system were 10, 28 and 36, respectively. For cotton-wheat rotation, the limits for 10 and 25 percent reduction in yield were 6 and 10 meq/l (Bajwa, Choudhary and Josan 1992). Some of the selected data on wheat yields with variable *kharif* crops from the experiments conducted by Bajwa and associates (Table 3) suggest that an increase in RSC keeping the adj.SAR (using pCa+Mg) in the range 30-40, decreased the wheat yield, especially when grown following rice. The decline in wheat yield was not appreciable even with waters having high adj.SAR (25-52) and RSC (8.0 meq/l) with millets/maize sequence particularly with waters having higher ECiw/concentration of Ca^{2+}.

From the above experimental evidences, it is evident that deterioration of soils and decline in wheat yields with use of alkali waters is higher when rice is grown in rotation. So it is usually feared that rice-wheat system may not be sustainable with the use of alkali waters. However, a critical evaluation of ion chemistry of alkali waters used in the most of micro-plot/pot experiments indicates that the proportions of cations, mainly the amounts of calcium present in these artificially prepared waters were much less than the naturally occurring alkali waters. These alkali waters belonged mostly to Type IV (RSC > 4 meq/l, 10 < SAR < 40, $M^{2+}/\Sigma M^{n+}$ ratio < 0.25) as defined by Minhas and Gupta (1992) and have been stated to pose serious problems to soils and crops. While evaluating the causes of opting for paddy-wheat as the most favored cropping system of the farmers using naturally occurring alkali waters in different villages of Kaithal, Minhas, Sharma and Sharma (1996), concluded that in addition to economics, farmers prefer this system because:

- Attainment of desired plant population in any other *kharif* crop is rarely feasible with alkali waters, even the established crops are prone to aeration problems caused by stagnation of water,
- The salt leaching is higher and uniform under paddy where soil is continuously submerged than with fallow or other upland crops where much of the rain water is wasted as run-off or removed by the farmers to avoid the aeration problems in upland crops,
- Soils irrigated with waters containing sufficient calcium, i.e., upto 4 meq/l, are less responsive to gypsum, and
- Canal waters are available to some extent and the rainfall of the areas is sufficient (50-55 cm), so the farmers usually go in for the conjunctive use of ground and canal waters and thus are able to sustain crop yields.

MANAGEMENT PRACTICES FOR SUSTAINED USE OF ALKALI WATERS

Consistent efforts have been made at different research centers in the country to devise ways for the safe utilization of alkali waters to raise agricultural crops. With scientific advances, the basic principles of soil-water-plant systems are now fairly well understood and advocate specialized soil, crop and irrigation management practices for preventing the deterioration of soil to levels which limit the crop productivity. Some such measures for control-

ling the build up of ESP and maintaining the physical and chemical properties of alkali water irrigated soils are being discussed below.

Use of Agro-Chemicals for Alleviating Sodicity/Alkalinity Effects

Amendment Needs

The adverse effects of alkali water irrigation on physico-chemical properties of soils can be mitigated by the application of calcium containing amendments such as gypsum or those inducing its release from the inherent mineral
sources in soils. The quantity of gypsum for neutralization of each meq/l of
RSC is 86 kg/ha per 1000 m^3 (or depth of irrigation = 10 cm) of water. The
agricultural grade gypsum is usually 70-80% pure. The need of gypsum
application for ameliorating the sodicity effects is of recurring nature. Application of gypsum has earlier been recommended when RSC of irrigation
water exceeded 2.5 meq/l (Bhumbla and Abrol, 1972). However later re-
searches have shown that factors such as the level of the existing deterioration of the soil, cropping intensity and the water requirements of the crops to
be raised will ultimately decide the amount of gypsum required. Field trials
have shown that gypsum helps in maintaining the yields of the crops irrigated
with alkali waters (RSC > 5 meq/l) especially when paddy is grown in
rotation and rainfall of the area is < 50 cm.

In wheat-fallow rotation no response to gypsum has been reported on well
drained light textured (sandy loam) soils when irrigated with waters having
RSC upto 10 meq/l (Gupta, 1980; AICRP-Saline Water, 1985). In an already
deteriorated soil (SARe 48.5) due to irrigation with an alkali water (ECiw 2.6
dS/m, SARiw 20.5 and RSC 9.5 meq/l), application of gypsum although did
not affect the rice yields but the yield of succeeding wheat crop increased
significantly (Sharma and Mondal 1982). Application of gypsum in fallow-
wheat system also improved the yield of wheat. Even a small dose of gypsum
(25% GR) improved the wheat yield from almost nil (0.06) to 2.67 Mg/ha in
a highly deteriorated sandy loam soil (pH 10, ESP 92 and infiltration rate < 2
mm/h) with the use of an alkali water (TSS 1000 ppm, RSC 10 meq/l). When
the gypsum dose was increased to 100% GR, the yield got enhanced to 6.33
Mg/ha (Manchanda et al., 1985). Similarly, sensitive crops like pearl-millet
and guar yielded 0.97 and 2.55 Mg/ha when gypsum @ 100% GR was added
to a soil (pH 9.5, ESP 45) yielding only 0.071 Mg/ha and nil, respectively,
due the use of a high RSC water (Manchanda et al., 1985). Later, Sharma and
Manchanda (1989) concluded that guar/pearl millet-wheat crops can be successfully grown in rotation with alkali waters provided the ESP of the soils is
maintained below 15 and 20 with the addition of gypsum @ 100% GR of the
soil. Joshi and Dhir (1991) studied the response of crops to the application of
gypsum on an abandoned land in arid climate of Rajasthan as a result of

irrigation with high RSC waters (7.1-8.9 meq/l). Application of gypsum (equaling 100% GR) plus that required to neutralize RSC in applied irrigation water during two years resulted in a moderate production of wheat (2.61 Mg/ha) and mustard (2.0 Mg/ha) in the second year. Yadav et al. (1991) reported that addition of gypsum @ 50% GR to a loamy sand soil (pH 9.6-9.7) irrigated with alkali water (EC 1.93 dS/m, RSC 12 meq/l) was appropriate for growing *kharif* crops like pearl-millet, urd-bean, mung-bean, cowpea and pigeon-pea whereas cluster-bean responded to gypsum upto 100% GR of soil. Amongst *rabi* crops, response of mustard to gypsum was more than wheat and barley. Gypsum to supply 2.5 and 5.0 meq/l to alkali irrigation water for wheat and rice, respectively, was sufficient for the maintenance of higher yields (Bajwa and Josan, 1989). Results of field trials conducted by CSSRI researchers involving alkali waters (Table 4) have so far shown that response to the use of gypsum to mitigate the adverse effects of high RSC waters has been controversial (Minhas, 1995).

It was concluded that:

- Under the rice-wheat system, pH and sodicity determine the wheat yields and it responded to the application of gypsum in almost all the experiments except when RSC < 5 meq/l,
- Wheat yields under fallow-wheat system were mainly governed by soil salinity and the response to gypsum was erratic, and
- Response of wheat to gypsum in sorghum-wheat rotation is masked by the interactive effects of ECe, SARe and pH of soils as is evident from the following multiple regression:

$$\text{Yield (Mg/ha)} = 10.66 - 0.081(\text{pH})^2 - (0.018\,\text{ECe} \times \text{SARe})$$
$$+ 0.88 \times 10^{-3}(\text{SARe})^2; \ R^2 = 0.78, \ n = 90$$

This equation shows that the ECe values for similar yields should decrease with pH and vice versa. Also at a given pH as SARe increases, ECe should be lower. Thus, in saline-alkali soils developed with the use of alkali waters, high levels of ECe, SARe and pHs together affect plant growth and need to be considered simultaneously for evaluating the salinity and sodicity tolerance in plants.

The need to add gypsum for sustained crop production especially of rice-wheat when irrigated with waters having high RSC is clearly evident from the above studies. Once the role of amendments is established for raising crops with alkali waters, questions regarding its mode, amount and time of application have to be answered. Bajwa, Hira and N.T. Singh (1983), observed that gypsum applied at each irrigation was more effective for increasing maize yields in maize-wheat sequence irrigated with RSC water (8 meq/l) as compared to its single dose applied annually. Later, Bajwa and Josan

TABLE 4. Crop responses to gypsum application in alkali water irrigated soils

Water quality			Soil properties			Grain yield		Gypsum applied	Grain yield	
EC (dS/m)	RSC (meq/l)	SAR	pH	SAR	ECe (dS/m)	Kharif ---- (Mg/ha) ----	Rabi	(Mg/ha)	Kharif ---- (Mg/ha) ----	Rabi
Rice-Wheat										
2.6	9.5	20.5	8.3	58	14.9	4.4	2.2	5.6-22.4	5.9-6.1	3.8-4.4
NA	12.5	NA	9.9*	NA	0.7*	0.6	1.0	2-4	2.0	2.1-3.7
3.1	9.5	15.4	9.6*	NA	NA	4.0	1.5	5	4.3	2.4
1.2-2.2	6.7-8.0	7.3-10.6	8.7*	19**	2.6	2.2	2.4	2.5	3.0	2.9
Sorghum-Wheat #										
3.1	9.5	15.4	8.7	25	7.1	2.8	2.0	1-3.5	2.2-2.6	2.1-3.1
Fallow-Wheat										
2.6	9.5	20.5	8.3	58	14.9	--	0.5	2.2-8.8	--	1.5-3.2

* 1:2 soil:water, ** ESP, # Gypsum @ 5 Mg/ha applied initially.
Adapted from Minhas (1994).

(1989c) reported that gypsum improved the soil properties and significantly increased the yields of rice and wheat crops irrigated with water of RSC 6.8 meq/l, ECiw 0.85 dS/m. Response to gypsum, either applied annually as one dose or at each irrigation was the same. With higher RSC (10.3 meq/l) water, although, the improvement in wheat yields was similar for the two modes of gypsum application, rice responded better to gypsum when applied with irrigation. This was because more water was applied to rice which lead to appreciable increase in soil sodicity during the season affecting rice yields. The depth of irrigation water applied for wheat being smaller, the increase in sodium saturation was not sufficient to adversely affect the wheat yields. While comparing the time of application of gypsum, Yadav et al. (1994) observed that its application before the onset of monsoons was better than its application before pre-sowing irrigation of the *rabi* crops and at each irrigation. Pyrite has also been used for amending the deleterious effects of high RSC waters. Pyrite application once before the sowing of wheat has proved better than its split application at each irrigation or mixing it with irrigation water (Chauhan et al., 1986).

Gypsum Beds

Results of the above studies indicate that application of gypsum at each irrigation proved better or at least equal in alleviating the deleterious effects of alkali waters in rice-wheat system. Translation of these results in practical terms requires some mechanism for dissolution of gypsum in the irrigation water itself. Such a practice will also eliminate the costs involved in powdering, bagging and proper storage before its actual use. In view of the costs involved, the dissolution of gypsum directly in water through the use of gypsum beds or its application to the irrigation channels appears to be an economical preposition. Alkalinity in irrigation waters (ECiw 1.83 dS/m, RSC 15 meq/l) could be decreased by passing it through a bed containing gypsum in assorted sized (2-50 mm) clods (Pal and Poonia, 1979). Dissolution of gypsum is affected by factors such as size distribution of gypsum fragments, flow velocity, salt content and chemical composition of water (Bashir et al., 1979; Kemper, Olsen and De Mooy, 1975; Singh, Poonia and Pal, 1986). For flowing water to pick up calcium through dissolution of gypsum, special gypsum bed has been designed (Singh, Poonia and Pal, 1986) but it may be mentioned here that gypsum bed water quality improvement technique may not dissolve > 8 meq/l of Ca^{2+}. The response of paddy and wheat to the application of equivalent amounts of gypsum, either by passing the water (RSC 9 meq/l) through gypsum beds where the thickness of bed was maintained at 7 and 15 cm, or the soil application of gypsum, is presented in Table 5 (AICRP-Saline Water, 1998-99). Though the crops responded to the application of gypsum through either of the methods, response of paddy was more

TABLE 5. Average yields (Mg/ha) of paddy and wheat and soil properties* as affected by equivalent doses of gypsum applied either to soil or passing alkali water through gypsum beds

Treatment	Paddy (1993-98)	Wheat (1995-99)	pH	ECe (dS/m)	ESP	Infil. rate (mm/d)
Control (T_1)	3.18	2.74	9.4	3.3	65	5.6
Gypsum Through Beds						
3.3 meq/l (T_2)	4.56	3.91	8.1	2.9	20	9.1
5.2 meq/l (T_3)	4.82	4.14	8.1	1.7	18	9.6
Equivalent Soil Application						
as in T_2 (T_4)	4.31	3.93	8.3	2.0	22	8.9
as in T_3 (T_5)	4.52	4.13	8.2	1.8	21	9.3

*At wheat harvest (1998-99).
Source: AICRP-Saline Water (1994-99).

in case of alkali water (RSC 9.0 meq/l) which was ameliorated (3-5 meq/l) after passing through gypsum beds. Thus it seems that gypsum bed technique can help in efficient utilization of gypsum.

Organic Materials

It is generally accepted that additions of organic materials improve sodic soils through mobilization of inherent Ca^{2+} from $CaCO_3$ and other minerals by organic acids (formed during its break down) and increased pCO_2 in soils. The solublized Ca^{2+} in soil replaces Na^+ from the exchange complex. Reclamation of barren alkali soils by addition of organic materials has been widely reported (Rao, 1996), there is some disagreement in literature concerning short-term effects of organic matter on the dispersion of sodic soil particles. Poonia and Pal (1979) studied the Na-(Ca+Mg) exchange equilibrium on a sandy loam soil treated with or without farm yard manure (FYM) and reported that variations in the proportions of Ca:Mg in the equilibrium solutions only slightly improved the Na^+ selectivity of the soils over the soils treated with FYM. In another study, Poonia et al. (1980) observed that the applied organic matter "apparently" had a greater preference for divalent cations than that present in natural forms in the soils. However, Gupta, Bhumbla and Abrol (1984) cautioned against the use of organic manure on the soils undergoing sodication process through irrigation with alkali waters. Organic matter was shown to enhance dispersion of soils due to greater inter-particle interactive forces at high pH. Sharma and Manchanda (1989) studied the effect of irrigation with alkali water (ECiw 4 dS/m, SAR 26 and RSC 15 meq/l) on the growth of pearl-millet and sorghum crops with and without gypsum and farmyard manure on a non-calcareous sandy clay loam soil. The soil was previously deteriorated due to irrigation with alkali water. Six year results with fallow-wheat rotation showed that the use of FYM alone further decreased the crop yields and the permeability of the soils. In a long-term experiment on a soil that received alkali waters (RSC 2.4-16 meq/l) without additions of FYM, the infiltration rate, pH and wheat yields were 5.2 mm/h, 10.34 and 2.7 Mg/ha, respectively. These values improved to 8.1 mm/h, 9.7 and 3.14 Mg/ha, respectively, for soils receiving FYM (Dhankar et al., 1990). The response to FYM, however, decreased with increase in RSC of irrigation waters. Thus it may be opined that the addition of organic materials for use of alkali waters should be preceded by gypsum application when upland *kharif* crops are taken. Nevertheless, short-term reduction in permeability may rather be beneficial for paddy that requires submerged conditions for its growth. As the additions of FYM decreased soil pH and sodicity and improved soil fertility, the yields of rice and wheat improved by 8-10 percent on a soil that received irrigation with an alkali water (ECw 3.2 dS/m, RSC 5.6 meq/l, SAR 11.3) (Minhas, Sharma and Singh, 1995). Similar increases in yields of crops with the application of FYM alone or along with gypsum have earlier been reported by Puntamkar, Mehta and Seth (1972); Manchanda and Sharma (1988) and ACRIP-Saline water, Agra (1989). Sekhon and Bajwa (1993) reported the salt balances in soil under rice-wheat-maize system irrigated with alkali waters (RSC 6.0 and 10.6 meq/l) from a green house experiment. Incorporation of organic materials decreased the precipitation of Ca^{2+} and carbonates, increased removal of Na in drainage waters, decreased soil pH and ESP and improved crop yields. The effectiveness followed the order: paddy straw > green manure > FYM. We can therefore conclude that with the mobilization of Ca^{2+} during decomposition of organic materials, the quantity of gypsum required for controlling the harmful effects of alkali water irrigation can be considerably decreased. Thus, occasional application of organic materials should help in sustaining yields of rice-wheat system receiving alkali waters.

Fertilizers

Indian soils are almost universally deficient in nitrogen that needs to be supplemented through fertilizer. Urea, which is by far the most widely used N source, is first hydrolyzed to NH_3 and CO_2 by the enzyme urease. Under alkaline conditions, NH^+_4 ions are converted to volatile NH_3 (NH^+_4 + HCO_3^- → ↑ NH_3 + CO_2 + H_2O) and can escape to the atmosphere resulting in a net loss of applied N. Losses of N due to ammonia volatilization have been reported to the extent of 40% of the added fertilizer from alkali water irrigated soil (Singh and Bajwa, 1985). Therefore, about 25% higher level of fertilizer N is recommended for sodic soils as compared to the normal soils

(Singh, 1986, Bajwa and Singh, 1992). Efforts have also been made to evolve practices that decrease losses and increase N-use efficiency. Amongst these, splitting of fertilizer N doses, placement of urea at some depth, application of gypsum and use of organic-N sources seem most practical (Bajwa and Singh, 1992, AICRP Saline Waters, 1996). Appropriate machinery may be required for placing urea at some depth in the soil. Experiment with wheat carried out on an alkali water irrigated soil at Kaithal (Minhas et al., 1994-6) demonstrated that 30-40 kg N/ha can be saved through its application before pre-sowing irrigation especially when the soil was deep tilled (chisel ploughed).

Irrigation Management

Conjunctive Use of Alkali and Canal Waters

In many situations where ground waters contain high concentrations of salts, limited canal water supplies may also be available. The two options for making the combined use of poor quality waters and the canal water are; (1) blend the two sources to bring down the alkalinity below tolerable limit of crops, and (2) apply them separately in cyclic mode. In addition to operational advantages, definite evidences exist in favor of the latter where use of canal waters is advocated during germination and seedling establishment stages and saline irrigation is postponed to later growth stages (Minhas and Gupta, 1992; Minhas, 1996). However, in case of alkali waters, the strategy that would either minimize the precipitation of calcium or maximize the dissolution of precipitated calcium can be expected to be better. Usually both canal and ground waters are in equilibrium with calcite, former at the P_{CO_2} of the atmosphere whereas the latter at a much higher P_{CO_2}. The relation between concentration of Ca^{2+} and P_{CO_2} is not linear and is governed by the relation:

$$mCa^{2+} = \frac{Kh\, P_{CO_2}\, K_1\, K_{cal}}{4\, K_2\, \lambda Ca^{2+}\, \lambda HCO_3}$$

In the above equation, mCa^{2+} refers to concentration of Ca^{2+} (g/L). K_1 and K_2 represent first and second dissociation constant of carbonic acid and Kh is Henry's gas constant. λ_{Ca} and λHCO_3 are the activity coefficient of ions while P_{CO_2} is the partial pressure of CO_2. Therefore, it seems that mixing of surface waters with ground waters of higher alkalinity and low calcium would result in under-saturation with respect to calcite. Consequently, the blended water will have tendency to pick up calcium through dissolution of native calcium. Benefits, those can be accrued from such a preposition are, however, yet to be quantified. Bajwa and Josan (1989) reported that irrigation of a sandy loam soil (18-26.8% clay) with alkali water (ECw 1.35 dS/m, RSC 10.1 meq/l, SAR 13.5, adj.SAR 26.7) increased the pH and ESP of the surface layers and reduced its infiltration rate to 14% (Table 6). The yields of rice and wheat decreased progressively with time and were 62 and 57%, respectively, of the potential yield, i.e., that obtained under canal irrigation during 6 years. However, when the alkali water was used in cyclic modes with canal water, yields of both the crops were maintained at par with canal water except in the CW-2AW mode. Cyclic use of two waters decreased sodication of soils. Interestingly after accounting for rainfall and canal water in estimating the adj.SAR, ESP of the surface soil was 1.2-1.5 times the adj.SAR for the cases where cyclic modes were adopted for irrigation compared with a factor of 1.8 observed with alkali waters alone (Table 4). In another experiment where alkali water (ECw 1.8 dS/m, RSC 10.5, Ca+Mg 5.6 meq/l, SAR 9.5) was used for last four years (Minhas et al., 1999), rice-wheat and sorghum-wheat were yet to respond to either mixing or cyclic use (irrigation-wise/seasonally). However, the yield of sorghum has started declining with continuous alkali irrigation. In a micro-plot experiment where an alkali water (EC 2.3 dS/m, RSC 11.3) and good quality tube well water (EC 0.5 dS/m, RSC nil) were used in cyclic modes (2TW:1 AW, 1TW:1AW, 1TW:2AW) with decline in yield in the range of 8-12 and 9-23% in case of paddy and wheat, respectively, performed better than their counter mixing modes where the decline ranged between 14-15 and 16-28%, respectively.

TABLE 6. Effect of cyclic use of alkali and canal waters on soil properties and yields of rice and wheat

Water quality/mode	adj.SAR*	pH	ESP	RIR	Average yield (Mg/ha)	
			(%)		Rice	Wheat
Canal water (CW)	0.3	8.2	4	100	6.78	5.43
Alkali water (AW)	22.0	9.7	46	14	4.17	3.08
2 CW-1 AW	8.9	8.8	13	72	6.67	5.22
1 CW-1 AW	12.8	9.2	18	59	6.30	5.72
1 CW-2 AW	18.5	9.3	22	34	5.72	4.85
	ECw	Ca	Ca + Mg	RSC	SAR	adj.SAR
	(dS/m)	-------(meq/l) -------				
CW	0.25	1.6	2.1	nil	0.3	0.4
AW	1.35	0.4	0.9	10.1	13.5	26.7

*After accounting for 828 and 434 cm of irrigation and rain water, respectively.
Compiled from Bajwa and Josan (1989c).

Thus, alternating alkali and canal waters can be considered to be a practical way to alleviate sodicity problems caused by the use of alkali waters. Field observations in Kaithal area further point out that farmers who are usually getting some canal water supplies are able to sustain yields of rice-wheat crops whereas yields of these crops decline on farmer's fields who do not have any access to canal water supplies (Minhas, Sharma and Sharma, 1996).

Leaching Requirements/Irrigation Schedules

Traditional brackish water management approach (US Salinity Lab, 1954; Rhoades, Kandiah and Mashali, 1992) assumes that steady state conditions exist in the long run, which implies that the economic way to control the salts is to ensure net downward flow of water through the root zone. As presented earlier, various indices in vogue predict lower build up of ESP with increase in leaching fraction (LF). However, the concept of leaching requirement (LR) has been shown to be inappropriate when growing season for post-monsoon winter crops starts with surface leached soil profile and thus attempts to meet LR do not pay off as of increased salt load with saline irrigation (Minhas, 1996). Similarly under alkali water irrigation, salinity control could not be achieved by applying 50% extra water (ECiw 3.2 dS/m, SAR 21, RSC 4 meq/l) to meet the leaching requirement in rice-wheat and maize/pearl millet-wheat systems (Bajwa, Hira and Singh, 1986). Rather such a practice resulted in 30-50% higher salinity build up which lowered the crop yields.

A general recommendation under sodic conditions is to apply light and frequent irrigation for overcoming the effects of poor hydraulic properties of soils. But with alkali waters, such an option would also mean an enhanced input of salt to soil and crops having higher water requirement (rice, sugar-cane, etc.) can result in greater deterioration of soil properties. Bajwa, Josan and Chaudhary (1993) reported that crop response to shorter irrigation intervals when using alkali waters depended upon the season in which crop was grown and its relative salt and Na tolerance. No effect of varying irrigation intervals was observed in wheat grown during winter season and maize grown for fodder during monsoon season. During summer months the shorter irrigation intervals lowered soil temperature and hence improved the yield of maize (fodder). Build up of salts and ESP for different irrigation frequencies were, however, similar.

Deep Tillage/Sub-Soiling

With the development of sodicity in the surface soil, the clay particles in alkali water irrigated soil become prone to dispersion and displacement and thus the possibility of formation of dense sub-soil layers (Plow sole) increases. Moreover, such soils become very hard and dense (hard setting soils) on drying. Both these factors retard root proliferation and poor crop yields are mainly ascribed to this. Therefore, deep plough/chiseling can be considered as a short-term measure to overcome physical hindrances in such soils. Wheat crop responded to deep tillage, and the average yield increase was of the order of 0.2-0.4 Mg/ha (Minhas, Sharma and Sharma, 1996). Such experiment needs to be continued to study the long-term effects of such a practice under different soils before some worthwhile conclusions can be arrived at.

Crop Tolerance/Varieties

Ability of the crops to perform under sodic/alkaline conditions vary a lot. Relative sodicity tolerances of important agricultural crops have been worked out in a number of field experiments at Karnal and other alkali sites undergoing reclamation through the use of gypsum. A summary of results as presented by Gupta and Abrol (1990) can serve as general guidelines for selection of crops (Table 7). However, when Minhas and Gupta (1992) compared the tolerance of wheat from experimental results under the two conditions of (i) alkali soils undergoing reclamation and (ii) soils being sodicated through irrigation with alkali waters, lower tolerance was observed under the latter. Though ESP of only surface 15 cm soil was considered for comparisons, the results could not even be explained on the basis of soil's ESP profiles. The differential availability of Ca^{2+} seemed to play an important role as during reclamation of alkali soils, calcium furnished through gypsum to reduce ESP also meets the calcium needs whereas in soils being sodicated, solution calcium continues to decrease. Use of alkali waters besides increasing sodicity and pH, reduces water infiltration with the result that salts added through irrigation concentrate in upper soil layers. Build up of these salts also influences

TABLE 7. Relative tolerance to alkalinity/sodicity of soils

ESP Range*	Crops
10-15	Safflower, mash, peas, lentil, pigeon-pea, urd bean
16-20	Bengal gram, soybean
20-25	Groundnut, cowpea, onion, pearl-millet
25-30	Linseed, garlic, guar
30-50	Mustard, wheat, sunflower
50-60	Barley, sesbania
60-70	Rice

*Relative crop yields are only 50% of the potential in respective sodicity ranges.

the performance of crops. In addition to crops, their cultivars also vary in tolerance to sodicity. Some of the indices established for higher tolerance of crop varieties have been avoidance of Na uptake and maintaining low Na/K ratio in shoot (Gill and Qadir, 1996) and the possession of penetrative root system (Choudhary, Bajwa and Josan, 1994). Therefore, breeding efforts have been going on for developing high yielding tolerant varieties of rice and wheat possessing these characters.

SOME RESEARCHABLE ISSUES

Keeping in view the inadequate availability of good quality water resources, the management of poor quality waters for sustained irrigation will perhaps be one of the most important determinants of the future of rice-wheat production system. Thus for ensuring the sustainability of rice-wheat system, water resources are required to be managed with minimum economic and environmental costs. The above discussions do indicate that considerable knowledge in the state-of-the art for the use of poor quality waters in rice-wheat and other cropping systems has become available from experiments conducted over prolonged periods on different soils and agro-climatic zones. Some of the selected technologies can now be transferred to the farmers for its adoption. However, certain issues demand continued research efforts through new experimentation or through modifications of the on-going research programs. Researchable issues that require attention at the farm and policy level are outlined below:

At the Farm Level

- The native calcium in soils and waters has a vital role in alleviating or minimizing the adverse effects of high RSC waters and its non-consideration often causes misleading inferences. Thus release of calcium in soils from both inherent and recent $CaCO_3$ needs to be quantified, its patterns defined and practical means to enhance dissolution of calcite be evolved so that appropriate modification can be made in the prevalent guidelines for the use of high RSC waters.
- Use of poor quality waters leads to lower fertilizer efficiencies, increased rates of fertilizer loss and decrease in the efficiency of *Rhizobium* nodulation. So the issues related to appropriate timing and placement of fertilizers, adjusting the timings of leaching treatment as well choice of slow release fertilizers require further research.
- Organic materials can be beneficial through increasing structural stability and infiltration rates, slow release of nutrient elements and some

lowering of pH and Ca release from $CaCO_3$. However, their role is still controversial for soils undergoing sodication and therefore requires detailed investigations by the researchers.

- There is a need to develop alternate tillage operation that can improve root growth, achieve improved water and nutrient use efficiencies through a better understanding of interactions of tillage and nutrients with respect to water management practices in soils irrigated with poor quality waters. As an example, long-term experimentation is required with respect to sub-soiling effects to combat hard setting and hard pan effects in alkali water irrigated soils.

- In many areas with poor quality underground waters, rice-wheat is practiced using traditional and wasteful water application schedules that aggravate the salinity problems. Therefore, there is a need to identify water saving but practical irrigation schedules both for rice and post-rice crops when poor quality waters are used.

- Because of higher deterioration of soils associated with its water requirements, doubts are often raised for the cultivation of rice with alkali waters whereas it has been shown to have ameliorative effects on alkali soils. Thus the site-specific guidelines with respect to ion chemistry of waters, rainfall and soil types require further refinements.

- Accurate information with regard to water quality impacts can only be perceived from long-term experimentation but the development of computer simulation models for determining the best options of water, crop and chemical management while minimizing the hazards from the usage of poor quality waters should be of great help. The models should include predictions on the changes in solute composition as a result of chemical interactions of the poor quality waters, effects of these changes on water transmission characteristics of soils vis-a-vis root water uptake and ultimately the performance of crops under such soil environments.

- Research has led to the recommendation of site-specific water quality guidelines for the successful utilization of poor quality waters using conventional (surface) irrigation method. However, the micro-irrigation systems, i.e., drip/trickle are considered to be the most efficient ways of utilization of saline waters but these systems have not been evaluated at large scale in India. Thus for a shift from conventional to newer, i.e., micro-irrigation systems, information on salinity/sodicity/toxicity limits of crops and horticultural species using sprinkler and drip irrigation systems is required. Also desired is the development of crop-water production functions with these micro-irrigation systems. The data generated can then be utilized for improving the existing guidelines for usage of poor quality waters.

- With liberalization of the economy, emphasis is being laid on the production of value added products with export potential. Economic studies should focus on the choice of feasible methods of using poor quality waters by evaluating the effects of these waters on productivity and profitability of crop production. Such studies will also be useful to compare the indigenous methods of managing poor quality waters. Thus, these may help in the development of location specific technologies for the varying socio-economic conditions and may help in reallocation of water resources for maximizing the benefits.
- Seasonal rainfall is a valuable water resource for maintaining the salt and water balances. Thus this needs to be utilized efficiently by planning for synchronizing the crop demands with rainfall, creation of on-farm storage reservoirs, etc.

System Level Policy Issues

At the system level, the following policy decisions require attention of the planners to solve water quality related problems.

- Predictions of yield vis-a-vis revenue losses accrued upon the use of poor quality waters.
- Re-defining of surface water policies with preferential allocation to areas having poor quality ground waters.
- Offering subsidies for installing tubewells in area with poor ground water zones, on the use of amendments like gypsum and creation of blending sites. Also provision for other incentives for rational use of poor quality waters should be made.
- Rain water harvesting and other water storage structures to ensure supplies during salt sensitive growth stages.
- Creation of options to dump/dispose extra salts by identifying sites for salt tolerant forestry plantations or halophytes, evaporation ponds or direct disposal to sea or possibly rivers during their full flow.

REFERENCES

Abrol, I.P. and D.R. Bhumbla. (1979). Crop responses to differential gypsum application in a highly sodic soil and the tolerance of several crops to exchangeable sodium under field conditions. *Soil Science* 127: 79-85.

Abrol, I.P., and M.S. Gill. (1994). Sustainability of rice-wheat system in India. In: *Sustainability of Rice-Wheat Systems in Asia*, Eds., R.S. Paroda, T. Woodhead and R.S. Singh. Regional Office of Asia and Pecific, FAO, Bangkok, pp. 112.

AICRP-Saline Water. (1972-99). *Annual Progress Reports*. All India Co-Ordinated

Research Project on Management of Salt-Affected Soils and Use of Saline Water in Agriculture, CSSRI, Karnal.

Ayers, R.S. and D.W. Westcot. (1985). *Water Quality for Agriculture, Irrigation and Drainage* Paper No. 29, Rev. 1, FAO, Rome, pp. 174.

Bajwa, M.S., O.P. Choudhary and A.S. Josan. (1992). Effect of continuous irrigation with sodic and saline sodic water on soil properties and crop yields under cotton-wheat rotation in northern India. *Agricultural Water Management*, 22: 345-350.

Bajwa, M.S., G.S. Hira, and N.T. Singh. (1983). Effect of sodium and bicarbonate irrigation waters on sodium accumulation and maize and wheat yields in northern India. *Irrigation Science*, 4:91-99.

Bajwa, M.S., G.S. Hira and N.T. Singh. (1986). Effect of sustained saline irrigation on soil salinity and crop yields. *Irrigation Science*, 7: 27-34.

Bajwa, M.S. and A.S. Josan. (1989a). Prediction of sustained sodic irrigation effects on soil sodium saturation and crop yields. *Agricultural Water Management*, 16: 227-228.

Bajwa, M.S. and A.S. Josan. (1989b). Effect of gypsum and sodic water irrigation on soil and crop yields in a rice-wheat rotation. *Agricultural Water Management*, 16: 53-61.

Bajwa, M.S. and A.S. Josan. (1989c). Effect of alternating sodic and non-sodic irrigation on build up of sodium in soil and crop yield in northern India. *Experimental Agriculture*, 25: 199-205.

Bajwa, M.S., A.S. Josan and O.P. Chaudhary. (1993). Effect of frequency of sodic and saline-sodic irrigation and gypsum on the build up of sodium in soil and crop yields. *Irrigation Science*, 13: 21-26.

Bajwa, M.S. and B. Singh. (1992). Fertilizer nitrogen management for rice and wheat crops grown in alkali soils. *Fertilizer News*, 37(8): 47-59.

Bajwa, M.S., N.T. Singh, N.S. Randhawa and S.P.S. Brar. (1975). Underground water quality map of Punjab state. *Journal of Research (PAU)*, 12: 117-122.

Bashir, A., W.D. Kemper, G. Haider and M.A. Niazi. (1979). Use of gypsum to lower the sodium adsorption ratio of irrigation waters. *Soil Science Society of American Journal*, 43: 698-702.

Bhu Dayal. (1985). *Studies on residual sodium carbonate of irrigation waters and its effect on sodium characteristics and crop growth*. Ph D. thesis, Agra Univ., Agra. p. 126

Bhumbla, D.R. and I.P. Abrol. (1972). Is your water suitable for irrigation–get it tested. *Indian Farming*, 21: 15-16.

Bower, C.A., G. Ogata and J.M. Tucker. (1968). Sodium hazards of irrigation waters as influenced by leaching fraction and by precipitating or selection of calcium carbonates. *Soil Science*, 106: 29-34.

Chauhan, R.P.S., C.P.S. Chauhan and V.P. Singh. (1986). Use of pyrites in minimising the adverse effects of sodic waters. *Indian Journal of Agricultural Science*, 56: 717-721.

Choudhary, O.P., M.S. Bajwa and A.S. Josan. (1994). Characteristics of different wheat cultivars for tolerance to soil sodium saturation. *Trans. World Soil Science Congres*, 3(b): 368-369.

Dhanker, O.P., H.D. Yadav and O.P. Yadav. (1990). *Long term effect of sodic water* on soil deterioration and crop yields in loamy sand soil of semiarid regions. National Symposium of Water Resource Conservation Recycling and Reuse, Nagpur. Feb. 3-5.pp. 57-60.

Dhir, R.P., B.K. Sharma and N. Singh. (1980). *Sodic characteristics of highly saline water irrigated soils and the importance of sulphate ions.* Int. Symp. Salt Affected Soils, CSSRI, Karnal, India, pp. 369-375.

Dubey, D.D., S.C. Tiwari, R.K. Gupta and O.P. Sharma. (1988). Effect of chloride and bicarbonates in irrigation waters on dissolution and precipitation of soil minerals. *Journal of Indian Society of Soil Science*, 37: 268-273.

Eaton, F.M. (1950). Significance of carbonates in irrigation waters. *Soil Science*, 69: 123-133.

Gill, K.S. and A. Qadir. (1997). Physiological aspects of salt tolerance, pp. 243-260. In: *Agricultural Salinity Management in India*, eds. N.K. Tyagi and P.S. Minhas, CSSRI, Karnal.

Girdhar, I.K. and J.S.P. Yadav. (1980). *Effect of different Ca/Mg ratios, SAR values and electrolytic concentration in leaching water on the dispersion and hydraulic conductivity of soils.* Proceedings of International Symposium of Salt Affected Soils, CSSRI, Karnal, pp. 210-218.

Gupta, I.C. (1979). *Use of Saline Water in Agriculture in Arid and Semiarid Zones of India.* Oxford and IBH Publ. Co., New Delhi, p. 210.

Gupta, I.C. (1980). *Effect of irrigation with high sodium waters on soil properties and growth of wheat.* Proceedings of International Symposium of Salt Affected Soils, CSSRI, Karnal, pp. 382-388.

Gupta, I.C. (1983). Concept of residual sodium carbonate in irrigation water in relation to sodium hazard of irrigated soils. *Current Agriculture*, 7: 97-113.

Gupta, R.K. and I.P. Abrol. (1990). Salt affected soils–their reclamation and management for crop production. *Advances in Soil Science*, 12: 223-275.

Gupta, R.K., D.R. Bhumbla and I.P. Abrol. (1984). Effect of soil pH, organic matter and calcium carbonate on the dispersion behaviour of alkali soils. *Soil Science*, 137: 245-251.

Joshi, D.C. and R.P. Dhir. (1991). Rehabilitation of degraded sodic soil in an arid environment by using residual Na-carbonate water for irrigation. *Arid Soil Research and Rehabilitation*, 5: 175-1185.

Kanwar, B.S. and J.S. Kanwar. (1971). Effect of residual sodium carbonate in irrigation waters on plant and soil. *Indian Journal of Agricultural Science*, 41: 54-66.

Kemper, W.D., J. Olsen and C.J. De Mooy. (1975). Predicting salinization and sodication of a bare sandy-loam soil after number of irrigations with poor quality water interspersed with rain water. *Soil Science*, 39: 458-463.

Manchnada, H.R. (1993). Long term use of sodic waters in North India and the reliability of equations for predicting their sodium hazard. In: *Towards the Rational Use of High Salinity Tolerant Plants*, eds. H. Lieth and A. Al. Masoom, Vol. 2:433-438, Kluwer Academic Publishers, The Netherlands.

Manchanda, H.R., R.N. Garg, S.K. Sharma and J.P. Singh. (1985). Effect of continuous use of sodium and bicarbonate rich irrigation water with gypsum and farm yard manure on soil properties and yield of wheat in a fine loamy soil. *Journal of Indian Society of Soil Science*, 33: 876-883.

Manchanda, H.R., I.C. Gupta and B.L. Jain. (1989). Use of poor quality waters. In: *Review of Research on Sandy Soils in India*, Int. Symp. Managing Sandy Soils, CAZRI, Jodhpur. Feb. 6-10, 1989, pp. 362-383

Manchanda, H.R., S.K. Sharma, H.R. Malik and B.K. Suneja. (1985). Technology for using sodic water for guar and bajra. *Indian Farming*, 35: 11.

Manchanda, H.R., S.K. Sharma, S.S. Yadav, B.S. Buttar and B.K. Suneja. (1985). Sodic water for wheat. *Indian Farming*, 35: 9-11.

Manchanda, H.R., S.K. Sharma and J.P. Singh. (1985). Effect of increasing levels of residual sodium carbonate in irrigation waters on exchangeable sodium percentage of a sandy loam soil and crop yields. *Journal of Indian Society of Soil Science*, 33: 366-371.

Manchanda, H.R., S.L. Verma and S.S. Khanna. (1982). Identification of some factors for use of sodic waters with high residual sodium carbonate. *Journal of Indian Society of Soil Science*, 30: 353-360.

Minhas, P.S. (1996). Saline water management for irrigation in India. *Agricultral Water Management*, 30: 1-24.

Minhas, P.S. (1994). Use of saline waters. In: *Salinity Management for Sustainable Agriculture*, eds. Rao, D.L.N. et al. CSSRI, Karnal, pp. 201-225.

Minhas, P.S. and R.K. Gupta. (1992). *Quality of Irrigation Water–Assessment and Management*, ICAR Pub., New Delhi, p. 123

Minhas, P.S., R..K. Naresh, C.P.S. Chauhan and Raj K. Gupta. (1994). Field determined hydraulic properties of a sandy loam soil irrigated with various salinity and SAR waters. *Agricultural and Water Management*, 24: 93-104.

Minhas, P.S. and D.R. Sharma. (1986). Hydraulic conductivity and clay dispersion as affected by the application sequence of saline and simulated rain water. *Irrigation Science*, 7: 159-167.

Minhas, P.S., S.K. Dubey and D.R. Sharma. (1998-99). *Salt and water dynamics in soils irrigated with multiquality waters and their influence on crop performance*. Annual Reports. CSSRI, Karnal.

Minhas, P.S., D.R. Sharma and D.K. Sharma. (1996). Perspective of sodic water management for paddy-wheat cropping system. *Journal of Indian Water Resources Society*, 2(1): 57-61.

Minhas, P.S., D.R. Sharma and Y.P. Singh. (1995). Response of paddy and wheat to applied gypsum and FYM on an alkali water irrigated soil. *Journal of Indian Society of Soil Science*, 43: 452-455.

Minhas, P.S., Y.P. Singh, D.S. Chhabba and V.K. Sharma. (1999). Changes in hydraulic conductivity of a highly calcareous and non-calcareous soil under cycles of irrigation with saline and simulated rainwater. *Irrigation Science*, 18: 199-203.

Pal, R. and S.R. Poonia. (1979). Dimension of gypsum bed in relation to residual sodium carbonate of irrigation water, size of gypsum fragments and flow velocity. *Journal of Indian Society of Soil Science*, 27: 5-10.

Pal, R., S. Singh and S.R. Poonia. (1980). *Effect of water quality on some water transmission parameters of soils and their evaluation using different methods*. Proceedings of International Symposium on Salt-Affected Soils. CSSRI, Karnal, pp. 210-219.

Poonia, S.R., S.C. Mehta and R. Pal. (1980). *Calcium-sodium, magnesium-sodium exchange equilibria in relation to organic matter in soils.* Proceedings of International Symposium on Salt-Affected Soils. CSSRI, Karnal, pp. 135-142.

Poonia, S.R. and R. Pal. (1979). Effect of organic manuring and water quality on water transmission parameters and sodication of a sandy loam soil. *Agricultural Water Management*, 2: 163-175.

Puntamkar, S.S., P.C. Mehta and S.P. Seth. (1972). Effect of gypsum and manure on the growth of wheat irrigated with bicarbonate rich waters. *Journal of Indian Society of Soil Science*, 20: 281-285.

Rao, D.L.N. (1998). Microbiological processes, pp. 125-144. In: *Agricultural Salinity Management in India*, eds. N.K. Tyagi and P.S. Minhas, CSSRI, Karnal.

Rhoades, J.D. (1968). Mineral weathering correction for estimating sodium hazard of irrigation waters. *Proceedings of Soil Science Society of America*, 32: 648-652.

Rhoades, J.D., A. Kandiah and A.M. Mashali. (1992). *The Use of Saline Water for Crop Production.* Irrig. & Drainage Paper No. 48. FAO, Rome, 133 p.

Saxena, R.K., M.L. Sharma and H.R. Jodha. (1966). Quality of ground waters for irrigation in Ahor development block, Jalore. *Annals of Arid Zone*, 5: 204-218.

Sharma, D.R. (1978). Testing of a model for predicting sodium hazard of irrigation waters. *Journal of Indian Society of Soil Science*, 27: 204-208.

Sharma, D.R. and P.S. Minhas. (1986). *Effect of water quality on hydraulic properties of different textured soils.* Annual Report, CSSRI, Karnal, India, p. 73.

Sharma, D.R. and R.C. Mondal. (1981). Case study on sodic hazard of irrigation waters. *Journal of Indian Society of Soil Science*, 29: 270-273.

Sharma, D.R. and R.C. Mondal. (1982). *Effect of irrigation of sodic water and gypsum application on soil properties and crop yields.* 12th Intnl. Congr. of Soil Sci. Feb. 8-12, 1982, New Delhi. Abst. No. 604, p. 170.

Sharma, S.K. and H.R. Manchanda. (1989). Using sodic water with gypsum for some crops in relation to soil ESP. *Journal of Indian Society of Soil Science*, 37: 135-139.

Shekhon, M.S. and M.S. Bajwa. (1993). Effect of incorporation of organic materials and gypsum in controlling sodic irrigation effects on soil properties under rice-wheat-maize system. *Agricultural Water Management*, 24: 15-25.

Singh, A. and N.T. Singh. (1971). *Effect of quality of water on soil properties.* Proc. All India Symp. Soil Salinity, Kanpur, pp. 132-137.

Singh, B. and M.S. Bajwa. (1989). Effect of poor quality irrigation waters on ammonia volatilization losses. *Journal of Indian Society of Soil Science*, 39: 779-780.

Singh, B., R.S. Rana and M.S. Bajwa. (1977). Salinity and sodium hazard of underground irrigation waters in Bhatinda district (Punjab). *Indian Journal of Ecology*, 4: 32-41.

Singh, Hargopal and M.S. Bajwa. (1991). Effect of sodic irrigation and gypsum on reclamation of sodic soil and growth of rice and wheat plants. *Agricultural Water Management*, 20: 163-171.

Singh, K.S. and R.P. Sharma. (1971). Studies on the effect of saline irrigation waters on physico-chemical properties of some soils of Rajasthan. *Journal of Indian Society of Soil Science*, 18: 345-356.

Singh, M., S.R. Poonia and R. Pal. (1986). Improvement of irrigation waters by gypsum beds. *Agricultural Water Management*, 11: 293-301.

Siyag, R.S., M.S. Lamba, R. Pal and S.R. Poonia. (1988). Predicting sodication of calcium saturated soil columns on leaching with sodic waters. *Journal of Agricultural Science* (Camb.), 111: 159-163.

Suarez, D.L. (1981). Relationship between pHc and sodium adsorption ration (SAR) and an alternative method of estimating SAR of soil or drainage waters. *Soil Science Society of America Journal*, 45: 469-474.

US Salinity Laboratory Staff. (1954). *Diagnosis and Improvement of Saline and Alkali Soils*, USDA Handbk No. 60, p 160.

Verma, S.K., R.K. Gupta and R.A. Sharma. (1987). Hydraulic properties of sodic clay as modified by the quality of irrigation waters. *Journal of Indian Society of Soil Science, 35*: 1-4.

Yadav, H.D. and V. Kumar. (1994). Management of sodic water in light textured soils. In. Proc. Sem. *Reclamation and Management of Waterlogged Saline Soils*. April 5-8, 1994, CSSRI, Karnal, pp. 226-241.

Yadav, H.D., V. Kumar, S. Singh and O.P. Yadva. (1991). Effect of gypsum on some kharif crops in sodic soils. *Journal of Research (HAU)*, 22: 170-173.

Index